CAMBRIDGE LIBRARY COLLECTION

Books of enduring scholarly value

Technology

The focus of this series is engineering, broadly construed. It covers technological innovation from a range of periods and cultures, but centres on the technological achievements of the industrial era in the West, particularly in the nineteenth century, as understood by their contemporaries. Infrastructure is one major focus, covering the building of railways and canals, bridges and tunnels, land drainage, the laying of submarine cables, and the construction of docks and lighthouses. Other key topics include developments in industrial and manufacturing fields such as mining technology, the production of iron and steel, the use of steam power, and chemical processes such as photography and textile dyes.

Dynamo-Electricity Machinery

Silvanus P. Thompson (1851–1916) was a physicist and electrical engineer. A professor by the age of 27, he taught at University College, Bristol, and the City and Guilds Finsbury Technical College in London, and was a leading expert on the newly emerging subject of electrical lighting. This work, first published in 1884, is considered a classic in the field. In this third edition (1888), Thompson explains that he has updated much of the work, and made an important amendment in Chapter XIV about the introduction of magnetic circuits into theoretical arguments about energy production. The book begins with an explanation of how dynamos turn mechanical power into electricity, and moves on to discuss some historical background and theoretical aspects before giving detailed descriptions and illustrations of the many types of dynamo. It is an important source document for the field of electrical engineering at the end of the nineteenth century.

Cambridge University Press has long been a pioneer in the reissuing of out-of-print titles from its own backlist, producing digital reprints of books that are still sought after by scholars and students but could not be reprinted economically using traditional technology. The Cambridge Library Collection extends this activity to a wider range of books which are still of importance to researchers and professionals, either for the source material they contain, or as landmarks in the history of their academic discipline.

Drawing from the world-renowned collections in the Cambridge University Library, and guided by the advice of experts in each subject area, Cambridge University Press is using state-of-the-art scanning machines in its own Printing House to capture the content of each book selected for inclusion. The files are processed to give a consistently clear, crisp image, and the books finished to the high quality standard for which the Press is recognised around the world. The latest print-on-demand technology ensures that the books will remain available indefinitely, and that orders for single or multiple copies can quickly be supplied.

The Cambridge Library Collection will bring back to life books of enduring scholarly value (including out-of-copyright works originally issued by other publishers) across a wide range of disciplines in the humanities and social sciences and in science and technology.

Dynamo-Electricity Machinery

A Manual for Students of Electrotechnics

SILVANUS PHILLIPS THOMPSON

CAMBRIDGE
UNIVERSITY PRESS

CAMBRIDGE UNIVERSITY PRESS

Cambridge, New York, Melbourne, Madrid, Cape Town, Singapore,
São Paolo, Delhi, Dubai, Tokyo, Mexico City

Published in the United States of America by Cambridge University Press, New York

www.cambridge.org
Information on this title: www.cambridge.org/9781108026871

© in this compilation Cambridge University Press 2011

This edition first published 1888
This digitally printed version 2011

ISBN 978-1-108-02687-1 Paperback

DYNAMO-ELECTRIC
MACHINERY.

DYNAMO-ELECTRIC MACHINERY:

A MANUAL
FOR STUDENTS OF ELECTROTECHNICS.

BY

SILVANUS P. THOMPSON, D.Sc., B.A.;

PRINCIPAL OF, AND PROFESSOR OF PHYSICS IN, THE CITY AND GUILDS OF LONDON TECHNICAL
COLLEGE, FINSBURY ;
LATE PROFESSOR OF EXPERIMENTAL PHYSICS IN UNIVERSITY COLLEGE, BRISTOL ;
MEMBER OF THE SOCIETY OF TELEGRAPH-ENGINEERS AND ELECTRICIANS ;
MEMBER OF THE PHYSICAL SOCIETY OF LONDON ;
MEMBRE DE LA SOCIÉTÉ DE PHYSIQUE DE PARIS ;
HONORARY MEMBER OF THE PHYSICAL SOCIETY OF FRANKFORT-ON-THE-MAIN ;
FELLOW OF THE ROYAL ASTRONOMICAL SOCIETY.

THIRD EDITION, ENLARGED AND REVISED.

E. & F. N. SPON, 125, STRAND, LONDON.
NEW YORK: 12, CORTLANDT STREET.

1888.

PREFACE

TO THE THIRD EDITION.

———•———

MOST of this treatise has been rewritten for this edition, and much new matter has been added. In order to render the work more complete, some historical notes and a brief account of the system of electric units have been added to the introductory chapter. In that part of the book which is descriptive, the machines for furnishing continuous currents have been considered separately from those for furnishing alternate currents. The mathematical parts of the work have been entirely rewritten and curtailed.

The algebraic and graphic methods of treating the various problems connected with dynamo-electric machinery were in the former editions kept apart : in this edition they are now combined. Some additional data useful in electro-metallurgy have been inserted in Chapter XIII. A chapter has been added upon transformers for distributing alternating and continuous currents. The most important change, however, is the introduction (hinted at in the preface to the first edition of this work) of the conception of the magnetic circuit into the theoretical arguments. This conception, which during the last four years has revolutionised the practice of dynamo design, is developed in Chapter XIV., and is made the basis of the theoretical treatment which follows. The essential part of the investigations of Rowland, Hopkinson, and Kapp in this direction is embodied in this chapter. The author has also followed Kapp in the method of reckoning armature-quantities by counting the number of conductors at the periphery, instead of the number of convolutions, so as to make the formulæ applicable equally to ring armatures and to drum armatures. An important error in the treatment of constant-current motors has been rectified. The author's thanks are due to numerous correspondents and friends for corrections, hints, and suggestions, and for valuable information respecting new machines.

To the Appendices have been added an abstract of the theoretical researches of Clausius, an account of some recent investigations respecting certain points in dynamo design, an enquiry into the conditions of self-excitation, and some references to the use of the dynamo in telegraphy.

FINSBURY TECHNICAL COLLEGE,
LONDON, *May* 1888.

(vii)

CONTENTS

Contents.

Contents.

APPENDICES.

ERRATA.

Page 24, line 19 from top, *dele* "and in Appendix XI."

,, 58 ,, 6 ,, bottom, *for* $\dfrac{\text{``R}^2\text{''}}{r}$ *read* $\dfrac{\text{``R}^2\text{''}}{r_a}$.

,, 67 ,, 19 ,, bottom, *for* Fig. 35, *read* Fig. 34.

,, 97 ,, 8 ,, top, *for* Fig. 66, *read* Fig. 62.

,, 101 ,, 2 ,, top, *for* $r\,i$, *read* $r\,i^2$.

,, 159 ,, 2 ,, top, *for* "brushes" *read* "bushes."

DYNAMO-ELECTRIC MACHINERY.

CHAPTER I.

INTRODUCTORY. HISTORICAL NOTES AND ELECTRICAL UNITS.

A dynamo-electric machine is a machine for converting energy in the form of mechanical power into energy in the form of electric currents, or vice versâ, by the operation of setting conductors (usually in the form of coils of copper wire) *to rotate in a magnetic field, or by varying a magnetic field in the presence of conductors.* This definition is framed to include all machines, the action of which is dependent on the principle of magneto-electric induction, discovered by Faraday in 1831.

Every dynamo-electric machine is, indeed, capable of serving two distinct functions, the converse of one another. When supplied with mechanical power from some external source of power, such as a steam-engine, it furnishes electric currents. When supplied with electric currents from some external source such as a voltaic battery, it furnishes mechanical power. On the one hand the dynamo serves as a *generator*, on the other hand as a *motor*. In general every dynamo, whether intended for use as a generator or as a motor, consists of two essential parts, a *field-magnet*, usually a massive, stationary structure of iron surrounded by coils of insulated copper wires, and an *armature*, a peculiarly arranged system of copper conductors, usually wound upon the periphery of a ring, drum or disk, fixed upon a shaft whereby rotation can be imparted mechanically. For receiving the

B

electric currents from the armature and imparting them to the electric circuit, or *vice versâ*, there are special devices, known as *collectors* or *commutators*, attached to the armature and rotating with it, and there are collecting *brushes*, constituting sliding circuit-connections, which press upon the moving surface of the collector or commutator.

The function of the field-magnet is to provide a *magnetic field* of great extent and intensity ; that is to say, to provide an enormous number of lines of magnetic force in the space wherein the armature-conductors are to revolve. It must consequently consist of a large, and powerful, and therefore well-designed, magnet or electromagnet, having its poles so shaped that the magnetic lines that issue from them shall be utilised in the armature-space. The magnetic field and the magnetic lines are dealt with in Chapter II.; the fundamental principles of magnetism and of the magnetic circuit, including the designing of field-magnets, are dealt with in Chapter XIV.

The function of the armature is to rotate in the magnetic field, whilst carrying electric currents in its copper coils or conductors. It must be remembered that there is a two-fold action between a conducting wire (forming part of a circuit) and a magnetic field. *Firstly*, if the conducting wire is forcibly moved across the magnetic field (so as to cut across the magnetic lines), electric currents are generated in the conductor, and a mechanical effort is required to move the conductor. This action, discovered by Faraday, is termed " magneto-electric induction." In every case the induction or generation of currents necessitates the application of mechanical power and the expenditure of energy. This is the principle of the dynamo used as a generator. *Secondly*, if the conducting wire, while situated in the magnetic field, is actually conveying an electric current (from whatever source) it experiences a lateral thrust, tending to move it forcibly, parallel to itself, across the magnetic lines, and so enables it to exert power and to do work. This action, which is the converse of the former, is the principle of the dynamo used as a motor. In the first case, power is required to drive the armature; in the second, the armature rotating becomes a source of power. If we have the magnetic field and supply power to drive the rotating conductor, we get the electric currents; if we have the magnetic field and supply the electric currents to the conductor, it rotates and furnishes power. Whether the machine be used as generator or as motor, the magnetic field must be present: hence the most important point in theory is the theory of the magnetic field. As every dynamo will work (at least theoretically) either as generator or as motor, it should be possible to frame a general theory for any machine serving either of these two converse functions. For the sake of simplicity, however, these two functions will be separately considered in the present work.

The mathematical theory of the dynamo is, indeed, complex, and takes different forms for its expression in the various classes of machine now included under the one name of "dynamo." The progress recently made in the theoretical treatment of magnetic problems has simplified matters so much that it is now possible to predict from the construction and dimensions of a dynamo its electrical output under given conditions of speed and load. The theory of alternate-current machines is different in many points from that of machines which are to furnish continuous currents. The theory of the dynamo, then, which will be developed in the present work, will not be a general mathematical theory. The aim will be to deal with physical and experimental rather than mathematical ideas, though of necessity mathematical symbols must be used here as in every kind of engineering work. A physical theory of the dynamo is not new, though none of any great completeness had been given[1] prior to the appearance of the author's lectures in 1882.

There are, in fact, three distinct methods of dealing with the principles of the dynamo: (1) a physical method, dealing

[1] See J. M. Gaugain, *Annales de Chimie et de Physique*, 1873; Antoine Breguet, *Annales de Chimie et de Physique*, 1879; Du Moncel, *Exposé des Applications de l'Électricité*, vol. ii.; Niaudet, *Machines Électriques*; Dredge's *Electric Illumination*; Schellen, *Die Magneto- und Dynamo-elektrischen Maschinen* (3rd edition, 1883); Cunynghame, *The Law of Electric Lighting*, 1883.

with the lines of magnetic force and with currents, in which these quantities are made, without further inquiry into their why or how, the subjects of the arguments; (2) an algebraical method, founded upon the mathematical laws of electric induction and of theoretical mechanics; and (3) a graphical method, based upon the possibility of representing the action of a dynamo by a so-called "characteristic" curve, in the manner originally devised by Dr. Hopkinson, and subsequently developed by Frölich, Deprez, and others.

These three methods are really three aspects of the theory. The number of lines of magnetic force, with which we deal in the next chapter, may be expressed by a certain length of line geometrically, or by the symbol N algebraically, or they may be represented optically by a mere pictorial demonstration. What some people write N for, other people indicate by drawing a line of a certain length in a certain direction. We approximate, in fact, toward understanding the true theory by various processes : sometimes by algebra ; sometimes by geometry ; sometimes by magnetic diagrams ; and each of these processes is of value in its turn.

It will be our aim first to develop a general physical theory, applicable to all the varied types of dynamo-electric machines, and to trace it out into a number of corollaries bearing upon the construction of such machines. Having recited these consequences, which we shall deduce from theory, it will then remain to see how they are verified and embodied in the various forms assumed by the dynamo in practice. After that come chapters on the magnetic circuit and on the algebraic and geometrical methods of treating the theoretical parts of the subject. The last section of the book deals with the dynamo in its functions as a mechanical motor.

Before, however, proceeding to the physical theory of the dynamo, it will be expedient to introduce a few explanatory historical notes, and also to give some account of the system of electric and magnetic units universally adopted by electricians.

HISTORICAL NOTES.

Faraday's discovery of the magneto-electric induction of currents was made in the autumn of 1831, and communicated, on Nov. 24th, to the Royal Society in a paper printed in the *Philosophical Transactions*, and reprinted in the beginning of the first volume of Faraday's *Experimental Researches in Electricity*. His first experiments related to the production of induced currents in a coil by means of currents started or stopped in a neighbouring coil; from these he went on to currents generated in a coil moved in front of the poles of a powerful steel magnet. Upon thus obtaining electricity from magnets he attempted to construct "a new electrical machine." A disk of copper, 12 inches in diameter, and about one-fifth of an inch in thickness, fixed upon a brass axle, was mounted in frames, so as to allow of revolution, its edge being at the same time introduced between the magnetic poles of a large compound permanent magnet, the poles being about half an inch apart. The edge of the plate was well amalgamated, for the purpose of obtaining a good but movable contact, and a part round the axle was also prepared in a similar manner. Conducting strips of copper and lead, to serve as electric collectors, were prepared, so as to be placed in contact with the edge of the copper disk; one of these was held by hand to touch the edge of the disk between the magnet poles. The wires from a galvanometer were connected the one to the collecting strip, the other to the brass axle; then on revolving the disk a deflexion of the galvanometer was obtained, which was reversed in direction when the direction of the rotation was reversed. "Here, therefore, was demonstrated the production of a permanent current of electricity by ordinary magnets." These effects were also obtained from the poles of electro-magnets, and from copper helices without iron cores. Two other forms of magneto-electric machines were tried by Faraday.

Within a few months machines on the principle of magneto-induction had been devised by Dal Negro,[1] and by Pixii.[2] In the latter's apparatus a steel horseshoe magnet, with its poles upwards, was caused to rotate about a vertical shaft, inducing alternate currents in a pair of bobbins fixed above it, and provided with a horseshoe core of soft iron. Later, in 1832, Pixii produced, at the suggestion of Ampère,[3] a second machine, provided with commutators

[1] *Phil. Mag.* [3] i. 45, July 1832 (an oscillatory apparatus).
[2] *Ann. Chim. Phys.* l. 322, 1832.
[3] *Ann. Chim. Phys.*, li. 76, 1832.

to rectify the alternating currents. Further improvements were made by Ritchie[1] and Watkins.[2] In 1833 appeared the machine of Saxton,[3] and two years later that of Clarke;[4] both having the steel horseshoe magnet a fixture, and having as a revolving armature an electromagnet consisting of a pair of bobbins wound upon a simple horseshoe of iron. Clarke's machine possessed many original details, including a special form of commutator for giving short, sharp currents for physiological purposes. In it the armature rotated, not opposite the ends, but in close proximity to the flat faces of the magnet. In Saxton's machine, which was shown to the British Association at Cambridge in 1883, the armature was rotated opposite the polar ends, and consisted of four coils. Von Ettingshausen,[5] in 1837, brought out a very similar alternate-current machine, with a special device by which the alternate currents could be cut out. Poggendorff,[6] in 1838, devised a special mercury-cup commutator for Saxton's machine, to make the currents less discontinuous.

Other improvements in detail were made by Petrina,[7] who improved the commutator; Jacobi,[8] who pointed out the importance of using short cores for the armatures; Sturgeon,[9] who placed a shuttle-wound coil longitudinally between the limbs of a horseshoe magnet, and who also invented the simple two-part commutator or "unio-directive discharger," as he termed it; Stöhrer,[10] who showed how to construct a six-pole machine with six bobbins in the armature; Ritchie,[11] who employed tubular cores and a double winding. Woolrich,[12] in 1841, devised a multipolar machine for electroplating, having twice as many rotating coils as magnet-poles. Wheatstone [13] began his improvements in 1841, with a machine in which for the

---

[1] *Phil. Mag.* [3] viii. 455 ; [3] x. 280, 1837 ; and *Phil. Trans.*, 1833, ii. 318.
[2] *Phil. Mag.* [3] vii. 107, 1835.
[3] *Phil. Mag.* [3] ix. 360, 1836.
[4] *Phil. Mag.* [3] ix. 262, 1836; x. 365, 455, 1837 ; and Sturgeon's *Annals of Electricity*, i. 145.
[5] Gehler's *Physikalisches Wörterbuch*, ix. 122, 1838.
[6] *Pogg. Ann.*, xlv. 385, 1838.
[7] *Pogg. Ann.*, lxiv. 58, 1845.
[8] *Pogg. Ann.*, lxix. 194, 1846.
[9] *Annals of Electricity*, ii. 1, 1838.   See also Sturgeon's *Scientific Researches*, p. 252; also *Phil. Mag.*, vii. 231, 1835.
[10] *Pogg. Ann.*, lxi. 417, 1844 ; lxxvii, 467, 1849.
[11] *Loc. cit.*
[12] See also Specification of Patent, 9431 of 1842.
[13] Specification of Patent, 9022 of 1841.

first time the armature coils were so grouped as to give a really continuous current. For this purpose five armatures, each consisting of a pair of short parallel cylindrical coils with iron cores, and each having a simple split-tube commutator, were arranged in a row along a single shaft, with six compound steel magnets between them, the five armatures being so set that they came successively into the position of greatest activity, no two of them being commuted at the same instant. They were connected in series with one another by wires, which joined the positive brush—a brass spring—of one to the negative brush of the next. In 1845 Wheatstone[1] and Cooke patented the use of electromagnets instead of steel permanent magnets in such machines. In 1848 Jacob Brett[2] made the important suggestion of causing the current developed in the armature by the permanent magnetism of the field-magnets to be transmitted through a coil of wire surrounding the magnet, so as to increase its action. This suggestion, which appears to be the first suggestion of the principle of the self-exciting dynamo, was independently made in 1851 by Sinsteden,[3] who appears to have had full knowledge of the fact, investigated by Müller, that steel is capable of receiving a temporary magnetisation not greatly inferior to that of wrought iron, and far in excess of that which it can permanently retain. Sinsteden's researches were numerous and important, relating to the best width of polar surface to employ, to the use of pole-pieces, and to the lamination of armature cores, for which purpose he employed, in 1849, iron wire bundles. A quite different type of machine was suggested independently by Ritchie,[4] by Page,[5] and by Dujardin,[6] in which neither field-magnet nor armature rotated; the coils in which the currents were to be induced were wound upon polar extensions of the field-magnets, and the induction was produced by rotating in front of them pieces oi soft iron, which set up rapid periodic variations in the magnetic field. Machines on this same principle were later devised by Holmes, Henley, Wheatstone, Wilde, and Sawyer.

Nollet,[7] in 1849, devised an alternate-current machine, in the

---

[1] Specification of Patent, 10655 of 1845.
[2] Specification of Patent, 12054 of 1848.
[3] *Pogg. Ann.* lxxxiv. 186, 1851. For Sinsteden's other researches see *Pogg. Ann.*, lxxvi. 29, 195 and 524, 1849 ; lxxxiv. 181, 1852 ; xcii. 1 and 220, 1854 ; xcvi. 353, 1855 ; cxxxvii. 290 and 483, 1869.
[4] *Phil. Mag.* [3] x. 280, 1837.
[5] *Annals of Electricity*, 489, 1839.
[6] *Comptes Rendus*, xviii. 837, 1844 ; xxi. 528, 892, 1181, 1845.
[7] See Specification of Patent, 13302 of 1850. See also Douglass in *Proc. Inst. Civil Engin.*, lvii. 1878–9.

construction of which he was joined by Van Malderen ; and after the death of Nollet this was developed, with the aid, first of Holmes, then of Masson and Du Moncel, into the " Alliance "[1] machine, which, from the year 1863, did good service in the lighthouses of France. Holmes continued to perfect his work, and produced a fine machine,[2] which in 1857 received high commendation from Faraday. The large machine of Holmes shown in the International Exhibition of 1862, was a continuous-current machine, with a large commutator and rotating rollers for brushes ; the bobbins, 160 in number, were arranged on the peripheries of two wheels, each about 9 ft. in diameter. There were sixty horseshoe magnets arranged in three circles, each presenting radially forty poles. In 1867 Holmes remodelled his machine, making the field-magnets more powerful in proportion, and leaving the induced currents uncommuted, and in 1869 he introduced the principle of diverting the current from a few of the armature coils, through a commutator, to excite the field-magnets. This period was one of great activity. In 1848 Hjorth[3] patented a remarkable machine, having for its field-magnets a compound arrangement of a permanent magnet to provide initial currents, and powerful electromagnets to be excited up by the currents generated by the machine itself.

C. W. Siemens[4] in 1856 provisionally patented the famous shuttle-wound longitudinal armature, invented by Werner Siemens. In 1859,[5] he made the suggestion that the core only need rotate, the coils being fixed in grooves in the pole-pieces of the field-magnets. Wilde,[6] of Manchester, embarked on a remarkable series of researches from 1861 to 1867. Beginning with small apparatus for telegraphic purposes, he was led in 1863 to devise an apparatus having a shuttle-wound Siemens armature between the poles of a powerful electromagnet, the coils of which were traversed by currents furnished by a small auxiliary machine—with shuttle-wound armature and permanent magnets—mounted upon its summit. In 1866 and

---

[1] See Du Moncel's *Exposé des Applications de l'Électricité*, i. 361. Also see Le Roux, *Bulletin de la Société d'Encouragement*, 1868.

[2] See Douglass, *loc. cit.* Also Specifications of Patents, 573 of 1856, 2060 of 1868, and 1744 of 1869.

[3] Specification of Patent, 12,295 of 1848.

[4] Specification of Patent, 2017 of 1856. See W. Siemens, *Pogg. Ann.* ci. 271, 1857.

[5] Specification of Patent, 512 of 1859.

[6] Specifications of Patents, 299, 858, 1994 and 2997 of 1861 ; 516 and 3006 of 1863, 1412 and 2753 of 1865 ; 3209 of 1866 ; and 824 of 1867.

1867 Wilde devised alternate-current machines, of which the latest had a number of bobbins mounted on the periphery of a disk rotating between two opposite crowns of alternately polarised field-magnets—a type which survives to the present day. These machines, originally separately excited by currents from a small magneto machine, were made self-exciting, in 1873, by diverting through a commutator the currents induced in one or more of the armature bobbins. The principle of using the whole or part of the machine's own currents to excite the requisite magnetism of its field-magnets was by this time becoming recognised. As mentioned above, Brett, Sinsteden, and Hjorth had all made use of this principle. In 1858, Johnson,[1] patent-agent for a foreign inventor, states : " It is proposed to employ the electromagnet in obtaining induced electricity, which supplies wholly or partially the electricity necessary for polarising the electromagnets, which electricity would otherwise be required to be obtained from batteries or other known sources." In July, 1866, Murray[2] stated that he had connected in series with the armature some coils wound on the field-magnets of his magneto machine, and recommended the adoption of this plan. In November, 1866, Baker[3] stated that the secondary currents from the revolving magnets might be applied to magnetise the fixed magnets. In December of the same year C. and S. A. Varley[4] filed a provisional specification for a machine having electromagnets only, which apparatus, however, required before using to have a small amount of permanent magnetism given to it by passing an electric current through the coils of the electromagnets ; a device which reappears in another machine patented by the same inventors in June, 1867, and again in another by O. and F. H. Varley in 1869. The electro-magnets of the 1867 machine were wound with two separate circuits, supplied alternately with currents from two commutators which received the currents from two separate pairs of coils. Mr. S. A. Varley continued, in 1868 and 1871, to patent magneto-electric generators. In 1876 he returned to the self-exciting method, employing a multiple armature in which the principle was applied of cutting out each coil in succession during the rotation. In this machine also there were two windings on the field-magnets,

[1] Specification of Patent, 2670 of 1858.
[2] See *Engineer*, July 20, 1866, p. 42
[3] Specification of Patent, 3039 of 1866.
[4] Specification of Patent, 3394 of 1867. Other Varley Specifications are 1755 of 1867, 315 of 1868, 131 and 1150 of 1871, 4905 of 1876, 270 and 4435 of 1877, 4100 of 1878.

one of greater resistance than the other, the circuit of greater resistance being always closed. It is not, however, clear that this method of double winding was what is now understood as " compound winding."[1] Returning to the self-exciting principle, we find that on January 17th, 1867, Dr. Werner Siemens[2] described to the Berlin Academy a machine for generating electric currents by the application of mechanical power, the currents being induced in the coils of a rotating armature by the action of electromagnets, which were themselves excited by the currents so generated. To mark the importance of this departure Siemens coined the name *dynamo-electric machine*, which now, in the shortened form of *dynamo*, has become the familiar term for all these electric machines driven by mechanical power, whether self-excited or not. On the same day that this discovery was announced to the Royal Society, February 14th, 1867, a paper was read by Sir C. Wheatstone,[3] making an almost identical suggestion ; but with this difference, that whilst Siemens proposed that the exciting coils should be in the main circuit in series with the armature coils, Wheatstone proposed that they should be connected as a shunt. A self-exciting machine without permanent magnets had been constructed for Wheatstone by Mr. Stroh in the summer of 1866. In 1867 Ladd[4] exhibited a self-exciting machine having two shuttle-wound armatures, a small one to excite the common field-magnet, a large one to supply currents for electric light.

Meantime the question of procuring continuous currents, with less fluctuation in their strength, had come up, and had received from Pacinotti[5] an answer, which, though it fell into temporary oblivion, is now recognised as of great merit. He devised a machine, first described in 1864, having as its armature an electromagnet in the form of a ring, the core consisting of a toothed iron wheel, between the teeth of which the coils were wound in sixteen separate sections. He denominated this a " transversal electromagnet." The coils being joined up in a closed circuit, if at any point a current was introduced, it flowed both ways through the coils to some other point where it was taken off by a return wire. By the device of leading down connections, at sixteen different points around the ring, to sixteen insulated pieces of

---

[1] See *Phil. Mag.* [4] xlv. 439, 1873.
[2] *Berliner Berichte*, Jan. 1867 ; *Proc. Roy. Soc.*, Feb. 14, 1867. Specification of patent, 261 of 1867 ; and *Pogg. Ann.*, cxxx. 332, 1867.
[3] *Proc. Roy. Soc.*, Feb. 14, 1867.
[4] *Phil. Mag.* [4] xxxiii. 544, 1867.
[5] *Nuovo Cimento*, xix. 378, 1865.

metal arranged as a commutator, it was possible to cause magnetic poles to appear in the ring at any desired points. The principle of winding a continuous coil in separate symmetrical sections around a ring, or other figure of revolution, was independently invented, in 1870, by Gramme[1], whose ring had no teeth, and was entirely overwound with wire. By winding an armature with a number of such symmetrically grouped coils which pass successively through the magnetic field, currents can be obtained that are practically steady. The introduction of the Gramme armature was at once recognised as marking an important step, and it gave a fresh impetus to invention. In 1873 von Hefner Alteneck[2] modified the longitudinal armature of Siemens by covering it with windings spaced out at symmetrical angles to secure the same advantage of continuity, and Lontin[3] in 1874 sought to perform a like transformation upon an armature with radiating poles. Gramme and Siemens both devised many special forms of machines, some furnishing alternating currents[4], others continuous currents. Bertin in 1875, Brush in 1879, and Siemens,[5] in 1880, revived the method of shunt-winding.

In 1878 Pacinotti[6] devised a form of armature in which the conductors took the form of a flat disk or fly-wheel. Brush[7] also introduced his famous dynamo embodying the principle of open-coil working. He also introduced the simultaneous use of a shunt and a series winding for the purpose of enabling the machine to do either a large or a small amount of work. Another open-coil machine was introduced in 1880 by Elihu Thomson and Houston,[8] of Philadelphia. About the same time Weston[9] devised several forms of dynamo, and in particular developed shunt-wound machines. Many other American inventors produced dynamos, amongst them Edison,[10] who began in 1878, with a machine in which the motion was oscillatory instead of rotatory, a device which had

[1] *Comptes Rendus*, lxxiii. 175, 1871, and lxxv., 1497, 1872; and Specification of Patent, 1668 of 1870.
[2] Specification of Patent, 2006 of 1873.
[3] Specifications of Patent, 473 of 1875, 386 and 3264 of 1876.
[4] Specification of Patents, Gramme, 953 of 1878; Siemens, 3134 of 1878.
[5] *Phil. Trans.*, March 1880.
[6] *Nuovo Cimento* [3] i. 1881.
[7] Specification of Patent, 2003 of 1878.
[8] Specification of Patent, 315 of 1880.
[9] Specifications of Patent, 4280 of 1876, 1614 and 2194 of 1882.
[10] Specifications of Patents, 4226 of 1878, 2402 of 1879, 1240 and 2954 of 1881, and 2052 of 1882.

been tried by Dujardin[1], in 1856, by Siemens[2], in 1859, by Wilde,[3] in 1861, and abandoned. Edison himself abandoned it in 1879 for a form of machine having a modified Hefner-Alteneck armature and an elongated shunt-wound electro-magnet. In 1881, he produced a disk dynamo on the same lines as Pacinotti's disk. The same year saw a revival of alternate-current machines in the forms devised by Sir W. Thomson[4] (and independently by Ferranti) and Gordon.[5]

About this time multipolar dynamos began to come into favour, the multipolar drum armature introduced by Lord Elphinstone[6] and Mr. Vincent, and the multipolar ring, independently, by Schuckert, Gramme, Gülcher, and Mordey.[7] Lord Elphinstone in particular drew attention to the importance of perfecting the magnetic circuit, though, for purely mechanical reasons, his machine soon became obsolete. Hopkinson[8] showed how greatly the performance of a dynamo was improved by improving and making more compact its magnetic circuit, whilst Crompton,[9] amidst a number of improvements in detail, showed the advantage of increasing the cross-section of iron in the armature core. Meantime theoretical considerations had led Marcel Deprez,[10] in 1881, to the conclusion that a dynamo driven at a certain critical speed ought to be able to distribute currents at a constant potential if its field-magnets were provided with a second coil to furnish from a battery or other source an independent and constant auxiliary excitation. This was almost immediately followed by the general adoption of the so-called compound-winding, for the purpose of obtaining a self-regulating dynamo, this advance being the subject of conflicting rival claims. Since 1883 the chief progress made has been in details of design and mechanical construction.

The other branch of the subject, that of the electric motor, goes back to the discovery by Faraday[11] in 1821 of electromagnetic rotation, and the invention, in 1823, by Barlow,[12] of his rotating wheel.

---

[1] See Du Moncel's *Exposé des Applications*, i. p. 372.
[2] Specification of Patent, 512 of 1859.
[3] Specification of Patent, 924 of 1861.
[4] Specification of Patent, 5668 of 1881.
[5] Specifications of Patent, 5536 of 1881, and 2871 of 1882.
[6] Specifications of Patent, 332 of 1879, and 2893 of 1880.
[7] Specification of Patent, 400 of 1883.
[8] Specification of Patent, 973 of 1883.
[9] Specifications of Patent, 2618 and 4810 of 1882, and 4302 of 1884.
[10] *La Lumière Électrique*, December 3, 1881, and January 5, 1884.
[11] Journal of *Royal Institution*, September 1821.
[12] Barlow, *On Magnetic Attraction* (1823), p. 279, and *Encyclopædia Metropolitana* (1824) iv. Art. *Electromagnetism*, p. 36.

The earliest electric motors in which the principle of attraction by an electromagnet was applied were those of Henry,[1] in 1831, and of Dal Negro,[2] in 1832, and these were followed in 1833 and 1834 by the motors of Ritchie[3] and of Jacobi,[4] and in 1837 by that of Davenport.[5] Many other inventors devised machines of this kind, some of the most famous being Page[6] in the United States, Wheatstone[7] in England, Froment[8] in France, and Pacinotti[9] in Italy. The discovery that the action of a dynamo is the simple converse of that of the motor, and that the same machine can serve either function, appears to have been gradually made, and was more or less unconsciously recognised before its explicit announcement. It was certainly known in 1852, for in the fourth edition of Davis's *Magnetism*, published at Boston, an apparatus, described as a "revolving electromagnet" (a slight modification of Ritchie's motor) is shown, on page 212, as a motor, and the same apparatus is again shown on page 268 as a generator, accompanied by the remark that "any of the electro-magnetic instruments in which motion is produced by the mutual action between a galvanic current and a steel magnet may be made to afford a magneto-electric current by producing the motion mechanically." Walenn[10] explicitly stated the same point in 1860 ; and it was also stated by Pacinotti in 1864. The principle of transmitting power from one dynamo used as a generator to another used as motor is claimed for Fontaine and Gramme, as a discovery made in 1873, when such an arrangement was shown at Vienna. It has been noisily claimed, but without the shadow of reason, for Marcel Deprez,[11] who did not, however, discover it until 1881. In 1882 Ayrton and Perry made the important discovery of the automatic regulation of motors, to run with constant velocity, by methods akin to, but the converse of, those adapted for making dynamos self-

---

[1] *Silliman's Journal*, xx. 340, 1831. Also Henry, *Scientific Writings* (1886), i. 54.
[2] *Annali delle Scienze Lombardo-Veneto*, March 1834.
[3] *Phil. Trans.*, 318 [2], 1833.
[4] *L'Institut*, lxxxii. Dec. 1834.
[5] See *Annals of Electricity*, ii., 1838, *Encyclopædia Britannica* (Ed. VII.), Art. *Voltaic Electricity*, p. 687.
[6] *Silliman's Journal*, xxxiii., 1838, and [2] x. 344 and 473, 1850.
[7] Specification of Patent, 9022 of 1841.
[8] See *Cosmos*, x. 495, 1857, and *La Lumière Électrique*, ix. 193, June 1883.
[9] *Nuovo Cimento*, xix. 378, 1865.
[10] Specification of Patent, 2587 of 1860.
[11] Specification of Patent, 2830 of 1882. See *Journ. Soc. Telegr. Engineers*, xii. 301, 1883.

regulating. Since that date the improvements made, though great,
have been in mechanical perfection of design and detail.

The theory of the dynamo dates back to the investigations of
Weber[1] and of Neumann[2] respecting the general laws of magneto-
electric induction, followed by Jacobi's[3] calculations and experiments
respecting the performance of an electric motor, by Poggendorff's[4] and
Koosen's[5] investigations of the theory of the Saxton magneto-machine,
and by the researches of Lenz,[6] Joule,[7] Le Roux[8], and of Sinsteden.
These researches were followed at a long interval by those of Favre,[10]
followed by silence for twenty years, broken only by the pregnant,
but almost totally forgotten, little paper in which Clerk-Maxwell[11] laid
down a theory for self-exciting machines. On the revival of electric
lighting the theory of the dynamo was again studied, important contributions being made by Mascart,[12] Hagenbach,[13] von Waltenhofen,[14]
Hopkinson,[15] Herwig,[16] Meyer and Auerbach,[17] and Joubert.[18] The
latter founded the modern theory of alternate-current machines.
Hopkinson[19] devised the method of representing graphically by a
curve the relation between the current and the working electromotive force of the machine : such curves, under the name of
" characteristics," subsequently formed the basis of the theoretical
researches of Marcel Deprez.[20] In 1880 Frölich[21] began a series of

[1] Elektrodynamische Maasbestimmungen (1846).
[2] *Berliner Berichte*, 1845, p. 1, and 1847, p. 1.
[3] *Pogg. Ann.*, li. 370, 1840; lxix. 181, 1846; and *Krönig's Journal*, iii. 377,
1851. Also *Ann. Chim. Phys.* [3] xxxiv. 451, 1852.
[4] *Pogg. Ann.* xlv. 390, 1838.
[5] *Pogg. Ann.* lxxxv. 226, and lxxxvii. 386, 1852.
[6] *Pogg. Ann.* xxxi. 483, 1834, xxxiv. 385, 1835. and xcii. 128, 1854.
[7] *Annals of Electricity*, iv. v. 1839 40, *Phil. Mag.* [3] xxiii. 263, 347, and 435,
Ann. Chim. Phys. [3] l. 463, 1857.   [1843
[9] *Pogg. Ann.* lxxxiv. 181, 1851.
[10] *Comptes Rendus*, xxxvi. 342, 1853 ; xxxix. 1212, 1854 ; xlvi. 337, 658, 1858.
[11] *Proc. Roy. Soc.*, Mar. 14, 1867, and *Phil. Mag.* [4] xxxiii. [474], 1867.
[12] *Journal de Physique*, vi. 204, 297, 1877 ; and vii. 89, 1878.
[13] *Archives des Sciences Physiques*, lv. 255, March 1876, and *Pogg Ann.*,
clviii. 599, 1876.
[14] *Wiener Berichte*, lxxx. 601, 1879.
[15] *Proc. Inst. Mech. Engineers*, 238, 1879, and 266, 1880.
[16] *Wied. Ann.*, viii. 494, 1880.
[17] *Wied. Ann.*, viii. 494, 1879.
[18] *Ann. de l'Ecole Normale*, x. 131, 1881, and *Journal de Physique* [2] ii. 293,
[19] *Proc. Inst. Mech. Engineers*, 238, 1879.   [1883.
[20] *Comptes Rendus*, xcii. 1152, 1881, and *La Lumière Électrique*, xv. 1, 1885.
[21] *Berl. Berichte*, 962, 1880 ; *Elektrotechnische Zeitschrift*, ii., 134, 170, 1881 ;
and vi. 128, etc., 1885.

investigations both experimental and theoretical, that led to equations of remarkable simplicity, if not of more than approximate value, and in 1883 Clausius,[1] adopting Frölich's fundamental expression for the law of the electromagnet, evolved with great elaboration a theory in which all the various secondary effects arising in generators were taken into account—a theory which he later extended to the case of motors. In 1886 John and Edward Hopkinson[2] published a remarkable paper, developing, from theoretical considerations respecting the induction of magnetism in a magnetic circuit of given form and materials, a theory of the dynamo, the perfection of which may be judged by the fact that its use enables the performance of a machine to be predicted with extraordinary accuracy from the design as laid down in the working drawings. Other contributions to the theory of dynamos have been made by Sir W. Thomson[3] (windings to secure maximum efficiency), Kapp[4] (pre-determination of characteristic curve), Rücker[5] (limits of self-regulation), and others.

ELECTRIC AND MAGNETIC UNITS.

The principal units employed by practical electricians, by international agreement, are :—

The *ampère*, or unit of current (formerly called the weber).
The *volt*, or unit of electromotive-force.
The *ohm*, or unit of electric resistance.

These three practical units are based upon certain abstract units, derived by mathematical reasoning and experimentally proven laws, from the three fundamental units :—

The *centimetre*, as unit of length.
The *gramme*, as unit of mass.
The *second*, as unit of time.

The system of "absolute" units derived from these is often denominated the "C.G.S" system of units,[6] to distinguish it from other systems based on other fundamental units.

[1] *Wied. Ann.*, xx. 353, 1883; xxi. 385, 1884; *Phil. Mag.* (5) xvii. 49 and 119, 1884.
[2] *Phil. Trans.*, 1886, i. 331.
[3] *Journal de Physique* (2) ii. 240, 1887, and *Comptes Rendus*, xciii, 474, 1881.
[4] *Journ. Soc. Telegr. Engineers*, xv. 518, 1887.
[5] *Phil. Mag.* (5) xix., 462, June. 1885.
[6] The reader who may desire fuller information about the C.G.S. system of units is referred to Professor Everett's *Units and Physical Constants*.

Every system of measurement is based upon some experimental fact or law. We can only measure an electric current by the effects it produces. An electric current can (1) cause a deposition of metals from their chemical solutions; (2) heat the wire that it flows through; (3) attract (or repel) a parallel neighbouring current; (4) accumulate as an electric charge that can repel (or attract) a neighbouring charge of electricity; (5) produce in its neighbourhood a magnetic field, that is to say can exert a force upon the pole of a magnet placed near it, as, for example, in galvanometers. Now any one of these effects *might have been* chosen as a basis for a system of units of measurement, and all of them have been proposed by one authority or another. As a matter of fact, the fifth of them is made the basis in the system now adopted by international agreement; and it is the best, because, firstly, it connects the electrical units with the magnetic ones, and, secondly, it is closely connected with the mechanical units, enabling the mechanical values of the electrical quantities to be readily calculated.

Taking then the experimental fact that an electric current flowing in a wire, can exert a force upon the pole of a magnet placed near it, we have next to define the conditions with the utmost precision. It is found by experiment that the force which is exerted upon the magnet-pole by the current, depends on several other things beside the strength of the current: the force is proportional (*ceteris paribus*) (1) to the length of the conducting wire, (2) to the inverse square of the distance between an element of the wire and the pole, (3) to the strength of the magnet-pole. To be very precise then we ought to take (1) a wire one unit in length, (2) bent into an arc of unit radius so that each element of the wire is at unit distance from the pole, (3) and take a magnetic pole of one unit strength. If these things were done, and there was made to flow through the wire a current so strong that it acted on the pole with one unit of force, then a current of such a strength might be taken as a standard of comparison; for a current that was twice as strong would exert two units of force on the pole, and so-forth. But in order to be exact we have yet to define what is meant by "one unit of force" and "a magnet-pole of one unit strength." Here again we have to go to experimental facts, and choose such as will best suit for the purpose of making a consistent system of units.

A force must be measured by one of its effects, such for example as these: that it can (1) raise a given mass against the downward pull of the earth; (2) elongate a spring; (3) impart motion to a given mass, or in other words accelerate it. The first of these, which would seem the most natural to select, is rejected, because the downward pull of the earth is different at different places, the second because it would require awkward definitions of the elastic properties of springs. So the third is chosen; and, to make the definition precise, it must be remembered that experiment proves that the velocity of motion which a force imparts to a mass is proportional (1) to the force, (2) to the time during which it is applied, (3) inversely, to the mass acted upon. If, therefore, one could get such a force that, if it lasted one second and was made to act on one gramme, it imparted to that mass a velocity of one centimetre per second, then such a force ought to be called the unit of force. This unit has received the name of "one *dyne.*" It may be remarked that the downward pull of the earth on a mass of one gramme is sufficient to give to it at the end of one second a velocity of about 32 feet per second, or, more exactly, 981 centimetres per second (in the latitude of London) ; hence it is clear that the pull of the earth on one gramme (what is commonly called the gramme's weight) is equal (at London) to 981 dynes. The pull of the earth on a pound (commonly called the pound's weight) is 444,971 dynes (at London). (A pound at the Pole would weigh 445,879, and at the Equator only 443,611 dynes.) One dyne is a pull equal to 0·0157, or about $\frac{1}{63}$ of the weight of a grain (at London). Now, as to the unit strength of the magnet pole or unit of magnetism : a magnet pole can (1) lift a piece of iron ; (2) repel (or attract) another magnet pole at a distance. The first of these two effects is rejected as a basis for a definition of a unit because the load that a magnet pole will lift does not depend only on the amount of magnetism at the pole, but on the shape and quality of the piece of iron lifted. For precise definition of the second effect, upon which the definition is based, it must be remembered that experiment has shown that the repulsion of one magnetpole by another is proportional (1) to the product of the strengths of the two poles, (2) inversely to the square of the distance between them. If, therefore, we choose two similar and equal poles of just such a strength that when placed at unit distance apart they repel each other with unit force, then such poles will possess that amount of magnetism that ought to be called the unit quantity of magnetism.

We may now retrace our steps and build up systematically the units of the C.G.S. system.

The absolute unit of force (" dyne ") is that force which, if it acts on one gramme for one second, gives to it a velocity of one centimetre per second.

c

The unit of magnetism, or unit magnet pole, is one of such a strength that when placed at a distance of one centimetre (in air) from a similar pole of equal strength it repels it with a force of one dyne.

The absolute unit of current is one of such a strength that when one centimetre length of its circuit is bent into an arc of one centimetre radius, the current in it exerts a force of one dyne on a unit magnet pole placed at the centre.

The last definition is difficult to realise in practice, and a complete circle of one centimetre radius is more easy to work with than an arc one centimetre long only. If the radius be more than one centimetre and there be more than one turn of wire, as in most tangent galvanators, then a formula is necessary. Writing r for the number of centimetres of the radius, the length of circumference will be equal to $2\pi r$. Then writing S for the number of turns of wire in the coil, and i for the strength in absolute units of the current, the formula connecting these with the force (in dynes) exerted by the current on a unit pole at the centre is :—

$$\frac{2\pi r\, S\, i}{r^2} = f;$$

whence

$$\frac{2\pi S\, i}{r} = f.$$

In the case of the tangent galvanometer, the force, instead of being measured directly, is ascertained indirectly, by knowing the value (at the place of observation) of the horizontal component of the magnetic field due to the earth's magnetism, commonly represented by symbol H, and measuring the tangent of the deflexion produced on a magnetic needle hung at the centre when the coil lies parallel to the magnetic meridian. In this case $f = $ H $\times \tan \delta$; whence

$$\frac{2\pi S\, i}{r} = \text{H} \tan \delta.$$

From this it follows that if S, r, H, and the tangent of deflexion are known, the strength of the current i will be reckoned by making the following calculation :—

$$i = \frac{r\,\text{H}}{2\pi S} \tan \delta.$$

(The value of H may be taken as o 18 at London, and of the

following values at other places: Glasgow ·17, Boston ·17, Montreal ·147, Niagara ·167, Halifax, N.S., ·159, New York, Cleveland, and Chicago, ·184, Philadelphia, ·194, Washington ·20, Berlin ·178, Paris ·188, Rome ·24, San Francisco ·255, New Orleans ·28, Mexico ·31, Bombay ·33.)

Now, the current that is so strong as to fulfil the above definition is far stronger than anything used in telegraphic work, being about as great in quantity as the current in an arc-light circuit. Accordingly *the practical unit* of current is fixed at one-tenth part of the absolute unit; and it is called " one *ampère.*" It follows that the above equation, when *i* is to be given in ampères, must be altered to

$$ i = \frac{10\,r\,H}{2\,\pi\,S} \tan \delta. $$

It follows that a simple tangent galvanometer to read as an *ampère-meter* can be made as follows. Take a piece of insulated copper wire, of a gauge not less than No 10 B.W.G., or, say, than three millimetres in diameter, and of this wire wind five turns only, so as to have a mean radius exactly as below ; then such a coil will, when traversed by 1 ampère, deflect the needle exactly to 45°, that is to the angle whose natural tangent is = 1, and the natural tangents of the deflexions will therefore read ampères directly. The radius has to be inversely proportional to the intensity of the horizontal component of the earth's magnetic force at the place where the ampèremeter is to be used. For use at London, where H is o·18, the radius of the coils must be 17·45 centimetres or 6⅞ inches. Other values are stated below.

Place.	Horizontal Component of Magnetic Intensity.	Radius of Coil.	
		Centimetres.	Inches.
Montreal	·147	21·37	8·41
Halifax, N.S.	·159	19·75	7·76
Glasgow and Boston	·170	18·50	7·28
Berlin	·178	17·05	6·95
New York, Cleveland, and Chicago	·184	17·07	6·72
Paris	·188	16·76	6·60
Philadelphia	·194	16·19	6·37
Washington	·200	15·70	6·18
San Francisco	·255	12·32	4·85
New Orleans	·280	11·22	4·42
Bombay	·330	9·52	3·75

It may further be noted that the current of one ampère strength will cause the deposition in 1 hour of 1·174 grammes, or 18·116 grains of copper in a copper electrolytic cell. It will in 1 hour deposit 4 024 grammes, or 60·52 grains, of silver in a silver cell.

The other electrical units also require definition. The *electro-motive-force* of a battery, or of a dynamo, is only another name for the power which it possesses to drive electricity through a circuit. (Formerly the electromotive-force of a battery was called its "intensity" as a distinction from the "quantity" of current it would furnish.) It is also sometimes called the electric "pressure." As a basis for a unit of electromotive-force any one of the following experimental facts might have been selected. The electromotive-force is proportional, (1) to the current that it sets up in a circuit of given resistance ; (2) to the quantity of electricity that it will force as a charge into a condenser of given capacity ; (3) to the number of lines of magnetic force cut per second by a conductor moving in a magnetic field. The first of these would do if the unit of resistance were given, but it is more convenient to make this fact the basis of definition of that unit rather than of the unit of electromotive-force : the second is useful for defining the unit of capacity : the third is selected for defining the unit of electromotive-force, and is extremely appropriate for the purpose, as it is the very principle of the dynamo machine. Clearly that electromotive-force ought to be reckoned as of unit value which is produced by the motion of a conductor cutting across one line of magnetic force in one second. But this involves the preliminary definition of the unit line of magnetic force. This is as follows. The so-called magnetic lines of force represent by their direction, the direction of the resultant magnetic force in the space through which they pass : the space traversed by magnetic forces, and lines of force, being called a magnetic "field." To make the number of magnetic lines represent *numerically*, as well as in mere direction, the intensity of the magnetic forces, the following device is adopted. Remembering that experiment shows that the pull (or push) which a magnetic pole experiences when placed in a magnetic field, is proportional to the intensity of that field, let there be drawn as many lines to the square centimetre as there are dynes of force exerted on the unit pole. For example, if at any point it was found that the magnetic pull on a unit pole was 40 dynes, then at that place we should draw, or imagine to be drawn, 40 magnetic lines all packed within one square centimetre of sectional area. As the earth's horizontal component at London is only 0·18 (dynes on the unit pole) it follows that there would be only 18 lines passing through an area of 100 square centimetres set up vertically east and west. Returning to the definition of electromotive-force, we see that if the moving conductor cuts one magnetic line in one second, the electromotive-force engendered will be of unit value in this absolute C.G.S. system of measurement. But such a unit would be ridiculously small—far too small for practical use. Measured in such units the electromotive-force of a single Daniell's cell would be represented by the enormous number of 110,000,000, and a Latimer-Clark standard cell by 143,400,000 units. Hence practical electricians adopt, as their working unit, an electromotive-force one hundred million times as great as the absolute C.G.S. unit; and they call the practical unit " one *volt.*" Hence the definition of " one volt " is that electromotive-force which would be generated by a conductor cutting across a hundred million (10^8) magnetic lines per second. The electro-motive force of a Daniell's cell is about 1·1 volts; that of Clark's standard cell 1·434 volts. The appropriate instrument for measuring volts is called a *volt-meter.*

We come then to the unit of electrical *resistance.* It is found by experiment that the current which is produced in a circuit by applying a given electromotive-force depends on the resistance offered by the circuit to the flow of electricity, the current being less as the resistance is greater, in accordance with the famous law discovered by Dr. Ohm.

Ohm's law in fact states that the current is directly proportional to the electromotive-force that is exerted in, and is inversely proportional to the resistance of, the circuit. If the symbol E stands for the number of units of electromotive-force, and R for the number of units of resistance of the circuit, and i for the current that results, then Ohm's law will be written :—

$$\frac{E}{R} = i$$

or, the resulting current can be calculated by dividing the number of units of electromotive-force by the number of units of resistance. Another way of writing Ohm's law, which is useful when it is desired to calculate the electromotive-force that will drive any prescribed current through a given resistance, is

$$E = Ri.$$

Now suppose we had an electromotive-force equal to one absolute C.G.S. unit, and we required to produce by its means a current of unit strength as previously defined in the absolute system, it would be requisite to adjust the resistance of the circuit to a definite value; and that value would be extremely small, otherwise such a minute electromotive-force could not maintain so large a current. Nevertheless this very minute resistance would be rightly taken as the unit in the absolute C.G.S. system, for then Ohm's law would be numerically fulfilled as,

$$\frac{\text{one unit of electromotive force}}{\text{one unit of resistance}} = \text{one unit of current.}$$

But as there are already practical units of electromotive-force and of current, so there is required a practical unit of resistance to correspond. And reflection will show that the practical unit must be a thousand million times as great as the absolute unit. For then, again, Ohm's law will be fulfilled as

$$\frac{\text{one hundred million C.G.S. units of electromotive-force}}{\text{one thousand million C.G.S. units of resistance}}$$
$$= \text{one-tenth C.G.S. unit of current.}$$

The name of "one *ohm*" is given to this practical unit of resistance; and many researches have been made to determine its working value. The British Association Committee produced standard wire coils, which were long accepted as being exact *ohms*, but they are now known to be all slightly too low in resistance. In 1882 the International Congress fixed the *legal value of the ohm* as being *a resistance equal to that of a column of mercury one square millimetre in cross-section, and* 106 *centimetres long* (measured at the freezing point of water). According to Lord Rayleigh's most careful measurements the true ohm ought to be, not 106, but 106·2 centimetres long.

The resistances of wires and circuits are measured in practice by comparing them with certain standard "resistance coils," sets of which are often employed arranged in "resistance boxes;" the particular instruments employed in making the comparison being of two kinds, namely, the differential galvanometers and the Wheatstone's bridge. For further information the reader must refer to the text-books on electric testing.

A rough and ready idea of the resistance called "one ohm," may be obtained by remembering that a mile of ordinary iron telegraph line offers from 13½ to 20 ohms resistance.

One other unit is required by electricians, namely, a unit of *power*, in which to express the quantity of work per second done in any electrical system.

To measure the work done by a current in a wire, or in a lamp, or other thing supplied with electric power, we must measure both the *ampères* of current that are running through it, and the *volts* of electromotive-force that are actually applied at that part of the circuit, and having found the two numbers we must multiply them together. For just as engineers express power mechanically as the number of "foot-pounds" expended in a given time, so the electrician expresses electric power as the number of "volt-ampères." The more convenient name of "one *watt*" is given to the unit of electric power. Calculation shows that one "watt" or "volt-ampère" is equal to one seven-hundred-and-forty-sixth part of a horse-power.

As an example of calculation of electric power the following may be taken. It was required to ascertain the power expended in maintaining a certain arc lamp. The voltmeter showed an electric pressure of 57 volts between the terminals of the lamp, and the ampère-meter showed a current of 10·5 ampères running through it. The product is 598·5 watts. Dividing by 746 to bring to horse-power, we get 0·80, or eight-tenths of a horse-power. In the same way the output of a dynamo may be measured and worked out. For example, suppose a dynamo to drive a current of 10 ampères through a circuit of arc lamps by exerting a potential or available electromotive-force of 800 volts at its terminals, the activity or output of electric energy per second will be 8000 *watts*. Under the provisions of the Electric Lighting Act, the Board of Trade has issued regulations in which 1000 watts of output are termed "one unit." The output of the above-mentioned machine would therefore be briefly expressed as 8 Board-of-Trade units, or more simply as 8 units. (Since 746 watts = one electric horse-power, it is clear that one Board-of-Trade-unit = 1·34 electric horse-power.) Now, for every dynamo there is a certain limit of output, determined by the mechanical and electrical conditions of its design and construction. It must not give a current of more than a certain strength, or it will become overheated. It must not be run too fast, or its bearings will heat. The maximum output at which a machine is able to be worked safely and continually is termed its capacity. If, in the instance given above, 8000 watts were the safe working maximum output of the machine, it would be known as an 8-*unit machine.* A machine capable of giving 100 ampères at a potential of 80 volts (for incandescent lighting), or one capable of giving 2000 ampères at 4 volts (for electrolysis), would equally be called an 8-unit machine. Taking incandescent lamps as they are now manufactured, a 20-candle lamp requires approximately 50 watts of output. A "one-unit" dynamo will therefore supply about 20 lamps of 20 candle-power each. Again, taking arc lamps of approximately 2000 candle power as requiring 10 ampères of current at 50 volts of potential, or 500 watts per arc, it will be seen that a "one-unit" dynamo will supply about two arc lamps.

It may be added that, for a given type of dynamo, the capacity is very nearly proportional to the weight. But an enormous difference is to be found between the relative capacities of dynamos of equal weight but of different types. A somewhat fairer basis of comparison between different types of machine than is afforded by a mere statement of their capacity, is the figure of merit obtained by dividing their number of watts of output by the number of revolutions per minute at which they are to be driven. Great strides have been made since 1882 in improving the capacity of machines, particularly with regard to the amount of copper employed. Some data as to the number of watts per pound weight of copper in the armatures of different machines will be found in the succeeding chapters and in Appendix XI.

CHAPTER II.

PHYSICAL THEORY OF DYNAMO-ELECTRIC MACHINES.

ALL dynamos are based upon the discovery made by Faraday in 1831, that electric currents are generated in conductors by moving them in a magnetic field. Faraday's principle may be enunciated as follows:—When a conductor is moved in a field of magnetic force in any way so as to cut the lines of force, there is an electromotive-force produced in the conductor, in a direction at right angles to the direction of the motion, and at right angles also to the direction of the lines of force, and to the right of the lines of force, as viewed from the point from which the motion originates.[1]

This induced electromotive-force is, as Faraday showed, proportional to the number of lines of magnetic force cut per second ; and is, therefore, proportional to the intensity of the magnetic " field," and to the length and velocity of the moving conductor. For steady currents, the flow of electricity in the conductor is, by Ohm's well-known law, directly proportional to this electromotive-force, and inversely proportional to the resistance of the conductor. For sudden currents, or currents whose strength is varying rapidly, this is no longer true. And it is one of the most important matters, though one too often overlooked in the construction of dynamo-electric machinery, that the " resistance " of a coil of wire, or of a circuit, is by no means the only obstacle offered to the generation of a momen-

---

[1] A more usual rule for remembering the direction of the induced currents is the following adaptation from Ampère's well-known rule :—Supposing a figure swimming in any conductor to turn so as to look along the (positive direction of the) lines of force. Then, if he and the conductor be moved towards his right-hand, he will be swimming with the current induced by this motion.

tary current in that coil or circuit ; but that, on the contrary, the " self-induction " exercised by one part of a coil or circuit upon another part or parts of the same, is a consideration, in many cases quite as important as, and in some cases more important than, the resistance.

To understand clearly Faraday's principle—that is to say how it is that the act of moving a wire so as to cut magnetic lines of force can generate a current of electricity in that wire —let us inquire what a current of electricity is.

A wire through which a current of electricity is flowing looks in no way different from any other wire. No man has ever yet seen the electricity running along in a wire, or knows precisely what is happening there. Indeed, it is still a disputed point which way the electricity flows, or whether or not there are two currents flowing simultaneously in opposite directions. Until we know with absolute certainty what electricity is, we cannot expect to know precisely what a current of electricity is. But no electrician is in any doubt as to one most vital matter, namely, that when an electric current flows through a wire, the magnetic forces with which that wire is thereby, for the time, endowed, reside not in the wire at all, but in the space surrounding it. Every one knows that the space or "field" surrounding a magnet is full of magnetic "lines of force," and that these lines run in tufts (Fig. 1[1]) from

FIG. 1.[1]

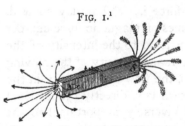

LINES OF FORCE OF BAR-MAGNET.

the N-pointing pole to the S-pointing pole of the magnet, invisible until, by dusting iron filings into the field, their presence is made known, though they are always in reality there (Fig. 2). A view of the magnetic field at the pole of a

[1] For the use of Figs. 1, 2, 3, 4, and 5, I am indebted to the kind courtesy of the editors and publishers of *Engineering*, who permit me to reproduce them from the volume, *Electric Illumination*, recently issued under the editorship of Mr. James Dredge, C.E.

bar-magnet, as seen end-on, would of course exhibit merely
radial lines, as in Fig. 3.

FIG. 2.

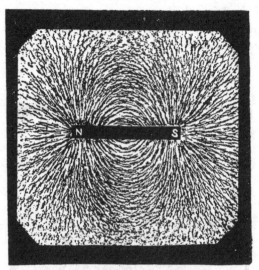

MAGNETIC FIELD OF BAR-MAGNET.

FIG. 3.

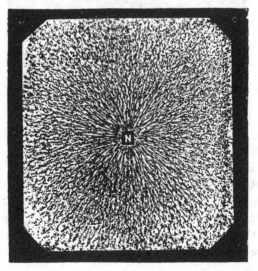

MAGNETIC FIELD ROUND ONE POLE, END-ON.

Now, every electric current (so-called) is surrounded by a magnetic field, the lines of which can be similarly revealed. To observe them, a hole is bored through a card or a piece of glass, and the wire which carries the current must be passed up through the hole. When iron filings are dusted into the field they assume the form of concentric circles (Fig. 4),

FIG. 4.

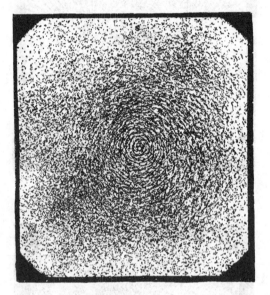

MAGNETIC FIELD SURROUNDING CURRENT. THE CONDUCTING WIRE SEEN END-ON.

showing that the lines of force run completely round the wire, and do not stand out in tufts. In fact, every conducting wire is surrounded by a sort of magnetic whirl, like that shown in Fig. 5. A great part of the energy of the so-called electric current in the wire consists in these external magnetic whirls. To set them up requires an expenditure of energy; and to maintain them requires also a constant expenditure of energy. It is these magnetic whirls which act on magnets, and cause them to set, as galvanometer needles do, at right angles to the conducting wire.

Now, Faraday's principle is nothing more or less than this :—That by moving a wire near a magnet, across a space in which there are magnetic lines, the motion of the wire, as it cuts across those magnetic lines, sets up magnetic whirls round the moving wire, or, in other language, generates a so-called current of electricity in that wire. Poking a magnet pole into a loop or circuit of wire also necessarily generates a momentary current in the wire loop, because it momentarily sets up magnetic whirls. In Faraday's language, this action increases the number of magnetic lines of force intercepted by the circuit.

It is, however, necessary that the moving conductor should, in its motion, so cut the lines of force as to alter the number of lines of force that pass through the circuit of which the moving conductor forms part. If a conducting circuit—a wire ring or single coil, for example—be moved along in a uniform magnetic field, as indicated in Fig. 6, so that only the same lines of force pass through it, no current will be generated. Or, if again, as in Fig. 7, the coil be moved by a motion of translation to another part of the

FIG. 5.

MAGNETIC WHIRL
SURROUNDING
WIRE CARRYING
CURRENT.

uniform field, as many lines of force will be left behind as are gained in advancing from its first to its second position, and there will be no current generated in the coil. If the coil be merely rotated on itself round a central axis, like the rim of a fly-wheel, it will not cut any more lines of force than before, and this motion will generate no current. But if, as in Fig. 8, the coil be tilted in its motion across the uniform field, or rotated round any axis in its own plane, then the number of lines of force that traverse it will be altered, and currents will be generated. These currents will flow round the ring coil in the positive[1] sense (as viewed from the point toward which the lines of force run), if the effect of the movement is to diminish

FIG. 6.

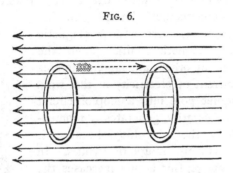

CIRCUIT MOVED WITHOUT CUTTING LINES OF FORCE OF UNIFORM MAGNETIC FIELD.

the number of lines of force that cross the coil ; they will flow round in the opposite sense, if the effect of the movement is to increase the number of intercepted lines of force.

FIG. 7.

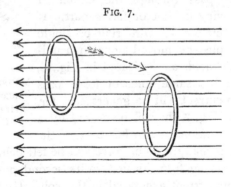

CIRCUIT MOVED WITHOUT CUTTING ANY MORE LINES OF FORCE.

If the field of force be not a uniform one, then the effect of taking the coil by a simple motion of translation from a place where the lines of force are dense to a place where they are

---

[1] The positive sense of motion round a circle is here taken as opposite to the sense in which the hands of a clock go round.

less dense, as from position 1 to position 2 in Fig. 9, will be to generate currents. Or, if the motion be to a place where

FIG. 8.

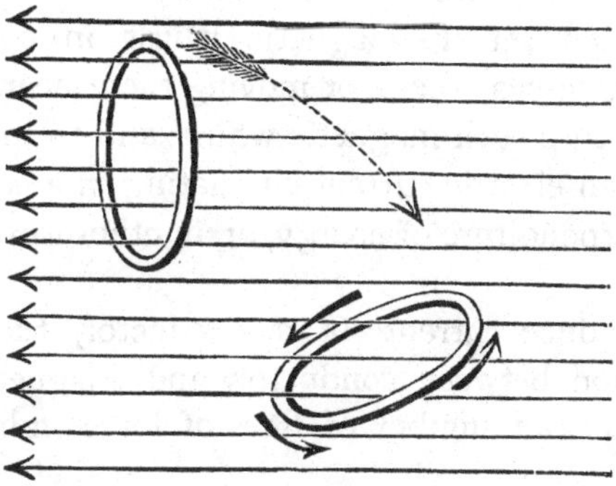

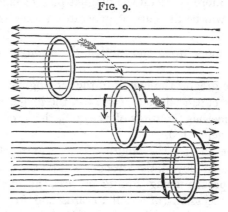

CIRCUIT MOVED SO AS TO ALTER NUMBER OF LINES OF FORCE THROUGH IT.

the lines of force run in the reverse direction, the effect will be the same, but even more powerful.

FIG. 9.

MOTION OF CIRCUIT IN NON-UNIFORM MAGNETIC FIELD.

We may now summarise the points under consideration and some of their immediate consequences, in the following manner:—

(1.) A part, at least, of the energy of an electric current

exists in the form of magnetic whirls in the space surrounding the conductor.

(2.) Currents can be generated in conductors by setting up magnetic whirls round them.

(3.) We can set up magnetic whirls in conductors by moving magnets near them, or moving them near magnets.

(4.) To set up such magnetic whirls, and to maintain them by means of an electric current circulating in a coil, requires a continuous expenditure of energy, or, in other words, consumes power.

(5.) To induce currents in a conductor, there must be relative motion between conductor and magnet, of such a kind as to alter the number of lines of force embraced in the circuit.

(6.) Increase in the number of lines of force embraced by the circuit produces a current in the opposite sense to decrease.

(7.) Approach induces an electromotive-force in the opposite direction to that induced by recession.

(8.) The more powerful the magnet-pole or magnetic field the stronger will be the current generated (other things being equal).

(9.) The more rapid the motion, the stronger will be the currents.

(10.) The greater the length of the moving conductor thus employed in cutting lines of force (*i.e.* the longer the bars, or the more numerous the turns of the coil), the stronger will be the currents generated.

(11.) The shorter the length of those parts of the conductor not so employed, the stronger will be the current.

(12.) Approach being a finite process, the method of approach and recession (of a coil towards and from a magnet pole) must necessarily yield currents alternating in direction.

(13.) By using a suitable commutator, all the currents, direct or inverse, produced during recession or approach, can be turned into the same direction in the wire that goes to supply currents to the external circuits, thereby yielding an almost uniform current.

(14.) In a circuit where the flow of currents is steady[1] it makes no difference what kind of magnets are used to procure the requisite magnetic field, whether permanent steel magnets or electromagnets, self-excited or otherwise.

(15.) Hence the current of the generator may be itself utilised to excite the magnetism of the field-magnets, by being caused, wholly or partially, to flow round the field-magnet coils.

A very large number of dynamo-electric machines have been constructed upon the foregoing principles. The variety is indeed so great, that classification is not altogether easy. Some have attempted to classify dynamos according to certain constructional points, such as whether the machine did or did not contain iron in its moving parts (which is a mere accident of manufacture, since almost all dynamos will work, though not equally well, either with or without iron in their armatures); or whether the currents generated were direct and continuous, or alternating (which is in many cases a mere question of arrangement of parts of the commutator or collectors); or what was the form of the rotating armature (which is, again, a matter of choice in construction, rather than of fundamental principle). A classification which I have adopted, is one which I have found more satisfactory and fundamental than any other. I distinguish three genera or main classes of dynamos.

Class I.—Dynamos in which there is rotation of a coil or coils in a uniform[2] field of force, such rotation being effected (as in the manner indicated in Fig. 8, p. 31), round an axis in the plane of the coil, or one parallel to such an axis.

---

[1] For currents that are not steady, there are other considerations to be taken into account, as will be shown hereafter.

[2] Or approximately uniform. A Gramme ring, or a Siemens drum armature, will work in a by no means uniform field, but is adapted to work in a field in which the lines of force run uniformly from one side to the other. But in such a field, a multipolar armature of many coils, such as that of Wilde, or such as is used in the Gramme alternate-current, or in the Siemens alternate-current machine, is useless and out of place. Indeed, the classification almost amounts to saying that in machines of Class I. there is one field of force, while in machines of Class II. there are many fields of force, or the whole field of force is complex.

D

EXAMPLES.—Gramme, Siemens (Alteneck), Edison, Lontin, Bürgin, Fein, Schuckert, Brush, Thomson-Houston, Crompton, &c.

Class II.—Dynamos in which there is translation [1] of coils to different parts of a complex field of varying strength, or of opposite sign. Most, but by no means all, of the machines of this class furnish alternate currents.

EXAMPLES.—Pixii, Clarke, Holmes, Niaudet, Wallace-Farmer, Wilde (alternate), Siemens (alternate), Hopkinson and Muirhead, Thomson-Ferranti (alternate), Gordon (alternate), Mechwart-Zipernowsky (alternate), Ayrton and Perry (Oblique-coiled Dynamo), Edison (Disk Dynamo), Mordey (alternate), &c.

Class III.—Dynamos having a conductor rotating so as to produce a continuous increase in the number of lines of force cut, by the device of sliding one part of the conductor on or round the magnet, or on some other part of the circuit.

EXAMPLES.—Faraday's Disc-machine, Siemens' " Unipolar " Dynamo, Forbes' " Non-polar " Dynamo.

There are a few nondescript machines, however, which do not fall exactly within any of these classes ; [2] one of these is the extraordinary tentative dynamo of Edison, in which the coils are waved to and fro at the ends of a gigantic tuning-fork, instead of being rotated on a spindle. In a few machines no conducting wires revolve, the only moving parts being of iron, revolving so as to vary the magnetic field.

---

[1] The motion by which the individual coils are carried round on such armatures as those of Niaudet, Wallace-Farmer, Siemens (alternate), &c., is, of course, not a pure translation. It may be regarded, however, as a combination of a motion of translation of the coil round the circumference of a circle, with a rotation of the coil round its own axis, which, as we have seen above, has *no* electrical effect. It is, of course, the translation of the coil to different parts of the field which is the effective motion.

[2] There are a few dynamos, including the four-pole Schuckert, the four-pole Gülcher, and the four-pole Elwell-Parker, which, though really belonging to the first class, are not named above, because they are, in reality, multiple machines. The Gülcher with its double field-magnets and four collecting brushes, is really a double machine, though it has but one rotating ring. The same is the case with an octagonal-pattern Gramme which has four brushes. The Schuckert-Mordey, or "Victoria " dynamo, though it has only two brushes, belongs to the same category. The Elphinstone-Vincent machine, a remarkable one in many respects, was a triple machine, having six brushes, or, in the later machines, two brushes only ; and might, indeed, be used as three machines, to feed three separate circuits.

Suppose, then, it were determined to construct a dynamo upon any one of these plans—say the first—a very slight acquaintance with Faraday's principle and its corollaries would suggest that, to obtain powerful electric currents, the machine must be constructed upon the following guiding lines:—

(*a.*) The field magnets should be as strong as possible, and their poles as near to the armature as possible.

(*b.*) The armature should have the greatest possible length of wire upon its coils.

(*c.*) The wire of the armature coils should be as thick as possible, so as to offer little resistance.

(*d.*) A very powerful steam-engine should be used to turn the armature, because,

(*e.*) The speed of rotation should be as great as possible.

Unfortunately, it is impossible to realise all these conditions at once, as they are incompatible with one another; and, moreover, there are a great many additional conditions to be observed in the construction of a successful dynamo. We will deal with the various matters in order, beginning with the various organs or parts of the machine. Having discussed these, we take up the nature of the processes that go on in the machine when it is at work, the action of the magnetic field on the rotating armature, the reactions of the armature upon the field in which it rotates, and the various methods of exciting and governing the magnetism of the field-magnets. After that we shall be in a position to enter upon the various actual types of machines for generating continuous and alternating currents.

CHAPTER III.

ORGANS OF DYNAMO-ELECTRIC MACHINES.

To make more clear the considerations which will occupy us when discussing individual types of dynamo, we will first examine some fundamental points in the general mechanism and design of dynamo machines ; and in particular those of Class I., which includes the great majority of the actual machines in use.   This will lead directly to a closer scrutiny of the construction and design of the various organs composing the dynamo and essential to its action.

Ideal Simple Dynamo.

The simplest conceivable dynamo is that sketched in Fig. 10, consisting of a single rectangular loop of wire rotating in a

FIG. 10.

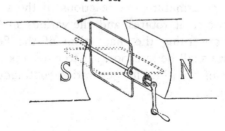

IDEAL SIMPLE DYNAMO.

simple and uniform magnetic field between the poles of a large magnet.   If the loop be placed at first in the vertical plane, the number of lines that pass through it from right to left will be a maximum, and as it is turned into the horizontal position the number diminishes to zero ; but on continuing the rotation the lines begin again to thread through the loop from the opposite side, so that there is a negative maximum when the loop has been turned through 180°. During the half-revolution, therefore, currents will have been induced in the loop, and these currents will be in the direction from back to front in the part of the loop which is rising on the left, and in the opposite direction, namely from front to back, in that part which is descending on the right. On passing the 180° position, there will begin an induction in the reverse sense, for now the number of negative lines of force is diminishing, which is equivalent to a positive increase in the number of lines of force ; and this increase would go on until the loop reached its original position, having made one complete turn. To commute these alternately-directed currents into one direction in the external circuit, there must be applied a commutator consisting of a metal tube slit into two parts, and mounted on a cylinder of hard wood or other

FIG. 11.

suitable insulating material ; each half being connected to one end of the loop, as indicated in Fig. 10. Against this commutator press a couple of metallic springs or " brushes " (Fig. 11), which lead away the currents to the main circuit. It is obvious that if the brushes are so set that the one part of the split tube slides out of contact, and the other part slides into contact with the brush, at the

TWO-PART COM-
MUTATOR OR
COLLECTOR.

moment when the loop passes through the positions when the induction reverses itself, the alternate currents induced in the loop will be " commuted " into one direction through the circuit. We should expect therefore the brushes to be set so that the commutation shall take place exactly as the loop passes through the vertical position. In practice, however, it is found that a slight forward lead must be given to the brushes, for reasons which will presently appear. In Fig. 12 are shown the brushes B B', displaced so as to touch the commutator not exactly at the highest and lowest points, but at points displaced in the direction of the line D D, which is called the "diameter of commutation." The argument is in nowise

changed if for the single ideal loop we substitute the simple
rectangular coil represented in Fig. 13, consisting of many
turns of wire in each of which a simultaneous inductive action

FIG. 12.   FIG. 13.

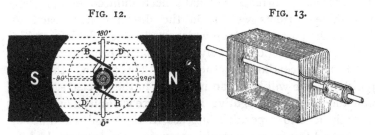

SIMPLE LOOP IN SIMPLE FIELD.   SIMPLE RECTANGULAR COIL.

is going on, making the total induced electromotive-force proportionally greater. This form, with the addition of an iron
core is, indeed, the
form given to ar-
matures in 1856
by Siemens, whose
shuttle - wound ar-

FIG. 15.

FIG. 14.

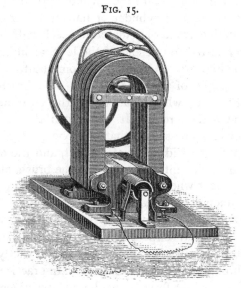

SECTION OF OLD SHUT-
TLE-WOUND SIEMENS
ARMATURE.

mature is repre-
sented in section in
Fig. 14. A small
magneto - electric
machine of the old   OLD SIEMENS MACHINE, WITH SHUTTLE-
pattern, having the   WOUND ARMATURE AND PERMANENT MAGNETS

shuttle-wound armature, is shown in Fig. 15. Though this
form has now for many years been abandoned, save for small
motors and similar work, it gave a great impetus to the

machines of its day ; but for all large work it has been entirely superseded by the ring armatures and drum armatures next to be described.

Armatures.

Returning to the ideal simple loop we may exhibit it in its relation to the 2-part commutator somewhat more clearly by referring to Fig. 16. The same split-tube or 2-part commutator will suffice if a loop of two or more turns be substituted, as shown in Fig. 17, for the single turn.

But we may substitute also for the one loop a small coil consisting of several turns wound upon an iron ring. This coil (Fig. 18), which may be considered as one section of a

FIG. 16.   FIG. 17.   FIG. 18.

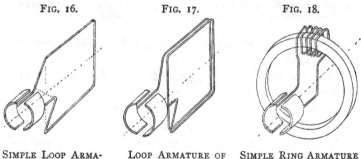

SIMPLE LOOP ARMA-   LOOP ARMATURE OF   SIMPLE RING ARMATURE
TURE.   Two Turns.   WITH ONE COIL.

Pacinotti or Gramme ring, will have lines of force induced through it as the loop had. In the position drawn, it occupies the highest point of its path and the induction of lines of force through it will be a maximum. As it turns, the number of lines of force threaded through it will diminish, and become zero when it is at 90° from its original position. But a little consideration of its action will suffice to show that if another coil be placed at the opposite side of the ring it will be undergoing exactly similar inductive action at the same moment, and may therefore be connected to the same commutator. If these two coils are united in parallel arc, as shown in Fig. 19, the joint electromotive-force will be the same as that due to either separately ; but the resistance offered to the current by the two jointly is half that of either. It is evident that we may connect two parallel loops in a similar fashion to one simple 2-part collector. If the two loops are of one turn each, we shall have the arrangement sketched in Fig. 20; but

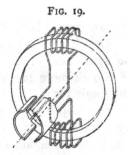

FIG. 19.

SIMPLE RING ARMATURE WITH
TWO COILS IN PARALLEL.

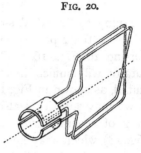

FIG. 20.

SIMPLE LOOP ARMATURE WITH
TWO COILS IN PARALLEL.

the method of connecting is equally good for loops consisting of many turns each.

Now with all these arrangements involving the use of a 2-part commutator, whether there be one circuit only or two circuits in parallel in the coils attached thereto, there is the disadvantage that the currents, though commuted into one direction, are not absolutely continuous. In any single coil without a commutator, there would be generated in successive revolutions, currents whose variations might, if the coil were destitute of self-induction, be graphically expressed by a recurring sinusoidal curve, such as Fig. 21. But if by

FIG. 21.

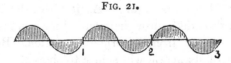

the addition of a simple split-tube commutator the alternate halves of these currents are reversed, so as to rectify their direction through the rest of the circuit, the resultant currents will not be continuous, but will be of one sign only, as shown in Fig. 22, there being two currents generated during each revolution of the coil. The currents are now "rectified," or, redressed " as our Continental neighbours say, but are not continuous. To give *continuity* to the currents, we must advance from the simple 2-part commutator to a form having a larger number of parts, and employ therewith a

FIG. 22.

larger number of coils. The coils must also be so arranged that one set comes into action while the other is going out of action. Accordingly if we fix upon our iron ring two sets of coils at right angles to each other's planes, as in Fig. 23, so that one comes into the position of best action while the other is in the position of least action (one being parallel to the lines of force when the other is normal to them), and their actions be superposed, the result will be, as shown in Fig. 24, to give a current which is continuous, but not steady, having four slight undulations per revolution. If any larger number of separate coils are used, and their effects, occurring at regular intervals, be superposed, a similar curve will be obtained, but with summits proportionately more numerous and less elevated. When the number of

FIG. 23.

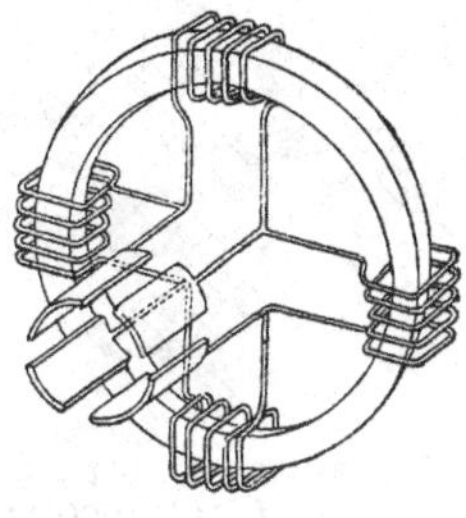

FOUR-PART RING
ARMATURE (CLOSED-COIL).

FIG. 24.

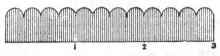

coils used is very great, and the overlappings of the curves are still more complete, the row of summits will form practically a straight line, or the whole current will be practically constant. As arranged in Fig. 23, the four coils are all united together in a *closed* circuit, the end of the first being united to the beginning of the second, and so forth all round, the last section closing in to the first. For perfect uniformity of effect the coils on the armature ought to be divided into a very large number of sections (see calculations, Chap. XV.), which come in regular succession into the position of maximum induction at regular intervals one after the other. In Fig. 25 a sketch is given of a drum armature wound with two pairs of coils at right angles one to

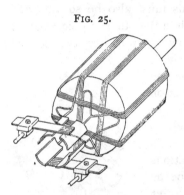

FIG. 25.

FOUR-PART DRUM ARMATURE (CLOSED-COIL).

the other, and connected to a 4-part collector. A little examination of Figs. 23 and 25 will show that each section of the coils is connected to the next in order to it; the whole of the windings constituting therefore a single closed coil. Also, the end of one section and the beginning of the next are both connected with a segment of the collector. In practice, the collector segments are not mere slices of metal tubing, but are built up

of a number of parallel bars of copper, gun metal, or phosphor-bronze, such as may be seen in Figs. 112 and 129, placed round the periphery of a cylinder of some insulating substance. It will also be noticed that, owing to the fact that there is a continuous circuit all round, there are two ways in which the current may flow through the armature from one brush to the other, as in all the ring and drum armatures; of which, indeed, Figs. 23 and 25 may be taken as simplified instances. The same reasoning now applied to 4-part armatures holds good for those having a still larger number of parts, such as is shown in Fig. 26. Of these more will be said in the subsequent chapters. Let it suffice to say here that in all closed-coil armatures, whether of the "ring" or the "drum" type, there must be just as many segments to the collector or commutator as there are sections in the coils of the armature. The special case of *open-coil armatures* is considered in Chapter X. In these machines the separate coils are not connected up together in series, and a special commutator is used instead of the usual collector.

As already explained, the "brushes" press against the commutator or collector, being usually held in position by a

FIG. 26.

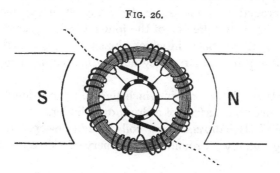

SIMPLE RING ARMATURE SHOWING CONNECTIONS OF CLOSED COIL.

spring. As the collector rotates, each of the bars passes successively under the brush, and makes contact with it. At one side—that towards which the two currents in the armature are flowing—the current flows from the collector-bar into the brush. At the other side the return current flows from the negative brush into the collector-bar in contact with it, and thence divides into two parts in the two circuits of the armature. If the brushes press strongly against the collector-bars, then when one collector-bar is leaving and the next coming up into contact with the brush, there will be contact made for a moment with two adjacent bars; and the coil, or section, whose two ends are united to these two bars, will, for an instant, be short-circuited. The effect of this will be considered in dealing with the reactions in armatures at a later stage.

So far, the only types of armature considered have been

the "drum" type, and the "ring" type; but these are not
the only possible cases. The object of all such combinations
of coils is to obtain the practical continuity and equability of
current explained above. To attain this end it is needful
that some of the individual coils should be moving through
the position of maximum action, whilst others are passing
through the neutral point, and are temporarily idle. Hence,
a symmetrical arrangement of the individual coils or groups
of coils around an axis is needed; and such symmetrical
arrangements may take one of the four following types :—

(1.) *Ring Armatures*, in which the coils are grouped upon
a ring whose principal axis of symmetry is its axis of rotation
also.

(2.) *Drum Armatures*, in which the coils are wound longitudinally over the surface of a drum or cylinder.

(3.) *Pole Armatures*, having coils wound on separate poles
projecting radially all round the periphery of a disk or central
hub.

(4.) *Disk Armatures*, in which the coils are flattened
against a disk. These armatures are mostly appropriate to
dynamos of Class II.

Ring armatures were adopted in practice in the dynamos
of Pacinotti, Gramme, Schuckert, Gülcher, Fein, Heinrichs,
De Meritens, Brush, Jürgensen, and others. Drum armatures
are found in the Siemens (Alteneck), Edison, Elphinstone-
Vincent, Weston, and other machines. Pole armatures were
used in the dynamos of Allan, Elmore, and of Lontin. There
are several intermediate forms. The Bürgin armature consists of eight or ten rings, side by side, so as to form a drum.
The Lontin (continuous-current dynamo) had the radial poles
affixed upon the surface of a cylinder. The Maxim armature
was a hollow elongated drum wound like a Gramme ring, and
a similar construction is used in numerous excellent recent
machines by Crompton, Cabella, Gramme, Paterson and
Cooper, Kapp, and Mather and Hopkinson. One early form
of the Weston armature had the drum surface cut up into
longitudinal poles; there was a similar armature by Jablochkoff,
in which the poles were oblique.

Ring armatures are found in many machines, but the ingenuity of inventors has been exercised chiefly in three directions :—The securing of practical continuity ; the avoidance of eddy currents in the cores ; and the reduction of useless resistance. In the greater part of these machines, the armatures are constructed with closed-coils : but there is no reason why open-coil armatures should not be constructed in the ring form, as indeed is the case with the well-known Brush dynamo. Most inventors have been content to secure approximate continuity by making the number of sections numerous. Pacinotti's early dynamo had the coils wound between projecting teeth upon an iron ring. Gramme rejected these cogs, preferring that the coils should be wound round the entire surface of the endless core. To prevent wasteful currents in the cores, Gramme employed for that portion a coil of varnished iron wire of many turns. In Gülcher's dynamo, the ring core was made up of thin flat rings cut out of sheet iron, furnished with projecting cogs, and laid upon one another. In the Schuckert-Mordey (Victoria) dynamo the core is built of hoops. The parts of the coils which pass through the interior of the ring (in spite of the late M. Antoine Breguet's ingenious proof that some of the lines of force of the field turned back into the core in this interior region) cut very few lines of force as they rotate, and therefore offer a certain amount of wasteful resistance. But this resistance in well-designed machines is insignificant compared with that of the external circuit, and the disadvantage is largely imaginary. Inventors have essayed to amend this, by either fitting projecting flanges to the pole-pieces (Fein and Heinrichs), or by using internal magnets (Jürgensen), or else by flattening the ring into a disk form, so as to reduce the interior parts of the ring coils into an insignificant amount (Schuckert and Gülcher). Indeed the flat-ring armatures may be said to present a distinct type from those in which the ring tends to the cylindrical form.

Drum armatures, as constructed by Siemens, had iron cores made of wire wound upon an internal non-magnetic nucleus. Weston substituted stamped washers of iron with teeth. Edison, iron washers without teeth. Special modes of winding or joining the copper conductors have been devised by these inventors, and more recently by Immisch and by Crompton and Swinburne.

Pole armatures, having the coils wound upon radially projecting poles have been devised by Allan, Lontin, and Weston. They are also used in some forms of alternate-current machine by Gramme, also by Elwell and Parker, and in the large Mechwart - Zipernowsky dynamo shown in 1883 at Vienna. The principle of Lontin's machine, in which the coils are connected like the sections of a Pacinotti or Gramme ring, is indicated in Fig. 27. Here the diameter of commutation

FIG. 27.

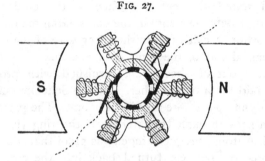

SIMPLE POLE ARMATURE SHOWING CONNEXIONS.

is parallel to the polar diameter, because the number of lines of force in this case is a maximum in the coils that are on the right and left positions. This armature is structurally a difficult one, because it is not mechanically strong unless the cores are solid ; and solid cores are electrically bad owing to their heating. It is impracticable, too, in this form to have very many sections, and the coils act, by reason of their position, prejudicially on one another. This form of armature is practically obsolete.

Disc armatures are now differentiated into two kinds : (1) those in which the coils are grouped on a number of small bobbins, side by side ; an arrangement suitable for alternate-current machines, such as those of Wilde, Holmes, Siemens, and Gordon ; (2) those in which the coils are made to overlap over a considerable angle of the periphery, as in the disk dynamos of Pacinotti, Edison, Ayrton and Perry, and of Rupp and Jehl, all of which are adapted to give continuous currents.

Armature Cores.

(*a.*) Theory dictates that if iron is employed in armatures, it must be slit or laminated, so as to prevent the generation of eddy currents. Such iron cores should be structurally divided in planes normal to the circuits round which electro-motive-force is induced ; or should be divided in planes parallel to the lines of force and to the direction of the motion. Thus, drum-armature cores should be built of discs of thin sheet iron. Ring armatures, if of the cylindrical or elongated type, should have cores made up of rings stamped out of sheet iron and clamped together side by side; but if of the flat-ring type they should be built of concentric hoops. Cores built up of varnished iron wire, or of thin discs of sheet iron separated by varnish, asbestos paper, or mica, partially realise the required condition.

(*b.*) The magnetic discontinuity of wire cores is, however, to a certain extent disadvantageous : it is better that the iron should be without discontinuity in the direction in which it is to be magnetised. It should therefore be laminated into sheets, rather than subdivided into wires. Pole armatures, however, ought to have wire cores. For this reason they are structurally unsatisfactory. Cores of solid iron are quite in-admissible, as currents are generated in them and heat them. The Wallace - Farmer armature, in which the coils were mounted on a solid disc of iron, used to grow very hot. Cores of solid metal other than iron—for example, of gun-metal, or of phosphor-bronze—should on no account be used in any armature.

(*c.*) In constant potential machines there ought to be so much iron in the armature that it is magnetised to about 16,000 magnetic lines to the square centimetre when the dynamo is generating its maximum working current. On the other hand, in constant current machines it is usual to work with a much higher degree of saturation.

Armature Coils.

(*d.*) All needless resistance should be avoided in armature coils, as hurtful to the efficiency of the machine. The wires should therefore be as short and as thick as is consistent with obtaining the requisite electromotive-force without requiring an undue speed of driving. In some dynamos—as, for example, some of Siemens' dynamos as used for electro-plating, and in those of Goolden and Trotter—several wires separately wound have their ends united in parallel arc. It is easier to wind four wires side by side than to coil up one very thick and unbendable wire.

(*e.*) Theoretically, since the function of armature coils is to enclose magnetic lines of force, the best form ought to be that which with minimum length of wire gives maximum area— namely a circle. In the Thomson-Houston machine the armature is spherical, and the coils circular ; and in some ring armatures, the section of the ring is circular. A drum arma-ture, whose longitudinal section is rectangular, ought to be as broad as long to preserve maximum area with minimum length of wire. Convenience of construction appears to dictate greater longitudinal dimensions.

(*f.*) The wire should be of the very best electric conduc-tivity. The conductivity of good copper is so nearly equal to that of silver (over 96 per cent.) that it is not worth while to use silver wires in the armature coils of dynamos.

(*g.*) In cases where rods or strips of copper are used instead of mere wires, care must be taken to avoid eddy currents by laminating such conductors, or slitting them in planes parallel to the electromotive-force, that is to say, in planes perpendicular to the lines of force and to the direction of the rotation.[1]

---

[1] It will be observed that the rule for eliminating the eddy-currents is different, in the three cases, for magnets and their pole-pieces, for moving iron armature cores, and for moving conductors in the armature.

(*h.*) We have seen that in order to obviate fluctuations in the current, the rotating armature coils must be divided into a large number of sections, each coming in regular succession into the position of best action. If these sections, or coils, are independent of each other, each coil, or diametral pair of coils must have its own commutator (as in the Brush machine). If they are not independent, but are wound on in continuous connection all round the armature, a collector is needed consisting of parallel metallic bars as numerous as the sections, or pairs of sections, each bar communicating with the end of one section and the beginning of the next.

(*i.*) In any case, the connections of such sections and of the commutators, or collectors, should be symmetrical round the axis ; for, if not symmetrical, the induction will be unequal in the parts that successively occupy the same positions with respect to the field-magnets, giving rise to inequalities in the electromotive-force, sparking at the commutator or collector, and other irregularities.

(*j.*) In certain cases it has been considered advantageous to arrange the commutator to cut out the coil that is in the position of least action, as the circuit is thereby relieved of the resistance of an idle coil. But no such coil should be short-circuited to cut it out, otherwise the harmful effects of induction will soon make themselves apparent by heating. Armatures so arranged are known as " open-coil " armatures.

(*k.*) In the case of pole armatures, the coils should be wound on the poles rather than on the middles of the projecting cores ; since the variations in the induced magnetism are most effective at or near the poles.

(*l.*) Since it is impossible to reduce the resistance of the armature coils to zero, it is impossible to prevent heat being developed in those coils during their rotation ; hence it is advisable that the coils should be wound with air spaces in some way between them, that they may be cooled by ventilation. In a well-ventilated armature a current-density of nearly 2500 ampères per square inch of the conductor may be attained.

(*m.*) The insulation of the armature coils should be en-

E

sured with particular care, and should be carried out as far as possible with mica and asbestos, or other materials not liable to be burned or melted if the armature coils become heated.

(*n.*) Care must be taken that the coils of armatures are so wound and held that they cannot fly out when rotating, Armatures have been known to become stripped of their insulating covering, and even to fly to pieces, for want of simple mechanical precautions.

(*o.*) Armatures ought to be very carefully balanced on their axles, otherwise when running at a high speed they will set up detrimental vibrations, and will tend to bend the axles. Great care should therefore be taken to secure perfect symmetry of construction.

Some further considerations respecting the reactions between the armature and the field-magnets are reserved for consideration until the rules for design of field-magnets have been considered.

Field-Magnets.

The coils of the field-magnets of a dynamo cannot be constructed so as to offer no resistance. They inevitably waste some of the energy of the currents in the form of heat. It has, therefore, been argued that it cannot be economical to use electromagnets instead of permanent magnets of steel, which have only to be magnetised once for all. Nevertheless, there are certain considerations which tell in favour of electromagnets. For equal power, their prime cost is less than that of steel magnets, which, moreover, are not permanent, but require remagnetising at intervals. Moreover, as we have seen, from the fact that there is a limiting velocity at which it is safe to run a machine, it is important, in order not to have machines of needlessly great size, to use the most powerful field-magnets possible. But if we do not get our magnetism for nothing, and find it more convenient to spend part of our current upon the electromagnets, economy dictates that we should so construct them that their magnetism may cost us as little as possible. To magnetise a piece of iron requires the expenditure of energy; but when once it is magnetised, it requires no further expenditure of energy (save the slight loss by heating in the coils, which may be reduced by making the resistance of the coils as little as possible) to keep it so magnetised, provided the magnet is doing no work. Even when doing no work, there will be loss if the current flowing round it be not steady. If the magnet does work in attracting a piece of iron to it, then there is an immediate and corresponding call upon the strength of the current in the coils, to provide the needful energy. This point may be illustrated by the following experiment :—Let a current from a steady source (see Fig. 28) pass through an incandescent lamp, and

FIG. 28.

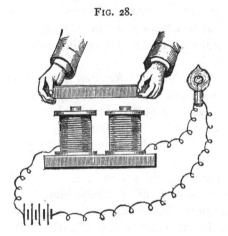

REACTION OF MOVING MASS OF IRON ON AN ELECTROMAGNET.

also through an electromagnet, whose cores it magnetises. When once established, the current is perfectly steady, and none of its energy is wasted on the magnet (save the negligible trifle due to the resistance of the coils) But if now the magnet is allowed to do work in attracting an iron bar towards itself, the light of the lamp is seen momentarily to fade. When the iron bar is snatched away, the light exhibits a momentary increase; in each case resuming its original intensity when the motion ceases. Now, in a dynamo there are, in many cases, revolving parts containing iron, and it is of importance that the approach or recession of the iron parts should not produce such reactions as these in the magnetism of the magnet. Large, slow-acting field-magnets are therefore advisable. The following points embody the conditions for attaining the end desired.

(*p.*) The body of the field-magnets should be solid. Even in the iron itself, currents are induced, and circulate round and round, whenever the strength of the magnetism is altered. These self-induced currents tend to retard all changes in the degree of magnetisation. They are stronger in proportion to the square of the diameter of the magnet, if cylindrical, or to its area of cross-section. A *thick* magnet will, therefore, be a slow-acting one, and will steady the current induced in its field.

(*q.*) Use magnets having in them *plenty of iron*. It is important to have a sufficient mass, that saturation may not be too soon attained.

(*r.*) Use the *softest possible iron* for field-magnets, not because soft iron magnetises and demagnetises quicker than other iron (that is here no advantage); but because soft iron has a higher magnetic susceptibility than other iron—is not so soon saturated. It is hardly possible to attach too great importance to this point. A small dynamo built of really good iron will do as much work as, and do it at a lower speed than, a much larger dynamo built of inferior iron.

(*s.*) So design the iron framework, cores, yokes and polar portions, that the magnetic circuit, or path of the flow of the magnetic lines, shall have the least magnetic resistance. To attain this the field-magnet cores and yokes, together with the iron core of the armature, should constitute as nearly as possible a closed circuit. And this closed circuit should be as short as is consistent with providing room for the requisite magnetising coils, and should have a large cross-section. The laws of the magnetic circuit, as bearing upon the design of dynamos, are considered in Chapter XIV Formerly it was considered good to use very long magnets as field-magnets. But with these it is found there is a greater "leakage" of the magnetic lines, and, further, with such elongated magnetic circuits a greater amount of electric energy must be spent in the magnetising coils to generate a given number of magnetic lines, than is the case with more compact magnetic circuits.

(*t.*) Employ iron cores of such a cross-section that there shall be as much iron as possible enclosed within the coils. The best cross-section might be expected to be circular ; as this requires less wire. Many constructors, however, prefer slabs of iron of rectangular section, but rounded at the edges.

(*u.*) The magnetism so obtained should be utilised as directly as possible ; therefore place the field-magnets or their pole-pieces as close to the rotating armature as is compatible with safety in running.

(*v.*) Avoid edges and corners on the magnets and pole-pieces if you want a uniform field. The laws of distribution of the magnetic lines of force round a pole are strikingly akin to those of the distribution of electrification over a conductor. We avoid edges and points in the latter case, and ought to do in the former. If the field-magnets or their pole-pieces have sharp edges, the field cannot be uniform, and some of the lines of force will run uselessly through the space outside the armature instead of going through it. Projecting edges invite leakage of the magnetic field.

(*w.*) It is of great importance that the magnetising effect of the field-magnets on the iron core of the armature should be much greater than the magnetising effect due to the current generated in the armature coils. If the latter is relatively powerful, there will be a great lead and much sparking at the brushes. The lead of the brushes, and its variation under different loads, can be reduced, and the tendency to spark can therefore be reduced, by making the field-magnets very powerful in proportion to the armature.

(*x.*) Reinforce the magnetic field by placing iron, or better still electromagnets, within the rotating armature. In most cases this is done by giving the armature coils iron cores which rotate with them ; in other cases, the iron cores or internal masses are stationary. In the former case there is loss by heating ; in the latter, there are structural difficulties to be overcome. Siemens has employed a stationary mass within his rotating drum armature.   Internal electromagnets serving the function of concentrating the magnetism of the field, have been used by Lord Elphinstone and Mr. Vincent. A similar device was tried in Sir W. Thomson's " mousemill " dynamo, and in Jürgensen's dynamo.

(*y*.) In cases where a uniform magnetic field is not desired, but where, as in dynamos of the second class, the field must have varying intensity at different points, it may be advisable specially to use field-magnets with edges or points, so as to concentrate the field at certain regions.

(*z*.) It is of great importance that the iron should present no physical discontinuity in structure in the direction in which it is to be magnetised.   The grain of the iron should lie in the direction of the lines of magnetic force that run through it. The maximum magnetic susceptibility of wrought iron is in the direction of its grain.   Further, at all surfaces of the field-magnets which are destined to be polar surfaces, where, therefore, the lines of force will run *out* of the iron *into* the air, the grain of the iron ought to be end-on.   This rule was not observed in the early form of Siemens dynamos, in which there were arched pieces of wrought iron as cores for the field-magnets, arranged as shown in Fig. 29, so as to be magnetised

FIG. 29.

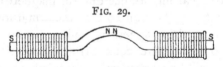

SIEMENS FIELD MAGNET.

from both ends, with a " consequent pole " in the middle of the arch.   Here the lines of force run along the grain to the middle, and then have to run out across the grain of the iron.

The student should contrast with Siemens' arrangement, that which is employed in the Edison dynamo (Fig. 140), which is more like that used by Wilde in 1865 (see Fig. 72). The ordinary Gramme has consequent poles at the middle of each of the electromagnets (see Figs. 87 and 88).   The modified double Gramme machine constructed by Marcel Deprez is sketched in Fig. 30. Other forms are considered in Chapter XIV.

(*aa.*) A great advantage is gained by thus working as nearly as possible with closed magnetic circuits ; that is to say, with a nearly continuous circuit of iron to conduct the lines of magnetic force round into themselves in closed curves. The enormous importance of this was pointed out so far back as 1879 by Lord Elphinstone and Mr. C. W. Vincent, whose dynamo embodied their idea. Every electrician knows that if a current of electricity has to pass through a circuit, part of which consists of copper and part of liquids—such as the acid

FIG. 30.

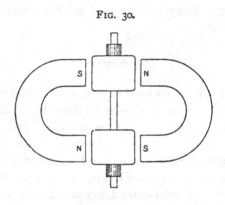

FIELD-MAGNETS OF DEPREZ'S DOUBLE GRAMME DYNAMO.

in a battery, or the solution in an electrolytic cell—the resistance of the liquid is, as a rule, much more serious than the resistance of the copper. Even with dilute sulphuric acid the resistance to the flow of the current by a thin stratum is 760,000 times as great as would be offered by an equally thick stratum of copper. And in the analogous case of using a field-magnet to magnetise the iron core of an armature, the stratum of air—or, it may be, of copper wire—in between the two pieces of iron, offers what we may term a relatively enormous resistance to the magnetic induction. If we take the magnetic permeability of air as 1, then the permeability of iron will be represented by values lying between 5 and 20,000 according to the quality of the iron. The permeability of copper is not very different from that of air. Or, in other words, a stratum of air or copper offers from 5 to 20,000 times as much resistance to the magnetic induction as if the space were filled up with soft iron. Obviously, then, it would be a gain to diminish as much as possible the gaps between the portions of iron in the circuit. The values of the magnetic permeability for iron, air, and copper, have been known for years, yet this simple deduction from theory was long set at defiance in the vast majority of cases. Further, all unnecessary *joints* in the iron cores of the field-magnets should be avoided ; and, where inevitable, should be tooled up to fit with the greatest nicety. Many makers now get their forgings made all in one piece.

Pole-pieces.

(*ab.*) If pole-pieces are used they should be massive, and of the softest iron, for reasons similar to those urged above. In many cases no specific pole-pieces are employed, the polar surfaces being cut out in the face of the solid iron core.

(*ac.*) The pole-pieces should be of shapes really adapted to their functions. If intended to form a single approximately uniform field, they should extend, but not too far, on each side. In the case of dynamos with flat-ring armatures, it is found, as demonstrated later on, that a narrow pole-piece is more advantageous than a very broad one. The distribution of the electromotive-force in the various sections of the coils on the armature depends very greatly on the shape of the pole-pieces.

(*ad.*) Pole-pieces should be constructed so as to avoid, if possible, the generation in them of useless eddy currents. The only way of diminishing loss from this source is to construct them of laminæ built up so that the mass of iron is divided by planes in a direction perpendicular to the direction of the currents, or of the electromotive-forces tending to start such currents.

(*ae.*) If the bed-plates of dynamos are of cast iron, care should be taken that these bed-plates do not short-circuit the magnetic lines of force from pole to pole of the field magnets. Masses of brass, zinc, or other non-magnetic metal may be interposed ; but are at best a poor resource. In a well-designed dynamo there should be no need of such devices.

Field-Magnets in Practice.

In the classification of dynamos laid down in Chapter II., we found that those of the first class required a single approximately uniform field of force, whilst those of the second class required a complex field of force differing in intensity and sign at different parts. Accordingly, we find a corresponding general demarcation between the field magnets in the two classes of machine. In the first, we have usually two pole-pieces on opposite sides of a rotating armature. In the second, a couple of series of poles set alternately round a circumference or crown, the coils which rotate being set upon a frame between two such crowns of poles.

Further differences in detail are pointed out in Chapter XIV. on Field-Magnets.

Field-magnet Coils.

(*af.*) The function of the field-magnets is to force magnetic lines through the armature ; and since the function of the field-magnet coils is to magnetise the field-magnets, it might have been expected that the best way of arranging the magnetising coils would have been to place them as near as possible to the armature, close to, or even over the pole-pieces. Some inventors have even essayed to place the magnetising coils over the armature itself. This arrangement is, however, not successful; it is inconvenient for the rotating armature to be enclosed within stationary coils; it is, moreover, found that the same quantity of wire wound in turns of smaller diameter upon the iron core of the field-magnet produces more powerful results. Formerly, when dynamos

were designed without special regard to the goodness of the
magnetic circuit, it was a matter of some importance at what
portion of the ironwork the coils were placed.   Coils placed
on or near the poles of a long straight magnet produce
greater changes in the distribution of the magnetism at the
pole than do coils placed at the middle of the magnet.   It is
for this reason that in Hughes's magnets and in Bell's tele-
phones, where it is desired to make a magnet most sensitive
to variations in the strength of the current, the coils are fixed
on at the pole.   But where the iron cores and yokes constitute
a good magnetic circuit it makes very little difference whether
the coils are placed at one part of the magnetic circuit or
another ; each turn of wire produces (with a given current) an
equal magnetising effect.   In such cases, therefore, it is
obvious that the greatest magnetising effect will be obtained
from a given length of wire by not heaping up the coils at
any one point but by distributing the turns of wire nearly
uniformly along the iron circuit.

(*ag.*) The proper resistances to give to the field-magnet
coils of dynamos have been calculated by Sir Wm. Thomson,[1]
who has given the following results :—

For a " Series Dynamo," make the resistance of the field-
magnets a little less than that of the armature.   Both of them
should be small compared with the resistance of the external
circuit.

For a " Shunt Dynamo " the rule is different.   The best
proportions are when such that

$$R = \sqrt{r_s\, r_a},$$

or that

$$r_s = \frac{R^2}{r},$$

where the symbols R, r_s, and r_a, stand respectively for the
resistances of the external circuit, of the shunt coils, and of
the armature.   The proof of this rule, as deduced from the
economic coefficient, is given in due place in Chapter XVIII. on
the Algebraic Theory of the Shunt Dynamo.   The rule for

[1] *British Association Report*, 1881, p. 528.

shunt dynamos when worked out shows that the resistance of the shunt coils ought to be at least 324 times as great as that of the armature, otherwise an efficiency of 90 per cent. is quite unattainable.

Commutators, Collectors, and Brushes.

(*ah.*) Commutators and collectors being liable to be heated through imperfect contact, and liable to be corroded by sparking, should be made of very substantial pieces of copper, or else of gun metal ; or, better still, of phosphor-bronze. Some makers cast the metal in the form of a hollow cylinder and saw it up into parallel strips. More frequently the metal is cast or drawn in rods of the form of the bars, which are afterwards filed up true and fitted into their places. Collectors of substantially such type as is here described are common to all dynamos of the first class, except only the "open-coil" dynamos, which require a special pattern of commutator. The collector of Pacinotti's early machine differed only in having the separate bars alternately a little displaced longitudinally along the cylinder, but still so that the same brush could slip from bar to bar. Niaudet's modification, in which the bars are radially attached to a disk, is a mere variety in detail, and is only applicable for small machines. In the collector once used in Weston's dynamo, and in some forms of Schuckert's dynamo, the bars are oblique or curved, without, however, any other effect than that of prolonging the moment during which the brush, while slipping from contact with one bar to contact with the next, short-circuits one section of the coil.

(*ai.*) In the case of a collector made of parallel bars of copper, ranged upon the periphery of a cylinder, the separate bars should be capable of being removable singly, to admit of repairs and examination. In the latest Siemens' dynamos the collector bars are of iron, very massive, and insulated with air-gaps.

(*aj.*) The brushes should touch the commutator or the collector at the two points, the potentials of which are respectively the highest and the lowest of all the circumference, and close to the points of least sparking. In a properly and symmetrically built two-pole dynamo, these points (called *neutral points*) will be at opposite ends of a diameter. In multipolar dynamos there are more than two neutral points, and therefore more than one pair of brushes, unless the neutral points of the armature are cross-connected.

(*ak.*) In consequence of the armature itself, when traversed by the currents, acting as a magnet, the magnetic lines of force of the field will not run straight across, from pole to pole, of the field-magnets, but will take, on the whole, an angular position, being twisted a considerable number of degrees in the direction of the rotation. This reaction of the induced current will be more particularly dealt with presently. In consequence of this and other reactions, the diameter of commutation (which is at right angles to the resultant lines of force in machines of the Siemens and Gramme type, and parallel to the resultant lines of force in machines of the Lontin type) will be shifted forward. In other words, the brushes will have a certain *angular lead*. The amount of this lead depends upon the relation between the intensity of the magnetic field and the strength of the current in the armature. This relation varies in the four different types of field magnets. In the series dynamo, where the one depends directly on the other, the angle of lead is nearly constant whatever the external resistance. In other forms of dynamo, the lead will not be the same, because the variations of resistance in the external circuit do not produce a proportionate variation between the two variables which chiefly determine the angle of lead.

(*al.*) Hence, in all dynamos, it is advisable to have an adjustment, enabling the brushes to be rotated round the commutator or collector to the position of the diameter of commutation for the time being. Otherwise there will be sparking at the brushes, and in part of the coils at least the current will be wasting itself by running against an opposing electromotive-force.

(*am.*) The arrangements of the collector or commutator should be such that, as the brushes slip from one part to the next, no coil or section in which there is an electromotive-force should be short-circuited, for any appreciable time, otherwise work will be lost in heating that coil. For this reason, it is well so to shape the pole-pieces that the several sections of coils on either side of the ′neutral point should differ but very slightly in potential from one another.

(*an.*) The number of contact points between the brush and the collector-bar should be as numerous as possible, for by increasing the number of contacts, the energy wasted in sparks will be diminished inversely as the square of that number. The brushes might with advantage be laminated, or made of parallel loose strips of copper, each bearing edgeways on the collector.

(*ao.*) The segments of the collector or commutator should be efficiently insulated from one another and from the axle. Many makers place layers of vulcanised fibre, asbestos paper, or mica between and below them. Others leave air spaces. If care is not taken the insulation may be spoiled by copper-dust worn off the brushes, or by the formation of a film of charred oil. The latter accident sometimes occurs where asbestos is employed; and is the more annoying as it is often accompanied with an unaccountable falling-off in the power of the dynamo, due to an invisible short-circuiting through a charred carbonaceous mass beneath the surface. The only remedy is to remove the insulating material and replace with new. On this account some experienced engineers prefer to treat the collector with French chalk instead of oil. In Hochhausen's dynamo the collector-bars are L-shaped, and are firmly bolted to an end-plate of slate, leaving air-gaps between the bars. Some makers drive air from a fan between the separate collector-bars to prevent dust lodging there and to keep them cool. To keep the collector-bars from flying it is the general practice to furnish them with projecting lugs at the ends (see Figs. 112, 124, 129, and 148), so that they can be clamped between insulated rings.

The collector should never be allowed to wear into ruts or to run untrue, otherwise there will be an altogether unnecessary and detrimental amount of sparking, which will rapidly ruin the machine. Whenever needed, the collector should be

filed or turned down until it runs truly. Sometimes one or
more of the collector-bars wear down flat; this is usually
caused by a fault in some one of the armature coils, or by some
mechanical defect in the construction that causes a periodical
spark.

Brushes.

The kind of brush most frequently used for receiving the
currents from the collector, consists of a quantity of straight
copper wires laid side by side,
soldered together at one end,
and held in a suitable clamp.
The number of points of contact secured by this method is
advantageous in reducing sparking. Two layers of wires are
often thus united in a single
brush, as shown in Fig. 31.

FIGS. 31, 32, 33.

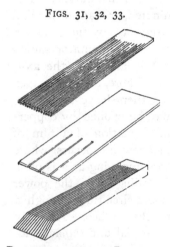

DIFFERENT KINDS OF BRUSHES.

Brushes are also made of
broad strips of springy copper,
slit for a short distance so as to
touch at several points, Fig. 32.
Such are used in the Brush and
Thomson - Houston dynamos.
This kind of brush is usually set
tangentially to the surface of the
commutator, not sloping to it at an angle as is the case with
the thicker kinds of brushes.

Edison used as brushes a number of copper strips placed
edgeways to the collector, and soldered flat against one another
at the end furthest from the collector, Fig. 33. Here, also, the
object in view was the subdividing of the spark at the contact.

In some of the Edison machines, a compound brush made
up alternately of layers of wire, like Fig. 31, and slit strips of
copper, like Fig. 32, has been adopted.

Rotating brushes in the form of metal disks have been
tried by Gramme, and others have been suggested by
Sir W. Thomson and Mr. C. F. Varley.

In those cases where, as in alternate-current machines and in dynamos of the "unipolar" type, a brush is used to obtain a sliding contact against an undivided collar or cylinder, it has been suggested by Professor G. Forbes to replace the brush by a slab of fine-grained and good conducting carbon.

The brushes of the dynamo should be properly trimmed before the machinery is set into motion. Neglect of this simple point will result in rapid deterioration.

Brush-holders and Rockers.

The brush-holders should be furnished with springs to bring a steady and even pressure to bear upon the points of contact. Spring brush-holders are shown in several of the figures of machines. If the pressure is too light the vibrations of the machine when running will cause a jumping of the brushes and consequent sparking and destruction of brushes and collector, and will probably cause one or more bars of the collector to become more worn away, and flattened at its surface, than the others. If the pressure of the brushes is too heavy the collector will be worn into ruts, and metal dust will be thrown off. Flimsy and rickety brush-holders always give trouble. Brush-holders in which the pressure is applied longitudinally, are preferable in those cases where the brush is a thick one set at a considerable angle to the surface of the collector.

The rockers which support the brush-holders should admit of a sufficient angular displacement being given to the brushes. Both rockers and brush-holders should be very substantial ; if they are not, trouble will arise from vibration and consequent sparking. In many dynamos the arrangement and construction of these parts is very defective. Good mechanical design is here all-essential ; but it must be combined with the two essential electrical conditions, (1) that the brush-holder shall be in perfect metallic contact, on the one hand with the brush, on the other with the conducting leads ; (2) that all these parts, brush, brush-holder, and leads shall be perfectly insulated from the bed-plate, brackets and bearings of the machine.

CHAPTER IV.

On the Induction of Currents in Armatures and the Distribution of Potentials around the Collector.

In considering the case of an ideal simple dynamo, it was shown that the induction in the rotating loop or coil was zero at the position where it lay in the diameter of commutation, and that the induction increased (as the sine of the angle) to its maximum value at about 90° (see Fig. 10, p. 36). In actual machines the induction at different points round the circle depends on the density of the magnetic lines of force in the gaps between the polar faces and the armature-core. In every well-designed dynamo it rises and falls according to a regular law of fluctuation. In fact Fig. 34, which re-

Fig. 34.

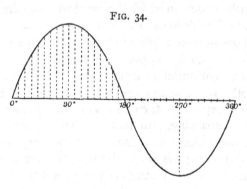

Curve of Induction.

presents a curve of *sines*, will serve to represent, by the height of the curve, the amount of induction going on in an armature at every 10° round the circle. If there are, for example, thirty-six sections in a ring armature, so that the sections are spaced out at 10° apart, the least active sections will be those at 0° and 180°, whilst the most active are those at 90° and 270°. But in all the ordinary "closed-coil" armatures, the separate sections are connected together so that any electromotive-force induced in the first section is added on to that induced in the second, and that in the third is added to these two, and so on all the way round to the brush at the other side. The separate electromotive-forces are added together just as are the separate electromotive-forces of a battery of voltaic cells united in series. A ring of battery cells united in series, but having one-half the cells set so

FIG. 35.

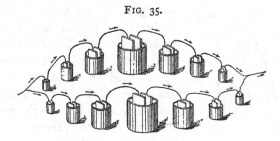

RING OF CELLS.

that the current in them tends to run the other way round the ring, forms a not inapt illustration of the inductions in the sections of a ring armature. If it could be indicated that those sections which are at 90° from the brushes are much more powerful in their inducing effect than those that occupy positions near the brushes, the analogy would be still more perfect.[1]

Now, knowing how the induction in individual coils or sections rises and falls round the ring, let us inquire what this will result in when we add up the separate electromotive-forces, so as to find the total effect. We shall have to add up the effects of all the sections round, from the negative brush at 0° on one side, to the positive brush at 180° on the other

[1] In Fig. 35 the middle cells of each row are drawn larger to suggest this ; only, unfortunately, *large* cells do not possess a higher electromotive-force than small ones, though they have less resistance internally.

F

side : and the result will be the same in each half of the ring, because of symmetry. Suppose we take the side from 0° by 90° to 180° (on the left in Figs. 11 and 12). If we look at the curve given above (Fig. 34), we shall see that as the heights of the dotted lines represent the amount of induction, the total effect will be got by adding up the lengths of all those from 0° to 180° ; and of course, the sum is equal to the sum of the negative lengths between 180° and 360°. But we may do the thing in another way, which beside giving us the final total, will show us how the sum grows as each length is successively added on. We should find that the sum grew slowly at first, then rapidly, then slowly again as it neared its highest value. The sum of the effects would grow, in fact, in a fashion represented on a reduced scale in the curve of Fig. 36. This process of adding up a continuously-varying set of values is called by mathematicians integrating. Fig. 36 is got by integrating the values of the curve Fig. 34 between the limits of 0° and 180°. Now in the actual dynamo this integration is effected by the very nature of things, in consequence of the fact that each section is united to those on either side of it.

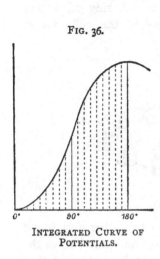

FIG. 36.

INTEGRATED CURVE OF POTENTIALS.

It is possible to investigate by experiment both these effects : the induction in the individual coils, and the total or integrated potential.

The electromotive-force induced in a single section as it passes any particular position, may be examined by means of a voltmeter or potential-galvanometer in the following way. Two small metal brushes are fixed to a piece of wood at a distance apart equal to the width between two consecutive bars of the collector. These brushes are united by wires to the voltmeter terminals, so that any difference of potentials between them will be indicated on the dial of the instrument. The two brushes are placed against the collector, as shown in Fig. 37, while it rotates ; and as they can be applied at any point, they will give on the voltmeter an indication which measures the amount of electro-motive-force in that section of the armature which is passing through the particular position in the field corresponding to the position of the contacts. This method was devised independently by the author and by Dr. Isenbeck. The author found, in the case of a small Siemens dynamo which he examined, that the difference of potential indicated was almost *nil* at the sections close to the proper brushes of the machine, and was a maximum about half-way between. In fact, the difference of potentials rose most markedly at 90° from the usual brushes, or precisely in the region where, as seen in Fig. 35, the induction is theoretically at its highest point, and where, as seen in Fig. 36, the slope of the curve of total potential is greatest.

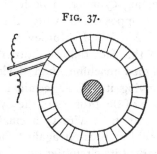

FIG. 37.

METHOD OF EXPERIMENTING
AT COLLECTOR OF DYNAMO.

After the experiment above detailed, the author experimented on his Siemens dynamo in another way. The machine was dismounted, and its field magnets separately excited. Two consecutive bars of the collector were then connected with a reflecting galvanometer having a moderately heavy and slow-moving needle. A small lever clamped to the collector allowed the armature to be rotated by hand through successive angles equal to 10°, there being thirty-six bars to the collector. The deflexions obtained, of course measured the intensity of the inductive effect at each position. The result confirmed those obtained by the method of the two wire brushes.

The rise of the totalised (*i. e.* "integrated") potential round the armature can be measured experimentally by a method first suggested by Mr. Mordey, and also involving the use of a voltmeter.

In a well-arranged dynamo of the first class, if one measures the difference of potential between the negative brush and the successive bars of the collector, one finds that the potential increases regularly all the way round the collecting cylinder, in both directions, becoming a maximum at the opposite side where the positive brush is.

Mr. W. M. Mordey, who first drew the author's attention to the fact that this distribution was irregular in badly designed machines, had devised the following method of observing it. One terminal of a voltmeter was connected to one of the brushes of the dynamo, and the other terminal was joined by a wire to a small metallic brush or spring, which could be pressed against the rotating collector at any desired part of its circumference. The author then made the suggestion that these indications might with advantage be plotted out round a circle corresponding to the circumference of the collector. Figs. 38 and 39, which are reproduced from his

Fig. 38.

Fig. 39.

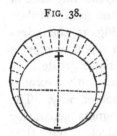

DIAGRAM OF POTENTIAL
ROUND THE COLLECTOR
OF GRAMME DYNAMO.

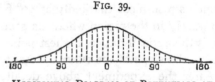

HORIZONTAL DIAGRAM OF POTENTIALS AT
COLLECTOR OF GRAMME DYNAMO.

Cantor Lectures, serve to show how the potential in a good Gramme machine rises gradually from its lowest to its highest value.

It will be seen that, taking the negative brush as the lowest point of the circle, the potential rises perfectly regularly to a maximum at the positive brush. The same values as are plotted round the circle in Fig. 38 are plotted out as vertical ordinates upon the level line in Fig. 39, which is nothing else than Fig. 36 completed for both halves of the collector. Fig. 36 is, however, a theoretical diagram of what the distribution ought to be, whilst Fig. 39 is an actual record taken from an "A" Gramme. If the magnetic field in which the armature rotated were uniform, this curve would be a true " sinusoid," or curve of sines ; and the steepness of the slope of the curve at different points will enable us to judge of the relative idleness or activity of coils in different parts of the field. The points marked + and − are close to the points of least sparking, or *neutral points.*

The rise of potential is not equal between each pair of bars, otherwise the curve would consist merely of two oblique straight lines, sloping right and left from the neutral point. On the contrary,ˑthere is very little difference of potential between the collector-bars immediately adjoining either of the neutral points. The greatest difference of potential occurs where the curve is steepest, at a position nearly 90° from the brushes, in fact, at that part of the circumference of the collector which is in connection with the coils that are passing through the position of best action. Were the field perfectly uniform, the number of lines of force that pass through a coil ought to be proportional to the sine of the angle which the plane of that coil makes with the resultant direction of the lines of force in the field, and the rate of cutting the lines of force should be proportional to the cosine of this angle. Now the cosine is a maximum when this angle = 0° ; hence, when the coil is parallel to the lines of force, or at 90° from the brushes, the rate of increase of potential should be at its greatest—as is very nearly realised in the diagram of Fig. 39, which, indeed, is very nearly a true "sinusoidal" curve. Such curves, plotted out from measurements of the distribution of potential at the collector, show not only where to place the brushes to get the best effect, but enable us to judge of the relative "idleness" or "activity" of coils in different parts of the field, and to gauge the actual intensity of different parts of the field while the machine is running. If the brushes are badly set, or if the pole-pieces are not judiciously shaped, the rise of potential will be irregular, and there will be maxima and minima of potential at other points. An actual diagram, taken from a dynamo in which these arrangements were faulty, is shown in Fig. 40, and again is plotted horizontally in Fig. 41 ; from which it will be seen, not only that the rise of potential was irregular, but that one part of the collector was more positive than the positive brush, and another part more negative than the negative. The brushes, therefore, were not getting their proper difference of potential ; and in part of the coils, the currents

FIG. 40.

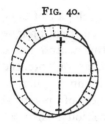

DIAGRAM OF POTENTIAL ROUND THE COLLECTOR OF A BADLY-ARRANGED DYNAMO.

FIG. 41.

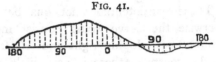

HORIZONTAL DIAGRAM OF POTENTIALS AT COLLECTOR OF FAULTY DYNAMO.

were actually being forced against an opposing electromotive-force.

This method of plotting the distribution of potential round the collector has proved very useful in practice, and elucidates various puzzling and anomalous results found by experimenters who have not known how to explain them. In a badly arranged dynamo, such as that giving a diagram like Fig. 40, a second pair of brushes, applied at the points showing maximum and minimum potential, could draw a good current without interfering greatly with the current flowing through the existing brushes !

Curves similar to those given can be obtained from the collectors of any dynamo of the first class—Gramme, Siemens, Edison, &c. — saving from the open-coil machines, which having no such collector, give diagrams of quite a different kind. It is, of course, not needful in taking such diagrams that the actual brushes of the machine should be in place, or that there should be any circuit between them, though in such cases the field-magnets must be separately excited. It should also be remembered that the presence of brushes, drawing a current at any point of the collector, will alter the distribution of potential in the collector ; and the manner and amount of such alteration will depend on the position of the brushes, and the resistance of the circuit between them.

The suggestion has several times been made to regulate

the available electromotive-force of a dynamo by shifting the brushes forward beyond the neutral point. Of course any desired fraction of the whole electromotive-force can thus be taken, but there is generally much sparking at brushes applied at any other point than the neutral point.

One immediate result of Mr. Mordey's observations on the distribution of potential, and of the author's method of mapping it, may be recorded. The author pointed out to Mr. Mordey that in a dynamo where the distribution was faulty, and where the curves of total potential showed irregularities, the fault was due to irregularities in the induction at different parts of the field ; and that the remedy must be sought in changing the distribution of the lines of force in the field by altering the shape of the pole-pieces. The author was able after the lapse of fifteen months, to congratulate Mr. Mordey on the entire and complete success with which he had followed out these suggestions. He entirely cured the Schuckert machine of its vice of sparking. The typical bad diagram given in Fig. 40, was taken from a Schuckert machine before it received from his hands the modifications which have been signally successful, and which are detailed in Chapter VII.

These methods have been dwelt on at some length here because of their great usefulness when applied to dynamos in which any such defect appears. They are also very closely related to the researches of Dr. August Isenbeck, which next claim attention.

Dr. Isenbeck described, in the *Elektrotechnische Zeitschrift* for August 1883, a beautiful little apparatus for investigating the induction in the coils of a Gramme ring, and for examining the influence exerted by pole-pieces of different form upon these actions.

Isenbeck's apparatus (Fig. 42) consists of a circular frame of wood placed between the poles of two small bar magnets of steel, each 25 centimetres long, lying 25 centimetres apart. On the frame, which is pivoted at the centre, is carried a ring of wood or iron, upon which is placed at one point a small coil of fine wire. This corresponds to a single section of the coils of a Pacinotti or of a Gramme ring, of which the ring of wood or iron constitutes the core. The coil can be adjusted to any desired position on the ring, and the ends communicate with a galvanometer. On vibrating it isochronously with the swing of the needle of the galvanometer, the latter is set in motion by the induced currents, and the deflection which results shows the relative amount of induction going on in the particular part of the field where the coil is situated. The vibrations of the frame are limited by stops to an angle of 7° 5'. Pole-pieces of soft iron, bent into arcs of about 160°, so as to embrace the ring on both sides, but not quite meeting, were constructed to fit upon the poles of the magnets. In some of the experiments a disk of iron was placed internally within the ring ; and in some other experiments a magnet was placed

FIG. 42.

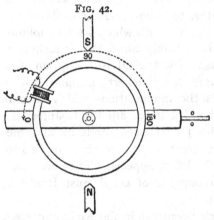

ISENBECK'S APPARATUS.

inside the ring, with its poles set so as either to reinforce or to oppose the action of the two external poles. In Dr. Isenbeck's hands this apparatus yielded some remarkable results. Using a wooden ring, and poles destitute of polar expansions, he observed a very remarkable inversion in the inductive action to take place at about 25° from the position nearest the poles.

Fig. 42 is a sketch of the main parts of Isenbeck's instrument, and shows the small coil mounted on the wooden ring, and capable of being vibrated to and fro between stops. When vibrated at 0°, or in a position on the diametral line at right angles to the polar diameter, there is no induction in the coil; but as the coil is moved into successive positions round the ring towards the poles, and vibrated there, the induction is observed first to increase, then die away, then begin again in a very powerful way, as it nears the pole, where the rate of cutting the lines of force is a maximum. This powerful induction near the poles is, however, confined to the narrow region within about 12° on each side of the pole. It is beyond these points that the false inductions occur, giving rise in the coil, as it passes through the regions beyond the 12°, to electromotive-forces opposing those which are generated in the regions which are close to the poles.

These inverse inductions were found by Isenbeck to be even worse when an iron disk, or an internal opposing magnet, was placed within the ring; but a reinforcing magnet slightly improved matters. Of course such an action in a Gramme armature going on in all the coils, except in those within 12° of the central line of the poles, would be most disastrous to the working of the machine; and the rise of potential round the collector would be anything but regular. In Fig. 43 I have copied out Isenbeck's curve of induction for the consecutive four quadrants. From 0° to 90° the exploring coil is supposed to be vibrated in successive positions from the place where, in the actual dynamo, the negative brush would be, round to a point opposite the S. pole of the pointed field magnet. From 90° to 180° it is passing round to the positive brush; from 180° to 270° it passes to a point opposite the N. pole; and from 270° to 360° returns to the negative brush. Now, since the height of this curve, at any point, measures the induction going on in a typical section as it moves through the corresponding region of the field, and since in the actual Pacinotti or Gramme ring the sections are connected all the way round the ring, it follows that the actual potential at any point in the series of sections will be got by adding up the total induced electromotive-force up to that point. In other words, we must integrate the curve to obtain the corresponding curve of potential, corresponding with the actual state of things round the collector of the machine. Fig. 44 gives the curve as integrated expressly for me from Fig. 43 by the aid of the very ingenious curve integrator of Mr. C. Vernon Boys. The height of the ordinate of this second curve at any point is proportional to the total area enclosed under the first curve up to the corresponding point. Thus the height at 90° in the second curve is proportional to the total area up to 90° below the first

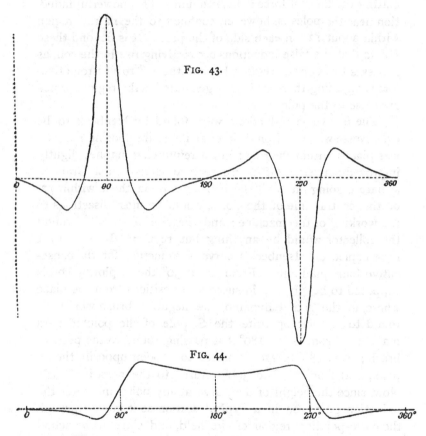

Fig. 43.

Fig. 44.

curve. And it will be noticed that though the induction (first curve, Fig. 43) decreases after 90°, and falls to zero at about 102°, the sum of the potentials (second curve, Fig. 44) goes on increasing up to 102°, where it is a maximum, and after that falls off, because, as the first curve shows, there is from that point onwards till 180° an opposing false induction. If this potential curve were actually observed on any dynamo, we might be sure that we could get a higher electromotive-force by moving the brush from 180° to 102°, or to 258°, where the potential is higher. Any dynamo in which the curve of potentials at the commutator presented such irregularities as Fig. 44 would be a very inefficient machine, and would probably spark excessively at the collector. It is evident that the induction in some of the coils is opposing that in some of the adjacent coils.

Two questions naturally arise—Why should such detrimental inductions arise in the ring; and how can they be obviated? The researches of Dr. Isenbeck supply the answer to both points. Dr. Isenbeck has calculated from the laws of magnetic potential the number of lines of force that will be cut at the various points of the path of the ring. He finds that the complicated mathematical expression for this case when examined, shows negative values for angles between 12° and 90°. The curves of values that satisfy his equations have minima exactly in those regions where his experiments revealed them. This is very satisfactory as far as it goes. But we may deduce a precisely similar conclusion in a much simpler manner, from considering the form and distribution of the lines of magnetic force in the field. These are shown in Fig. 45, together with the exploring coil situated as in Fig. 42 (p. 72). A simple inspection of the figure will show that at 0° a certain number of lines of force would thread themselves through the exploring coil. As the coil moved round towards the S. pole, the number would increase at first, then become for an instant stationary, with neither increase nor decrease; after that a very rapid decrease would set in, which, as the coil passed the 90° point, would result in there being no lines of force through the coil. But at the very same instant the lines of force would begin to crowd in on the other side of the coil and the number so threaded through negatively, would increase until the coil turned round to about the position marked T, where the lines of force are nearly tangential to its path, and here the inversion would occur, because, from that point onwards to 180°, the number of lines of force threaded through

the coil would decrease. We see, then, that such inversions
in the induction must occur of necessity to a *small* coil rotating in a magnetic field in which the lines of force are
distributed in the curved directions, and with unequal
density. The remedy is obvious ; either arrange a field, in

FIG. 45.

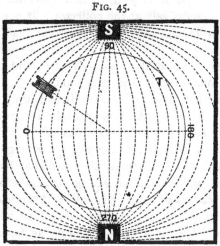

which the lines of force are more equally distributed, and are
straighter, or else use a coil of larger aperture. Another
remedy is to abandon the *ring* armature and use a *drum*
armature which gives no inversions.

If an iron core be substituted for the wooden core, the
useful induction is greater, and the false induction less; there
is still an inversion, but it takes place at about 25° from the
pole, and is quite trifling in amount. The introduction of iron
pole-pieces extending in two nearly semicircular arcs from
the magnets on either side has, if the wooden ring be still kept
as a core, the effect of completely changing the induction, so
that the curve, instead of showing a maximum at 90° from
starting, shows one at about 10°, and another at 170°. This is
given in Fig. 46.

The integrated curve of potential given in Fig. 47 is curious :
there are no reversals ; but the potentials rise and fall so very
suddenly on either side of the 0° and 180° points that any

small displacement of the brushes might produce disastrous inversions.

If, however, we make the double improvement of using the iron pole-pieces and the iron core at the same time, the effect

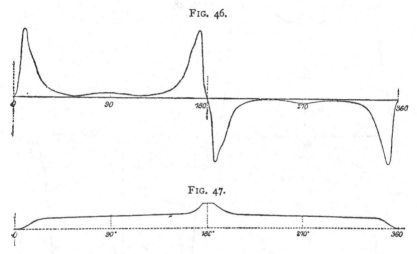

Fig. 46.

Fig. 47.

is at once changed. There are no longer any inversions, though the induction still shows some peculiarity. Fig. 48 shows the curve of induction adapted from Dr. Isenbeck's paper, and Fig. 49 the curve of potential which I have had integrated from it. Looking at Fig. 48, we see that on starting from 0°, induction soon mounts up, and becomes a maximum at about 20°, where the coil is getting well opposite the end of the encircling pole-piece. From this point on, though the induction is somewhat less, it still has a high value, showing a slight momentary increase as the coil passes the pole at 90°, and there is another maximum at about 160°, as the coil passes the other end of the pole-piece. My integrated curve (Fig. 49) tells us what would go on at the collectors if this were the action in the connected set of coils of a Pacinotti or Gramme ring. The potential rises from 0° all the way to close upon 180°. Still this is not perfect. In the perfect case the potential curve would rise in a perfect harmonic wave form, like that shown in Fig. 36 (p. 66). Fig. 49 departs widely from this, for it is

convex from 0° to 90°, and concave between 90° and 180°.
But there are no inversions.   The cause of the improvement is
easily told : the field—such as there is between the pole-piece

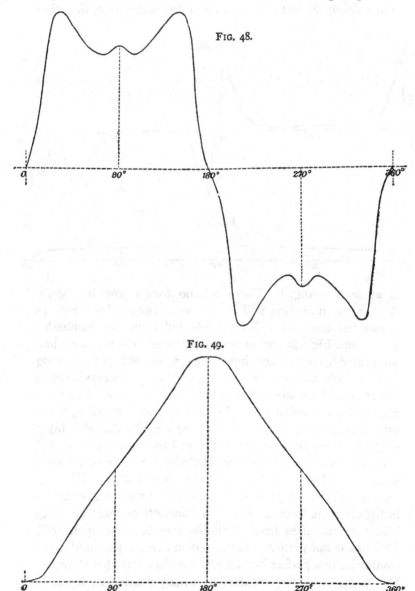

FIG. 48.

FIG. 49.

and the core—is "straighter," and the density of the lines of
force in it more uniform. I proved this experimentally in 1878,
by the simple process of examining the lines of force in such a
field by means of iron filings ; the actual filings, secured in
their places upon a sheet of gummed glass, were sent to the
late M. Alfred Niaudet, who had requested me to examine
the matter for him. Fig. 50 shows the actual lines of force
between the encircling pole-piece and the iron ring. It will

FIG. 50.

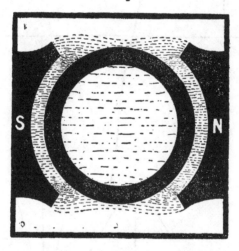

be seen that, though nearly straight in the narrow intervening
region, they are not equally distributed, being slightly denser
opposite the ends of the pole-pieces. One other case examined
by Dr. Isenbeck we will glance at. The effect of introducing
within the ring an interior magnet, having its S. pole opposite
the external S. pole, and its N. pole opposite the external N.
pole, was found to assist the action. The induction curve
is represented in Fig. 51. As will be seen, there are two
maxima at points a little beyond the end of the pole-pieces, as
before ; but in between them there is a still higher maximum,
right between the poles. This case also has been integrated
on Mr. Boys' machine, and shows the potential curve of
Fig. 52. This curve is a still nearer approach to the harmonic

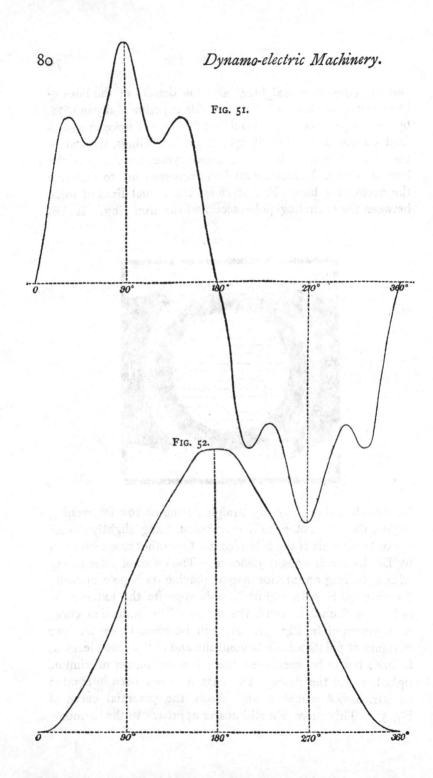

FIG. 51.

FIG. 52.

wave form, being concave from 0° to 90°, and convex from 90° to 180°.

I pass from Dr. Isenbeck's researches, and the integrated curves of potential which I have deduced from them, to some further researches of my own, which were undertaken with the view of throwing some light on the question whether the Pacinotti form of armature, with protruding iron teeth, or the Gramme form, in which the iron core is entirely over-wound with wire, is the better. It has been assumed, without, so far as I am aware, any reason having been assigned, that the Gramme ring was an improvement on that of Pacinotti. Pacinotti's was of solid iron, with teeth which projected both outwards and inwards, having the coils wound between. Gramme's was made " either out of one piece of iron, or of a bundle of iron wires," and had the coils wound " round the entire surface." Now the question whether the Gramme construction is better than the Pacinotti or not, can readily be tested by experiment ; and experiment alone can determine whether it is better to keep a thickness of wire always between pole-pieces and the core, or to intensify the field by giving to the lines of force the powerful reinforcement of protruding teeth of iron. The apparatus I constructed for determining this point is sketched in Fig. 53.

First, there are a couple of magnets set in a frame so as to give us a magnetic field, and there are pole-pieces that can be removed at will; in fact, there are three sets of pole-pieces for experimenting with different forms. Between the poles is set an axis of brass, upon which the armatures can be slid. These armatures are three in number. One is shown in Figs. 54 and 55, and consists of two coils of fine wire wound upon a wooden ring ; another armature is exactly like this, but is built up on a ring of iron wire ; a third (shown in its place in Fig. 53) is constructed upon a toothed ring made up of a number of plates of ferrotype iron cut out and placed flat upon one another. On each of the armatures are wound two coils at opposite ends of a diameter. The coils contain precisely equal lengths of silk-covered copper wire, cut from one piece. The cross-section of the core within each of these coils is in each case a square of 1 centimetre in the side, so

G

that the number of turns in each coil is as nearly equal as
possible.   I can slip any one of these armatures into the field

Fig. 53.

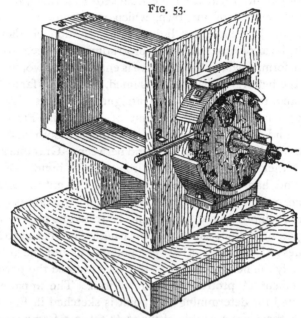

APPARATUS FOR INVESTIGATING INDUCTION OF ARMATURES.

Fig. 54.   Fig. 55.

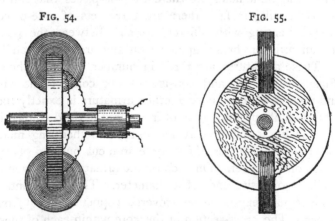

EXPERIMENTAL ARMATURE.

and connect it with a galvanometer. There is a lever handle
screwed to the armature, by means of which it can be moved.
I have used two methods of proceeding in order to compare
the coils. One of these methods is to turn the armature
suddenly through a quarter of a revolution, so that the coils
advance from 0° to 90°, when the "throw" of the needle of the
galvanometer—which is a slow-beat one—gives me a measure
of the total amount of induction in the armature. The results
are as follows :—

GRAMME.	GRAMME.	PACINOTTI.
Wooden Ring.	Iron Ring.	Iron Toothed Ring.
5	24	50

My second method of using these armatures consists in
jerking the coils through a distance equal to their own thickness,
the coils being successively placed at different positions in the
field, the throw of the galvanometer being observed as before.
Each of the coils occupies as nearly as possible 15° of angular
breadth. Accordingly, I have two stops set, limiting the
motion of the handle to that amount, and at the back there is
a graduated circle enabling me to set the armature with the
coils in any desired position. If we move the coils by six
such jerks, through their own angular breadth each time, then,
starting at 0°, the sixth jerk will bring us to 90°. The three
curves thus obtained are plotted out in Fig. 56, and the
corresponding numbers are given in the following table :—

	GRAMME.	GRAMME.	PACINOTTI.
	Wooden Ring.	Iron Ring.	Iron Toothed Ring.
0°–15°	− 5	25	30
15°–30°	− 10	60	70
30°–45°	0	120	140
45°–60°	45	195	320
60°–75°	40	200	380
75°–90°	30	220	360

These figures leave no doubt as to the question at issue. The Gramme pattern of ring armature, so far from being an

FIG. 56.

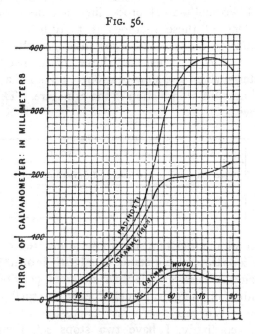

improvement on the Pacinotti, is theoretically a retrograde step; always supposing that the cost of construction, liability to heating, and other kindred matters be equal for the two.

In practice, however, there are two objections to the use of the toothed armature. The first is that the presence of these discontinuous projections causes heating in the pole-pieces of the field magnets. The second is that when there are projecting teeth it is less easy to secure durable insulation for the armature coils than when no such teeth project.

It ought not to be omitted that induction-curves, very similar to those of Isenbeck, were obtained in 1873 by M. J. M. Gaugain in an investigation into the action of the Gramme ring; an exploring coil being displaced through angles of 10°. Gaugain's work was published in the *Annales de Chimie et de Physique*, March 1873.

CHAPTER V.

REACTIONS IN THE ARMATURE AND MAGNETIC FIELD.

WHEN a dynamo is running, a set of entirely new phenomena arises in consequence of the magnetic and electric reactions set up between the armature and the field-magnets, and between the separate sections of the armature coils. These reactions manifest themselves by the "lead" which it is found necessary to give to the brushes, by sparking at the brushes, by variations of the lead and of the sparking when the speed or the number of lamps is altered, by heating of armature cores and coils, by heating of the pole-pieces of the field magnets, and by a discrepancy between the quantity of mechanical horse-power imparted to the shaft and the electric horse-power evolved in the electric circuit. The nature of these reactions demands careful attention.

Obliquity of Resultant Magnetisation.—We have seen (p. 65 and Fig. 35) that any closed-coil armature may be regarded as acting like a double voltaic battery, the two sets of coils acting like two rows of cells united "in parallel." We have now to show that a ring armature may be regarded also as a double magnet. Suppose a semi-ring of iron to be surrounded, as in Fig. 57, by a coil carrying a current, it will become, as every one knows, a magnet having a N. pole at one end, and a S. pole at the other. If a complete ring be similarly over-wound, but with an endless winding, and if then electric currents from a battery or other source are introduced into this coil at one point, flowing round the two halves of the ring to a point at the other side, and then leave the coil by an appropriate conductor, each half of the ring will be magnetised. There will be, if the currents circulate as represented by the arrows in Fig. 58, a double S pole at the

point where the currents enter, and a double N. pole at the
place where the currents leave.   The currents circulating in a
Gramme ring, will therefore tend to magnetise the ring in this

FIG. 57.

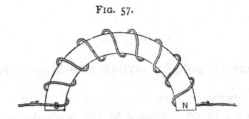

fashion.   Let us see how such a magnetisation is distributed
inside the iron itself.   Fig. 59 shows the general course of the
magnetic lines of force as they run through the iron ; where

FIG. 58.   FIG. 59.

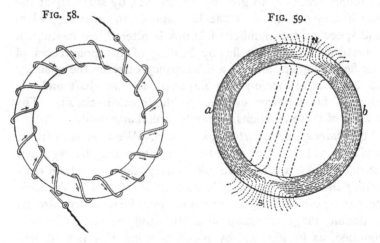

they emerge into the air are the effective poles of the ring
regarded as a magnet.   Fig. 59 should be very carefully
compared with Fig. 58.   It will be noticed that though the
majority of the lines of force pass externally into the air at
the outer circumference, a few of the lines of force find their
way across the interior of the ring, from its N. to its S.
pole.   This part of the magnetic field would in an actual
dynamo be deleterious if the number of lines of force were
not so few.

Now turn back to Fig. 50, p. 79, and observe the way in which the field-magnets tend to magnetise the ring when it is standing still, or at least when there is no current circulating in the armature coils. It is evident that when the dynamo is at work, if the brushes were set on a diameter at right angles to the line joining the poles, the ring will be subjected to two magnetising forces at once : the field magnets tending to magnetise it as in Fig. 50, the armature current tending to magnetise it as in Fig. 59. In consequence, the magnetisation which results is an oblique one, the lines of force in the field between the armature and the field-magnet being twisted round. But if the magnetic field is itself twisted round, the brushes must be correspondingly shifted, or else commutation will not take place at the moment when the number of lines of force is a maximum. This first reaction of the armature currents on the magnetisation of the ring, and on the magnetic lines of the field, shows that there ought to be in consequence a certain lead given to the brushes ; and that the lead will be greater as the armature currents are greater.

Neutral Points.—From the earliest time that dynamos have been used, engineers have found that, in order that the sparking may be a minimum, the brushes must be placed in certain positions, to be found by trial, called the *neutral points*. In ordinary two-pole dynamos the two neutral points lie at opposite ends of a diameter, which diameter is therefore called the *neutral line* or *diameter of commutation*. Experiment shows that in almost every case the neutral line is not exactly at right angles to the line joining the middles of the two pole-pieces, but lies obliquely across, being (in a generator) shifted round a few degrees in the direction of rotation. It was further found that in many machines the exact position of the neutral point was different according to the work that the dynamo was doing. If the brushes were set so as not to spark when a certain number of lamps were alight, then if the load of lamps was altered the machine sparked unless the brushes were adjusted to the corresponding neutral points. Hence arose the practice of mounting the brushes on *rockers*, by means of which their line of contact could be altered forward or backward to the neutral point. Great attention has naturally been paid by constructors to the practical problem how to get rid of variations in the angle of lead. On p. 69 it was stated that the neutral points lie close to the points of maximum positive and maximum negative potential on the commutator. But they are found not to coincide exactly with those points. At the point of maximum potential there is usually some sparking. The point of maximum potential has, as just shown, a forward lead, but the point of least sparking, the true *neutral* point, lies a little forward of this, this increased lead being due to another of the reactions now under consideration.

Lead of Brushes.—Formerly the fact that a lead must be given to the brushes was ascribed to a sluggishness in the demagnetisation of the iron of the armature, and even now Professors Ayrton and Perry take the view that part of the displacement of the pole is due to the sluggishness of demagnetisation of the iron. On the other hand no experimental proof has ever yet been given that the displacement is due to a true magnetic lag; the apparent magnetic sluggishness of thick masses of iron is demonstrably due to internal induced currents;[1] and no one uses solid iron in armature cores for this very reason. Neither has it been shown that thin iron plates or wires, such as are used in armature cores, are slower in demagnetising than magnetising. Indeed, the reverse is probably true; and, until further experimental evidence is forthcoming, it will be assumed that the alleged magnetic lag is negligibly small in its effects. For further discussion of this, see Appendix III. *On Movement of Neutral Point.*

In 1878, the late M. Antoine Breguet[2] suggested as a reason for the oblique position of the diameter of commutation, the influence of the actual current circulating round the armature coils, which would tend to produce in the iron of the armature a magnetisation at right angles to that due to the field-magnets. Breguet showed as above that there would

---

[1] The reader is referred to Appendix II. for further evidence on this point.

[2] Antoine Breguet: 'Théorie de la Machine de Gramme.' *Ann. de Chim. et de Phys.* [5], xvi., 1879.

be a resultant oblique direction of the lines of magnetism in the field, and therefore, since the "diameter of commutation" is at right angles to this direction, the brushes also must be displaced through an equal angle. Clausius accepts this view in his investigations, and adopts for the angle of the resultant field that whose tangent is the ratio of the two magnetising forces due to the current in the armature and field-magnets respectively. It will be shown presently that this rule is not correct, and that the sine, not the tangent of the angle of lead, represents the ratio of the two forces.

It may, however, be pointed out that, assuming as a first approximation that either the sine or the tangent of the angle of lead represents the ratio between the magnetising power of the field-magnets and of the armature coils, the lead may be diminished to a very small quantity, by increasing the relative power of the field-magnets, a course which is for many other reasons advisable. All practice confirms the rule that the magnetic moment of the field-magnets ought to be very great as compared with that of the armature.[1] Further than this, in all constant-potential machines there ought to be so much iron in the armature as only to be magnetised to near the critical point of saturation when the dynamo is working at its greatest activity. If there is less than this, there will be great alterations of lead, for after the armature core is well saturated then the magnetic effect due to the current in the armature will be of less importance relatively to that due to the field-magnets, and the machine will not regulate itself well. Every cause that tends to reduce the lead, makes the lead more constant, and therefore tends to reduce sparking at the brushes ; and the best means to secure this is obviously to use an unstinted quantity of iron—and that of the softest kind—both in the field-magnets and in the armature, for then

---

[1] The field-magnets ought to be so powerful as entirely to overpower the armature. I have seen a dynamo in which, on the contrary, the armature overpowered the field-magnets. When the brushes had a small lead there was a good electromotive-force, but it sparked excessively. With a large lead the sparks disappeared, but the electromotive-force also vanished !

the currents circulating in the armature will have less chance
of perturbing the field. In constant-current dynamos, however,
wherein usually the iron is less in quantity, and excited up to
a high degree of saturation even when the machine is not
lighting up its full load of lamps, there is much lead, and that
lead varies greatly with the load.

Diminution of Effective Magnetisation.—In relation to the
magnetisation of field-magnets, it may be pointed out that
the "characteristic" curves of dynamo machines (see Chap.
XVII.), which are used to show the rise of the electromotive-
force of the machine in relation to the corresponding
strength of the current, are sometimes assumed, though not
quite rightly, to represent the rise of magnetisation of the
field-magnets. Now, though the magnetisation of the magnet
may attain to practical saturation, it does not, under a still
more powerful current, show a diminished magnetisation.
But the characteristics of nearly all series-wound dynamos show
—at least for high speeds—a decided tendency to turn down
after attaining a maximum ; and in some machines, for example the older form of Brush with cast-iron ring (see Fig. 181),
this reaction is very marked. The electromotive-force diminishes, but the magnetism of the field magnets does not.
An explanation of this dip in the characteristic was put forward
by Dr. Hopkinson, in his lecture on "Electric Lighting,"
before the Institution of Civil Engineers, April 1883, attributing
this to the reaction of self-induction and mutual induction
between the sections of the armature. No doubt this cause
contributes to the effect, as all such reactions diminish the
effective electromotive-force. Part of the effect is due to the
shifting of the effective line of the field in consequence of
the iron of the armature becoming saturated at a different
rate from the iron of the field-magnets. It is at least significant that in the older form of Brush machine, where the
reduction of electromotive-force is very great, there is also
such a mass of iron in the armature, and so variable a lead
at the brushes. Machines with descending characteristics are
preferred for arc lighting.

Frolich, in certain experiments with an old-type Siemens

dynamo, found that the reaction of the armature current diminished the effective magnetism of the field-magnets nearly 25 per cent. For other researches on the effect of a cross magnetism in diminishing the magnetism of the core, see papers by Siemens[1] and Schültze in *Wiedemann's Annalen.* Schültze, in the course of twenty-four experiments, found that the cross-magnetisation of an iron core always diminished the longitudinal magnetisation: the diminution varied from 0 to 13·7 per cent. according to the relative strength of the two magnetising forces. In a more recent research Strömberg,[3] using a Schuckert dynamo, measured the demagnetising effect of the armature current, and expressed it as being equivalent to a certain number of negative ampere-turns. When the exciting power of the current in the field-magnet coils was maintained at 6250 ampere-turns, the demagnetising effect of the armature current was as follows :—

Armature Current.			Equivalent Negative Ampere-turns.
7·4	..	..	650
12·0	..	..	850
25·4	..	..	1450
44·0	..	..	2450
60·4	..	..	3650

With stronger constant excitation of the field-magnets the demagnetising effect of the armature currents was greater. With 13,400 ampère-turns, the demagnetising effect of sixty amperes in the armature was equivalent to 4400 ampere-turns. With 25,200 ampere-turns the demagnetising effect of only 30·5 amperes was equivalent to 3100 ampere-turns. This unexpected effect may probably arise from the iron of the field-magnet arriving earlier, by reason of inferior quality or insufficient section, at a higher stage of saturation than the armature core. It may also have been dependent on the

[1] Werner Siemens. *Wiedemann's Annalen,* XIV. p. 634, 1882.

[2] Schültze. *Wied. Ann.,* XXIV., p. 663, 1885. See also Oberbeck, Habili-tations-Schrift, 1878.

[3] Strömberg. *Centralblatt für Elektrotechnik,* IX., p. 283, 1887.

positions of the brushes, the lead of which is not specified by Stromberg. The greater the lead given to the brushes in a dynamo used as a generator, the greater is the demagnetising effect of the armature current ; the reverse is true in the case of the dynamo used as a motor, as will be seen in due course. If a negative lead (*i. e.* a displacement from the neutral line in the opposite direction to the sense of the rotation) is given to the brushes of a generator, the magnetising effect of the armature currents will tend to assist the magnetisation of the core. Drs. J. and E. Hopkinson [1] have even shown that with a backward lead a dynamo can excite itself by means of the armature currents only ; but in such cases of negative lead there is a destructive amount of sparking. The demagnetising effect is of course proportional to the number of effective ampere-turns of the armature circuit that surround the magnetic circuit ; and therefore to the actual number of ampereturns multiplied by the sine of the angle of lead.[2] Several expedients have been proposed to compensate the crossmagnetising tendency of the armature currents, and so obviate the variations of lead. In one due to Mather [3] a small bar electromagnet excited by the armature current is placed perpendicularly between the pole-pieces. Professor E. Thomson proposes to place a series coil on a movable frame over the armature and tilt it till it brings back the neutral point.

Theory of Angle of Lead.—A reference to Fig. 50, which represents the state of things when there is no current in the ring, shows that in the field between the ring and the polepieces the lines of force are nearly parallel and nearly uniform, with, however, a slight tendency to take a greater density near the outer extremities of the pole-pieces. Consider what will be the effect of inducing a current into the ring, thereby superimposing upon it the magnetisation which was shown in Fig. 59. The result, carefully thought out for

---

[1] *Phil Trans.*, 1886, Pt. I., p. 347.

[2] According to Peukert, who, however, does not specify the angle of lead, the demagnetising effect of the armature current is proportional to the $1 \cdot 3$ power of the armature current. See *Centralblatt für Elektrotechnik*, IX., 484, 1887.

[3] See *La Lumiere Electrique*, XIX., p. 404, 1885.

a particular case, is shown by the lines of Fig. 60, where, however, a slight further development has been introduced. Suppose that at first the brushes were set to touch at two points on the vertical diameter. The field-magnets tend to magnetise the ring, so that its extreme left point is a N. pole, and the currents tend to magnetise it so that its highest point, where the brush is, is a N. pole. The consequence of this will be a resultant magnetisation in an oblique direc-

FIG. 60.

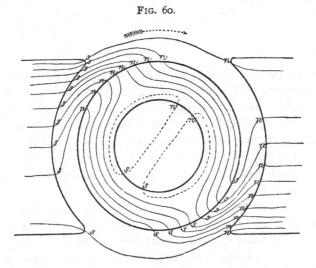

MAGNETIC REACTIONS BETWEEN FIELD-MAGNETS AND ARMATURE IN GENERATOR.

tion. Draw a line O F (Fig. 61) to represent the magnetising force due to the field magnets, and a line O C at right angles to represent the magnetising force due to the armature current, then the diagonal O R of the parallelogram will represent the direction of the resultant magnetisation. Draw a circle round O, and the point N will show how far the resultant induced pole is shifted round from the horizontal line. But the diameter of commutation where the brushes touch ought to be at right angles to the resultant poles in the ring, if the rise of potential round the commutator, as explained in the preceding (section) is to be regular. *There-fore we must give the brushes a lead,* and shift them round until the diameter of commutation is at right angles to O N. But shifting O C will itself alter N a little. We can find out

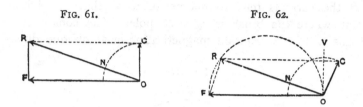

Fig. 61.   Fig. 62.

easily the new position. On O F (Fig. 62) describe a semi-circle, and set off F R, equal to the length that O C is to be, as a chord. Draw O C parallel and equal to F R ; and draw also the diagonal O R as before. The angle C O N is now a right angle, and N is very nearly where it was before. If O V be a vertical line, then angle V O C = angle F O R is the angle of lead. All this rearrangement of the lead is supposed to have been done in Fig. 60.

But a reference to Fig. 60 will also show that the mag-netism of the ring reacts on the magnetism of the pole-pieces. The lines of force in the iron of the left pole-piece are crowded up towards the top corner, and in the right pole-piece are crowded toward the bottom, as if the polarity had been attracted upwards on one side and downwards on the other. The density of the field is completely changed from what it was in Fig. 50. The lines of force at the upper left side are crowded together and are twisted across. The resultant N. pole of the ring—marked *n, n, n,* where the lines of force emerge from the ring—attracts the S. pole—marked *s, s, s,* where the lines of force enter the field magnet—and the steam-engine which drives the dynamo has to do hard work in dragging the armature round against these attractions. The stronger the current in the armature the stronger will be the poles in the ring, and in the field magnets the stronger will be the attraction of *n, n, n* toward *s, s, s,* and the steam-engine must work still harder to keep up the speed. It will also be noticed in this figure that a *few* of the lines of force due to the current in the armature—two of them are shown dotted in the figure—leak across internally and contribute nothing to the external field. The oblique direction of this internal field marks the angle of lead of the brushes. It will be remarked that the innermost layers of iron of the ring are magnetised differently from the outermost, for the "*n*" pole of the outer layers of iron occupies a region lying obliquely on the left, while the "*n*" pole of the inner layers lies to the right of the highest point. All these phenomena—the shifting of the field—its concentration under the pole-piece—the weak internal field—the discrepancy between the positions of the induced poles on the inner and outer sides of the ring, can all be observed in an actual dynamo. Fig. 63 shows the pattern produced experimen-

FIG. 63.

FIELD OF DYNAMO.

tally in iron filings by placing a magnetised ring between the poles S N of a field magnet, which would tend to induce in it poles *n'*, *s'*, and giving its own poles *n*, *s*, the proper lead. It should be compared with Figs. 50 and 60. It may perhaps be objected that in Fig. 60 the internal poles marked do not lie exactly at right angles to the external poles of the ring. Nor do they in actual dynamos. The position of the internal poles is determined by the lead given to the brushes, and *the brushes are so set that the diameter of commutation lies at the point of least sparking;* and this, as we shall see, is not exactly at right angles to the induced external poles in the ring, or to the diameter in which the density of the magnetic field is the greatest. Returning for a moment to Fig. 62, it will be seen that if O F or R C represents the intensity of the magnetic force due to field-magnets, and O C or F R that due to the armature current, then

$$\frac{O\,C}{R\,C} = sin\,C\,R\,O = sin\,F\,O\,R,$$

or the ratio of the two magnetic fields is proportional to the sine of the angle of lead—not to the tangent, as assumed by Breguet.

In the case of drum armatures, the phenomena, though of the same kind, are a little less easily traced. In consequence of the over-wrapping of the windings on the outside of the armature, the currents in some of the windings are partially neutralised in their magnetising effect on the core by those that lie across them, and consequently the polarity due to the current is not so well marked as with ring arma-tures. Neither can there be any internal field. But, with these exceptions, the same considerations apply as those we have traced out above. In fact drum armatures are less liable to induction troubles of all kinds than are ring-arma-tures.

That the obliquity of the diameter of commutation, and the resulting lead of the brushes is chiefly due to the obliquity of the effective magnetisation of the armature core, is finally demonstrated by the following facts. In those dynamos which are arranged to give a constant difference of potentials at the terminals, the current in the field-magnet rises very nearly *pari passu* with the current in the armature, and the lead in such machines is very constant, whether the currents are large or small. Moreover, in these dynamos if (as is the case in all good modern machines) the magnetism induced by the field-magnets in the armature core, is very much more powerful than that induced by the armature coils, there is practically no lead at all ; for the cross-magnetisation being negligible, *sin* F O R (Fig. 66) will be = 0° and the angle of lead V O C = 0° also. On the other hand, in those dynamos which are designed to yield a constant current (for arc-light circuits), there ought on this theory to be, and is in fact, a considerable lead, and one which varies with the resistance introduced into the circuit. For O C is constant, whilst O F must vary very nearly in proportion to the resistance of the circuit. The author has therefore suggested that in a constant current dynamo, there should be an automatic arrangement, to continually adjust the angle of lead, so that its sine shall be proportional inversely to the magnetism of the field-magnet.

Causes of Sparking.—Sparking at the brushes is due to several causes, chiefest amongst them being defective adjustment of the brushes ; long flashing sparks will inevitably be produced if the brushes are not set close to the neutral point. Hence the need of the adjustment mentioned on page 60. Another cause of sparking is want of symmetry in the winding of the armature: some of the older forms of the Siemens armature were defective in this respect. If the coils of one half of the armature are either more numerous or *nearer to the iron core*, on the average, than those of the other half, the two induced electromotive-forces in the two halves of the armature will be unequal, and, consequently, at every revolution, the neutral points will shift first forward, then backward, giving rise to sparks. Jumping of the brushes when the collector is untrue, or when the brush-holders are defective, is another prolific cause of sparking.

Reactions due to Induction.—We have next to consider certain effects due to self-induction, and to mutual induction.

In the armature when at work, half the current flows *up* the

H

coils on one half of the ring—say on the left—and the other
half of the current flows *up*[1] those in the other half of the ring
on the right. If the brush is near the top, the current flows
from left to right through the section on the left of the brush,
and from right to left through the section that is on the right
of the brush. But the section on the left of the brush is
gliding round (to the right in almost all these diagrams),
and will presently belong to the other half of the ring. It is
clear then that as each section passes the brush the current
in it is stopped and reversed. Of course, if things are properly arranged, it will itself not be actively inducing any
current at the moment when it passes the brush, but it is
receiving the current generated in the other sections. Moreover, there is just an instant when the two collector-bars, to
which the ends of this section of the coil are connected, are
both in contact with the brush; and therefore, just for an
instant, this section of the coil will be short-circuited.

   If, at the moment of being short-circuited, the coil is
cutting any lines of force, there will be an electromotive-
force acting in that coil. Such an electromotive-force, even
though small, may produce momentarily a large current,
because the short-circuited resistance is so small. If, as
in ring-armatures, there is an internal field (p. 95, top), then
the position of no induced electromotive-force—and this in a
well-designed dynamo (see p. 68) is the same point as the
point of maximum potential—will be at such a point of the
field that the algebraic sum of the lines of force which enter
and leave the ring at that point is zero. Even if the induction
at the moment of short-circuiting is zero, still short-circuiting
produces a reaction. The ideal arrangement is attained if
the brushes be shifted just so far beyond the point of maximum electromotive-force that while the sections pass under
the brush and are short-circuited they should be actually
beginning to cut the lines of force and have a small negative

---

[1] That is to say the currents flow *up* if the north pole is on the right, the ring
right-handedly wound and the rotation right-handed: but if any one of these
things is reversed the currents will flow *down*. If any two of them are reversed
the currents will flow up.

electromotive-force induced in them ; and this action should last just so long in each successive section as to stop the current that was circulating, and start a current in an opposite direction and let it grow exactly equal in strength to that which is circulating in the other half of the armature, which it is then ready to join. If this set of conditions could be attained there should be *no sparks*. This is the main reason why the true neutral point is found to be a little ahead (in generators) of the point of maximum potential.

If the brushes are not shifted forward as far as the point of maximum potential, there will be more waste of energy in the coils than if they are short-circuited at that point. If the brushes are shifted far beyond that point there will also be waste of energy.

We know that every electric current possesses a property sometimes called " electric inertia," sometimes called " self-induction," by virtue of which it tends to go on. Just as a fly-wheel once set in motion tends to go on spinning, so a current circulating round a coil tends to go on circulating, even though the connection with the source be cut off. True, the current lasts in most cases only for a small fraction of a second, but it *tends* to go on. It is also known that this quasi-inertia is connected with its magnetic properties (see the Introductory Remarks, p. 26), and that it is in the current's own magnetic field that this inertia of self-induction resides.

A current circulating round an iron core has a much greater electric inertia (or self-induction), because it has a more intense magnetic field, than one without an iron core. It requires an expenditure of energy to *start* a current because of this pro-perty ; and that energy may be considerable. We know that— to return to a mechanical analogy—it requires much energy to set a heavy grindstone spinning ; when once spinning it does not require much energy to keep it going, only enough in fact to overcome the friction of the pivots. Also, if we stop the spinning grindstone, say by holding a piece of wood against it as a "brake," it will give up the energy that has been put into it and will manifest this energy in the form of heat.

H 2

So also the electric current circulating in a coil possesses energy, and if we stop it by opening the circuit, that energy will show itself by a spark, the spark of the so-called (but mis-named) "extra-current." If we short-circuit the coil, its current—unless there is an electromotive-force to keep it up—will also be stopped by the internal quasi-friction which we commonly call the "resistance" of the wire, and the wire will be heated. A frequent accident to dynamos is the burning of the insulation, or even the fusion of the wire of one section of the armature which has become short-circuited.

Spurious Resistance.—Now all these things clearly have a bearing on that which happens as the sections of the coils pass the brushes. In each section the current tends to go on, and in fact does actually go on for a brief time after the brush has been reached. Then the energy of the current in that section is wasted in heating the copper wire during the interval when it is short-circuited; and as it passes on, energy must again be spent in starting a current in it in the inverse direction. All these reactions are of course detrimental to the output of current by the dynamo: especially the loss in short-circuiting. It has been shown by M. Joubert[1] that the loss of energy due to the mere reversals of the current in the sections of a ring armature is equal to $\dfrac{n\,L\,i_a^2}{4}$ per second, where n is the number of revolutions per second, L the coefficient of self-induction for the entire ring, and i_a the armature current. Professors Ayrton and Perry have more recently pointed out[2] that the matter may be conveniently expressed in another way.

---

[1] *Comptes Rendus*, June 23, 1880, January 9, 1882, March 5, 1883; and *L'Electricien*, April 1883. The proof of the above expression is simple. Electric work per second is expressed as product of an E. M. F. and a current. If $\bar{N}$ magnetic lines are cut n times per second the average E. M. F. is $n\,N$. If a ring circuit has coefficient of self-induction L, then current $\frac{1}{2}i^a$ on being sent through that circuit will create $\frac{1}{2}L\,i_a$ lines of force. Hence the current $\frac{1}{2}i_a$ is virtually caused n times per second to cut across $\frac{1}{2}L\,i_a$ lines of force; or the work done in stopping the half-armature current in all the sections one after the other for a second of time is $\frac{1}{4}\,n\,L\,i^2$.

[2] *Journ. Soc. Telegr. Eng. and Electr.*, vol. xii., No. 49, 1883, where, however, the letters n and L are used in a slightly different sense.

Since the energy per second conveyed by a current running through a resistance r is equal to ri , it is evident that the energy lost per second by self-induction is the same as if there were an additional resistance [1] in the armature of the value $r = \dfrac{n\,\mathrm{L}}{4}$. There is, therefore, in a rotating armature, an *apparent* increase of resistance proportional to the speed, and this apparent increase, due to self-induction, cannot be got rid of by increasing the number of sections of the armature. It can be diminished in degree by using in the armature more iron and fewer turns of wire, in other words by diminishing the magnetic moment of the coil while giving the field magnet an increased advantage. The existence of an apparent resistance varying with the speed was first pointed out by M. Cabanellas.[2]

Eddy-Currents.—There are two other inductive reactions in the armature to be considered. In the iron of the armature cores, internal eddy-currents (the so-called " Foucault currents ") may be set up, absorbing energy and producing detrimental heat ; and such currents may be even produced within the conductors which form the coil of the armature, if these are massive as are the bars in Edison's dynamo. Frölich, in 1880,[3] pointed out the effect of the presence of these currents ; and to them he attributed not only the otherwise unexplained deficit in the work transmitted electrically by a generator to a motor, but also the diminution in the effective magnetism (discussed above as a result of cross-magnetism, and found by Frölich to amount to 25 per cent. of the whole magnetism) observed with great currents and high speeds ; and further he attributed to this cause the apparent increase in the number of " dead-turns "[4] at high speeds. Doubtless such currents exist, and the energy they

---

[1] Professor O. J. Lodge has given the more accurate value of the spurious resistance as $r = n\dfrac{\mathrm{L}}{4} + \dfrac{(\pi n\,\mathrm{L})^2}{8\,\mathrm{R} + 2\,n\,\mathrm{L}}$; see *Electrician*, July 31, 1885.

[2] *Comptes Rendus*, January 9, 1882 ; see also Picou, *Manuel d'Electrometrie*, p. 123.

[3] Berlin Academy, *Berichte*, Nov. 18, 1880, and *Elektrotechnische Zeitschrift*, vol. ii., p. 174, May 1881.   [4] See Appendix VII.

waste will be nearly proportional to the square of the speed[1] : but they may be indefinitely diminished by proper lamination, insulation, and disposition of the structures of the armature. The new laminated armature of the Brush Machine (Fig. 181), when used in place of the old solid armature (Fig. 182), was found to diminish greatly the number of dead turns, as well as not wasting so much energy in heating.

Effects of Mutual Induction.—Some forms of armature are peculiarly defective in the matter of being so constructed as to allow of much induction between neighbouring sections or parts of the coil, causing the rise of the current in one section to exert an opposing induction on a neighbouring section, and thereby, though not necessarily wasting any energy, making the machine act as if it were a smaller machine. The Bürgin armature, which has six or eight rings side by side on one spindle, suffers from induction between each section, and those belonging to the rings on the right and left of it : and it is only by a careful alternation of positions that this defect has been mitigated. In armatures of the Niaudet and Wallace-Farmer type each of the parallel coils acts inductively on its neighbour. Beyond doubt the armature with least of this defect is the Siemens (Alteneck) drum armature as used in Siemens, Edison, and Weston machines. Clausius has shown[2] that after a coil has been short-circuited on passing a brush, it exercises a deleterious inductive effect on the neighbouring coil in advance of it, and that this effect is proportional to the number of turns in the section. It can therefore be diminished by increasing the number of sections, thereby diminishing the number of turns of wire in any one section of the armature.

Lag due to Self-Induction.—This electric inertia of the current which circulates in the sections affects slightly the lead that must be given to the brushes, and it also reacts on

---

[1] Clausius has introduced terms into his equations (*Wied. Ann.* xx., p. 354, 1883 ; and *Phil. Mag.*, Series 5, xvii., p. 46 and 119, 1883) to include the effects of the eddy-currents ; see Appendix viii. They have also been theoretically treated by H. Lorberg. (*Wied. Ann.* xx., p. 389, 1887.)

[2] *Wiedemann's Annalen*, Nov. and Dec. 1883, and *Phil. Mag.*, Jan. and Feb. 1884.

the neighbouring coils. Whenever a coil is short-circuited, the sudden rush of its own current round itself tends by mutual induction to stop the current in the coil behind it, and to accelerate the inverse current in the coil in front of it. These actions are diminished by increasing the number of sections and making the individual sections consequently smaller. The self-induction even extends to the iron of the cores. In every particle of the iron at the moment when it arrives at the position where its magnetism must be reversed, an internal current is set up which retards the reversal of the magnetism and makes it *apparently* lag in its magnetisation, as well as grow hot. This effect can also be diminished by properly laminating the core and arranging it so that its magnetism is reversed gradually instead of suddenly. Niaudet's arma-ture and that of the Wallace-Farmer machine are essentially defective from this latter point of view.

Induction is of enormous importance in alternate-current machines ; and indeed everywhere, throughout dynamos in general, but we cannot dwell longer on its effects at this point.

Remedy for Induction Troubles.—The one important way of diminishing these deleterious reactions is happily a very simple one. It is shown in Chapter XV. that the electro-motive-force of the dynamo is proportional to three things, the number n of revolutions per second, the total number N of magnetic lines in the effective field, and the number C of conductors around the armature. Now, for a given size of armature, the inductive reactions are proportional to C. If we can decrease C while increasing either of the other terms, we may thereby decrease the deleterious reactions and yet keep the same electromotive-force as before. Now it is inconvenient to increase the speed, and moreover some of the deleterious reactions, mechanical (such as friction) as well as electrical, increase when the speed increases. The only way then is to increase N, the number of magnetic lines in the effective magnetic field. This can be done by having enormously strong field-magnets which will entirely overpower the armature. If the field-magnets are large, and of wrought iron, and if there is plenty of iron in the armature core, then, without increasing the speed, we may get the same electro-motive-force while using fewer turns of wire in the armature. The ideal dynamo of the future has but one turn of wire to each section. It will have practically no lead at the brushes, will not spark, and its internal resistance will be practically *nil.*

It is also important to observe that distortion of the magnetic field and some of the resulting troubles can be partially obviated by so shaping the polar surfaces that they come nearer to the armature at the region at right angles to the diameter of commutation ; the pole-pieces being cut away so as to give a wider clearance at the outer edges. It is obviously possible by proper shaping to produce concentration of the magnetic lines at any desired region of the magnetic field.

Heating of Cores.—It is impossible to prevent the cores of armatures from heating ; and this is in every case detrimental to the action of the dynamo. Hot iron has a lower magnetic susceptibility than cold iron, and therefore requires a greater expenditure of current to magnetise it to an equal degree. This causes the output and efficiency of a dynamo to be less when hot than when cool. To say nothing of the risks of destruction by overheating, this is an additional reason for so designing dynamos as to secure proper ventilation of armatures.

Heating of Magnets.—All field-magnets are liable to heat : the cores by reason of eddy-currents induced in them ; the coils because even the purest copper offers resistance. The amount of heat developed per second in a coil is the product of the resistance into the square of the strength of the current. To avoid waste, therefore, no unnecessary resistance should be introduced into the coil. It is easy to show that with a coil of *given volume*, the heat-waste is the same for the same magnetising power, no matter whether the coil consist of few windings of thick wire or many windings of thin wire. The heat per second is $i^2 r$, and the magnetising power is $S i$; i being the current, r the resistance, and S the number of turns. But r varies as the square of S, if the volume occupied by the coils is constant: for suppose we double the number of coils, and halve the cross-sectional area of the wire. Each foot of the thinner wire will offer twice as much resistance as before; and there are twice as many feet of wire. The resistance is quadrupled therefore. The heat is then proportional to $i^2 S^2$: and therefore the heat is proportional to the square of the magnetising power. If, therefore, we apply the same magnetising power by means of the coil, the heat-waste is the same, however the coil is wound. To magnetise the field-magnets of a dynamo to the same degree of intensity requires the same expenditure of electric energy, whether they are series-wound or shunt-wound, provided the volume is the same. But if the *volume* of the coil (and the weight of copper in it) may be increased, then the heat-waste may be proportionally lessened. For example, suppose a shunt coil of resistance r has Z turns, if we wind on another Z turns in addition, the magnetising power will remain nearly the same, though the current will be cut down to one-half owing to the doubling of the resistance; and the heat loss will be halved, for $2\,r \times$ ($\frac{1}{2}\,i^2$) will be $\frac{1}{2}\,i^2 r$. In fact one ought to wind on so much copper wire that the annual interest on the prime cost is exactly equal to the annual cost of the electric energy spent in the inevitable heating. This law is not quite exact, because the magnetising effects of all the turns are not exactly equal to one another except when the core forms a closed circuit. It is also assumed in the foregoing argument that we get double the number of turns on if we halve the sectional area of the copper wire. This is not quite true, because the thickness of the insulating covering bears a greater ratio to the diameter of the wire for wires of small gauge than for wires of large gauge. In designing dynamos, moreover, one ought to be guided by the question of economy, not by the accident of there being only a certain volume left for winding. If there is insufficient space round the cores to wind on the amount of wire that economy dictates, new cores should be prepared having a sufficient length to receive the wire which is economically appropriate.

But there is another cause of heating in field-magnet

cores. Whenever, either from a change in the strength of the
exciting current, or from a change in the reactive influence of
the armature, any variation in the magnetisation of the core
occurs, such variation is inevitably accompanied by a generation of internal induced currents. Every one knows how, in
the ordinary induction-coil, the changes of magnetisation of
the core induce transient currents in the secondary wire. The
same is true of dynamos; any change in the magnetism
arouses transient currents in the surrounding coil (which acts
both as primary and secondary), or in the core, or in both.
If the core is solid it heats. Ought we then to laminate the
entire structure, or build our field-magnets of bundles of iron
wire? If we do they will certainly heat less, but any changes
in their magnetisation that occur, will occur much more suddenly; the momentary extra-currents induced in the external
coils will be fiercer and more dangerous. The necessity for
keeping the magnetism of the field-magnets steady, dictates
solidity (p. 52). In certain machines a copper envelope
is purposely placed around field-magnet cores to absorb the
induced extra-currents that arise from variations in the
magnetism of the core. They add, electrically speaking, to
the stability of the field-magnets, for the induced extra-
currents always circulate in such a sense as to oppose the
change of magnetisation which gives rise to them.

Heating in Pole-pieces.—If the masses of iron in the armature are so disposed that as it rotates, the distribution of the
lines of force in the narrow field between the armature and
the pole-piece is being continually altered, then, even though
the total amount of magnetism of the field-magnet remains
unchanged, eddy-currents will be set up in the pole-piece and
will heat it. This is shown by Figs. 64 to 68, which represent
the effect of a projecting tooth, such as that of a Pacinotti
ring, in changing the distribution of the magnetism of the
pole-piece. Figs. 67 and 68 (corresponding respectively to
Figs. 65 and 66) show the eddy-currents, grouped in pairs
of vortices. The strongest current flows between the vortices
and is situated just below the projecting tooth, where the
magnetism is most intense; it moves onward following the

tooth. Fig. 69 shows what occurs during the final retreat of the tooth from the pole-piece. These eddy-currents pene-

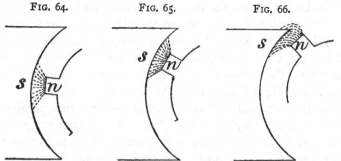

FIG. 64.   FIG. 65.   FIG. 66.

ALTERATION OF MAGNETIC FIELD DUE TO MOVEMENT OF MASS OF IRON IN ARMATURE.

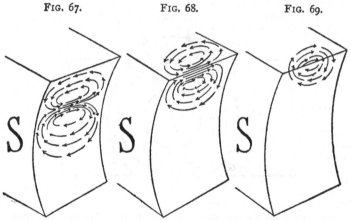

FIG. 67.   FIG. 68.   FIG. 69.

EDDY-CURRENTS INDUCED IN POLE-PIECES BY MOVEMENT OF MASSES OF IRON.

trate into the interior of the iron, although to no great depth. Clearly the greatest amount of such eddy-currents will be generated at that part of the pole-piece where the magnetic perturbations are greatest and most sudden. A glance at Figs. 60, 63, 68, and 69 will at once tell us that this should be at the leading corner or "horn" of the pole-piece of the generating dynamo. As a matter of fact, when any dynamo which has horned pole-pieces (such as the Gramme) has been running for some time as a generator this is found to be the case. The leading horns *a* and *c*, of Fig. 70, are found to be hot, whilst the following horns *b* and *d* are found to be comparatively cool. When the dynamo is used as a motor, the reverse is found to be the case : the leading horns *a* and *c* are cool, the following horns *b* and *d* are hot. A reference to the magnetic field of the motor, as drawn in Chap. XXVIII., will explain the latter case.

Closely connected with this effect is another, first pointed out to the author by M. Cabanellas. A Gramme magneto machine with permanent magnets is observed to lose power during its use as a motor ; the field-magnets decrease in strength. If, then, it is used as a generator, the field-magnets regain their magnetism. This seems at first sight impossible,

FIG. 70.

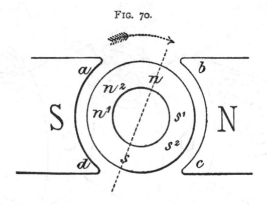

because the magnetic fields respectively due to the field-magnet and to the armature help one another in the motor (Chap. XXVIII.), whilst they oppose one another's actions (p. 90) in the generator. The effect is explicable[1] when the magnetising effect of the eddy-currents is taken into consideration.

---

[1] The following explanation was given by the author at the International Conference of Electricians at Philadelphia, 1884 (see report in *Electrical Review*, Dec. 13, 1884). "To explain these facts, and their mutual relation, I must relate one other observation which I have made, and which connects both sets

of facts. . . . . Suppose you take a horse-shoe magnet, having the usual armature or ' keeper ' of iron. You can purchase such an instrument of any optician,
who will probably give you instructions never to pull the armature off suddenly
for fear you injure the magnetism. He could not possibly give you worse directions. Take such a magnet and try what the effect really is. Fasten it down
upon a board with brass screws, and fix a magnetometer near it—a common
compass will answer—and notice how much the magnet pulls the needle round.
Then put on the armature, by placing it at the bend of the magnet; draw it
slowly to its usual position, and suddenly drag it off. You will find that by this
action your magnet will have grown stronger. Do this twenty times, and you
will make it considerably stronger. I have made a magnet 1·2 per cent. stronger
by putting on the armature very gently and pulling it off suddenly. If you reverse
the operation, by letting the armature slam suddenly against the poles and then
detaching it gently, you will find that the magnetism will go down. I have made
magnets lose 1·3 to 2·1 per cent. in this way. Why does this occur? How does
it explain the two phenomena noticed just now? If you suddenly take away a
piece of iron from a magnet, you do work against the magnetic attraction, and
the induced currents which are set up in the iron or steel of the magnet are
always (as we know from Lenz's Law) in such a direction as to oppose the motion ;
that is to say, they are in such a direction as will make the magnet pull more
strongly than before. By suddenly detaching the armature, we magnetise the
magnet more strongly than before, by means of currents circulating within its
own mass and within the mass of the armature. In the reverse motion, when
you allow the armature to slam up, there are induced currents which are in such
a direction as to oppose the motion of slamming ; they, therefore, decrease the
magnetism of the magnet. Apply this to the dynamo and to the motor. You
magnetise more highly by pulling off the armature. That is precisely what is
occurring in the field when the machine is being used as a generator. You are
dragging away the armature from the active horn a of the pole-piece, and the
effect is to generate induced currents in that horn. It therefore gets hot. So
does the other leading horn c, for the very same reason. In the case of the motor
the horns b and d are the active ones, and the armature is being continually
dragged up toward them, and they get hot from internally induced currents. It
is for this reason that in my Cantor Lectures (and also p. 56, ante) I recommended
that pole-pieces should always be laminated. The presence of these induced
currents explains the heating effect, and it also explains how it is that when a
magneto machine is used as a motor the magnet is weakened, and when used as
a generator the magnet is strengthened."

CHAPTER VI.

GOVERNMENT OF DYNAMOS.

Methods of Exciting the Field-Magnetism.

THE five simple methods of exciting the magnetism that is to be utilised in the magnetic field may be grouped under two heads, according to whether the armature of the machine supplies the machine's own magnetism, or whether the magnetism is provided for from some other source.

FIG. 71.

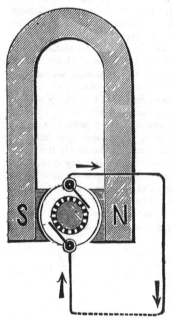

THE MAGNETO-DYNAMO.

Magneto - dynamo. — In the oldest machines there was no attempt to make the machine excite its own magnetism; this was provided for it once for all by the employment of a permanent magnet of steel. Unfortunately, the supposed permanent magnetism of steel magnets slowly decays, and is diminished by every mechanical shock or vibration to which the machine is subjected.

The *magneto-electric machine,* or *magneto-dynamo,* a diagrammatic drawing of which is given in Fig. 71, survives, indeed, in numerous small types of machines. It has the serious disadvantage of being both heavier and bulkier than other dynamos of equal capacity, because steel cannot be permanently mag-

netised to the same high degree as that to which wrought
iron, or cast iron, or even steel itself can be temporarily raised.

Separately-excited Dynamo.—It was an obvious step to
substitute for steel magnets, electromagnets excited by means
of currents from some independent source such as a voltaic
battery. The *separately-excited dynamo* (Fig. 72) comes there-

FIG. 72.

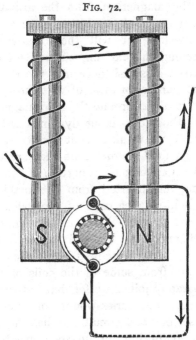

THE SEPARATELY-EXCITED DYNAMO.

fore second in the order of development. Though used by
Faraday, this method did not come into acceptance until, in
1866, Wilde employed a small auxiliary magneto machine to
furnish currents to excite the field-magnets of a larger dynamo.
The separately-excited dynamo, in common with the magneto
machine, possesses the property that, saving for reactions due
to the armature-current, the magnetism in its field, and therefore the electromotive-force of the machine, is independent of
changes of resistance going on in the working circuit.

The dynamos of either of the preceding kinds can be governed, either by altering the speed or by altering the amount of magnetism that passes across the armature. For long it has been the fashion to control the electromotive-force of magneto machines by the device of providing a movable piece of iron, which could be placed more or less over the poles of the field-magnet, serving as a magnetic shunt to divert some of the magnetism from the armature. In the case of separately - excited machines there are two other methods of diminishing at will the effective magnetism, namely by weakening the exciting current, for example, by introducing more or less resistance into the exciting circuit, or by altering the number of turns of wire through which the exciting current circulates around the field-magnet.

The simple methods of making dynamos self-exciting are three in number : (1) the whole current from the armature may be carried through field-magnet coils that are connected in series with the main circuit; (2) a portion of the current from the armature may be diverted from the main circuit and carried through field-magnet coils of somewhat high resistance connected as a shunt ; (3) the current required to excite the field-magnet may be procured either from a second armature revolving in the same field, or (if the armature consists of many coils) from some of the coils of the armature that may be separately joined up for that purpose.

Series Dynamo.—The series-wound, or ordinary dynamo (Fig. 73), possesses but one circuit. It has the disadvantage of not starting action until a certain speed has been attained, or, unless the resistance of the circuit is below a certain limit ; the machine refusing to magnetise its own magnets when there is too much resistance and too little speed. The speed that is critical for a given resistance in the circuit is discussed in Chapter XXII. Series-wound machines are also liable to become reversed in polarity, a serious disadvantage, and one that unfits this type of machine for employment in electroplating or for charging accumulators. Any increase in the resistance of the series-wound dynamo lessens its power to supply current, because it diminishes the current in the coils of the field-magnet, and therefore diminishes the amount of the effective magnetism. When lamps are in series (as in a 'Brush' arc-light circuit) in the circuit of a series-wound dynamo, the switching in of an additional lamp both adds to the resistance of the circuit and diminishes the power of the machine to supply current. On the other hand, when lamps are in parallel across a pair of mains fed by a dynamo, if that dynamo is series-wound, the switching on of additional lamps not only diminishes the resistance of the circuit, but causes the field-magnets to be further excited by the increased current, so that the more lamps are on, the greater becomes the risk of their getting too great a current.

FIG. 73. FIG. 74.

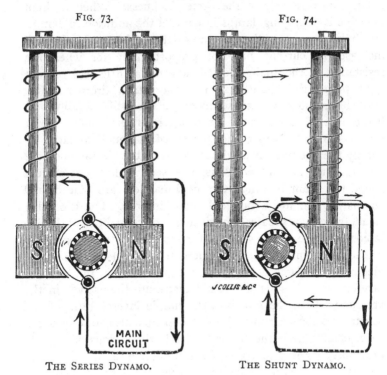

THE SERIES DYNAMO. THE SHUNT DYNAMO.

Shunt Dynamo.—In the shunt-wound machine (Fig. 74) the field-magnet is wound with many turns of fine wire to receive only a small portion of the whole current generated in

I

the armature. These coils are connected to the brushes of the machine, and constitute a by-pass circuit or shunt. Shunt machines are less liable to reverse their polarity than series machines. Owing to the somewhat greater cost of the fine wire of the shunt coil they are slightly dearer in prime cost than series machines of equal power ; but the expenditure of electric energy to keep up the magnetism is practically alike in both cases. It requires the same expenditure of electric energy to magnetise an electromagnet to the same degree, whether the coil consists of many turns of thin wire or of a few turns of thick wire, provided the *volume* occupied by the coil be alike in the two cases, and provided the insulation is relatively of the same thickness. When a shunt machine is supplying lamps in parallel the addition of lamps, which brings down the nett resistance of the circuit, will increase the current, but not proportionally, for when the resistance of the main circuit is lowered a little less current goes round the shunt and the magnetism drops a trifle ; nevertheless, such a machine may regulate itself tolerably well if the internal resistance of its armature is very small. For a set of lamps in series, the power of a shunt dynamo to supply the needful current increases with the demands of the circuit, since any added resistance sends additional current round the shunt in which the field-magnets are placed, and so makes the magnetic field more intense. On the other hand, there is a greater sensitiveness to inequalities of driving in consequence of the great self-induction in the shunt. As previously pointed out, when there are sudden changes in the electromotive-force acting in a complex circuit, the momentary currents thus set up do not distribute themselves in the various parts of the circuit in the simple inverse ratio of the resistances, for their distribution depends also, and in some cases chiefly, upon the self-induction in the various parts. As previously explained (p. 99) self-induction is an effect like inertia. It is more difficult to set up a sudden current in a circuit whose self-induction is great (or which, for example, consists of many turns wound closely together, so that they exercise great inductive action on each other, especially if they be wound about an iron core) than in one in which the self-induction is small. We cannot here follow further the mathematical law of the action of self-induction on momentary changes of electromotive-force ; but the application to the shunt-wound dynamo is too important to be passed over. The shunt part of the circuit in the present case consists of a fine wire of many turns wound upon iron cores. It therefore has a much higher coefficient of self-induction than the rest of the circuit ; and, consequently, any sudden variations in the speed of driving cannot but affect the current in the main circuit more than in the shunt. Briefly, the shunt-winding, though it steadies the current against perturbations due to changes of resistance in the circuit, does not steady the current against perturbations due to changes in speed of driving. In the series-wound dynamo, the converse holds good. The electromotive-force of the shunt-machine can be controlled by introducing a variable resistance into the shunt circuit.

Separate-circuit Self-exciting Dynamo.—There is yet the third species of self-exciting machine, in which the field-magnet coils are arranged to form part of a circuit entirely separate from the main circuit, but are supplied with currents from coils rotating in the field. There are two ways of carrying this into effect : (1) a second armature may be made to rotate between the same field-magnets in order to supply the exciting current. Machines on this plan were devised by Ladd[1] and by O. and F. H. Varley[2] ; (2) a few of the armature coils may be connected up separately to a special commutator to supply an exciting current. Such systems were devised by Wilde, Holmes, and Lontin, about the year 1868 or 1869, in order to make their alternate-current machines excite themselves. Holmes described a machine having twenty helices in the armature, ten of which supplied alternate currents to the lamps, whilst the remaining ten, or any part of them, could be so connected up through a special commutator

---

[1] *Phil. Mag.*, xxxiii., 544, 1867.
[2] Specification of Patent, 2525 of 1869.

as to supply the exciting current to the field-magnets. Ruhmkorff attained the same end by winding a second wire upon the Siemens (shuttle-wound) armature, which then was provided with a commutator at each end.

Any of these systems may be applied for direct-current machines. For alternate-current machines, neither series winding nor shunt winding is applicable. Each of these five systems of exciting the field-magnetism has its own merits for special cases, but none of them is perfect. Not one of these methods will ensure that, with a uniform speed of driving, either the potential at the terminals or the current shall be constant, however the resistances of the circuit are altered.

But though theory tells us that none of these systems is perfect, theory does not leave us without a guide. Thanks to M. Marcel Deprez, to Professor Perry, to Mr. Paget Higgs, to Mr. Bosanquet, to Messrs. Crompton and Kapp, to Herr Schuckert, to Messrs. Watson and Mordey, to Herr Zipernowsky and others, we have been taught how to combine these methods so as to secure in practice a machine which shall, when driven at a constant speed, give either a constant potential or a constant current. These methods are carefully developed in Chapters XX. and XXI. They are briefly described here also, so as to complete our summary of the methods of exciting the field-magnets.

COMBINATION METHODS.

The discovery of the method of rendering a dynamo machine automatically self-regulating when driven at a uniform speed, is due to M. Marcel Deprez, and is a result arising from the study of the diagrams of the characteristic curves of dynamos.[1] There are two distinct cases for which self-regulation is required.

As the first function of a dynamo in practice is to feed with

---

[1] See *La Lumiere Électrique*, December 3, 1881, and Jan. 5, 1884.

sufficiency and regularity a system of lamps, and as those lamps are always[1] in practice arranged either in parallel or in series, it is clear that in the former case *a constant difference of potentials*, and in the latter *a constant current* between the mains, is required.

Suppose a dynamo to have an armature without internal resistance, and to have its field-magnets excited from some independent constant source. At a constant speed it would give a constant potential at its terminals whatever the resistance in the circuit. But if it has internal resistance, the external potential will be less than the whole electromotive-force, and the discrepancy will be greater according as the internal resistance and the current are greater. Any resistance-less, separately-excited, or shunt dynamo would thus be self-regulating.

Now it is, we know, impossible to have an armature of no resistance. But if, knowing the resistance of the armature of our dynamo, we find out what additional magnetising power is necessary to increase the working electromotive-force of the dynamo, so that the nett electromotive-force (after discounting the part needed to overcome the internal resistance) shall be constant, and then, having found it out, provide for this variable part of the magnetisation by putting on coils in series, our dynamo thus reinforced will act as if it had no internal resistance, and will give, within certain limits, a constant difference of potentials at its terminals.

On the other hand, if a shunt dynamo were constructed, with an armature of considerable resistance, the electromotive-force which it would develop at a constant speed, would be nearly proportional to the external resistance, for doubling the external resistance would very nearly double the proportion of current thrown round the shunt, and therefore (always assuming the iron cores to be far from saturation) the magnetism of the field-magnet would be doubled; in other words

---

[1] I am aware that occasionally incandescent lamps have been arranged with two or three lamps, in series, in each parallel, or on a multiple series plan. In any case, distribution must fall under one or other of the two cases considered.

there would be an approximately constant current.[1] In this case, a high internal resistance in the armature would not be economical. But if we ascertain the internal resistance of the shunt dynamo, and make a similar calculation as to the amount of additional electromotive-force requisite in order that there shall always be enough current for the shunt circuit over and above that current which goes to the external circuit; and if we provide from some external constant source for this additional electromotive-force, either directly or by adding to the magnetisation, then the shunt dynamo so aided will theoretically give a constant current in the external circuit, no matter how great or how small the resistance of the circuit may be.

For distribution *at a constant potential*, we must have, therefore, dynamos in which there is a combination of *series* coils with some auxiliary independent constant excitation.

For distribution *with a constant current*, we must theoretically have dynamos in which there is a combination of *shunt* coils with some auxiliary independent constant excitation.

Combinations to give Constant Potential.

(1.) *Series and Separate (Deprez).*—This method, illustrated in Fig. 75, can be applied to any series dynamo, provided the coils are such that a separate current from an independent source can be passed through a part of them, so that there shall be an initial magnetic field, independent of the main-circuit current of the dynamo. When the machine is running, the electromotive-force producing the current will depend partly on this independent excitation, partly on the current's own excitation of the field-magnets. If the machine be run at such a speed that the quotient of the part

---

[1] In the first edition of this work it was not adequately explained at this point, as it is in Chapter XX. fully, how the conditions for obtaining either a constant potential or a constant current are related to the winding of the coils, and to the various resistances of the machine and circuit. These rules admit, as yet, of no practical solution for constant-current work. Brush arc lamps are run in series, and want constant current; yet no Brush machines are shunt-wound, and the expense of the fine wire for winding the shunts would probably be prohibitive: there would have to be nearly a mile of wire for every lamp supplied. For constant-current work other modes of regulation have to be used.

of the electromotive-force due to the self-excitation, divided by the strength of the current, is numerically equal to the internal resistance of the machine, then the electromotive-

FIG. 75.

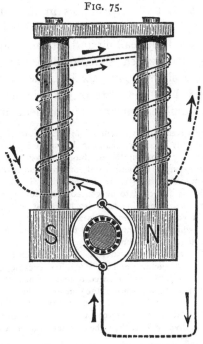

COMBINATION OF SERIES AND SEPARATE.

force in the circuit will be constant, however the external resistances are varied. M. Deprez has further shown that this velocity can be deduced from experiment, and that, when the critical velocity has once been determined, the machine can be adjusted to work at any desired electromotive-force by varying the strength of the separately-exciting current to the desired degree.

(2.) *Series and Magneto* (*Perry*).—The initial electromotive-force in the circuit required by Deprez's theory, need not necessarily consist in there being an initial magnetic field of independent origin. It is true that the addition of a permanent magnet, to give an initial partial magnetisation to the pole-pieces of the field-magnets, would meet the case to a certain extent; but Professor Perry has adopted the more general solution of introducing, into the circuit of a series dynamo, a separate magneto machine, also driven at a uniform speed, such that it produces in the circuit a constant electromotive-force equal to that which it is desired should exist between the leading and return mains.[1]

This arrangement, which is depicted in Fig. 76,[2] may be varied by using a shunt-wound exciter, the magnets being, as before, included in the part of the circuit outside the machine. The combination of a permanent magnet with

---

[1] Professor Perry gives, in his specification, the following numerical illustration, to which the only exception that can be taken is, that with so high a resistance as that of 3 ohms in the machine the system must be very uneconomical: "As an example, if there is only one dynamo machine, and if the resistance of the main cable, return cable, and machines, in fact of that part of the total circuit which is supposed to be constant, be, let us suppose, 3 ohms, then we find that the dynamo machine ought to be run at such a speed that the electromotive-force, in volts, produced in its moving parts, is three times the current, in ampères, which flows through the field-magnets; consequently this speed can readily be found by experiment. Suppose the constant electromotive-force of the magneto machine to be 50 volts, its resistance 0·3 ohms, and the resistance of the dynamo machine and of the other unchanging parts of the circuit 2·7 ohms; and suppose that the speed is that at which the electromotive-force produced by the rotating armature of the dynamo is three times the current. Now, let there be a consumers' resistance of 2 ohms, the total resistance is 5 ohms. Evidently the electromotive-force produced in the dynamo is 75 volts (for call this electromotive-force x, then the current will be $x \div 3$, whence it follows that $\dfrac{x + 50}{5} = \dfrac{x}{3}$, or $3x + 150 = 5x$, or $x = 75$), and the total electromotive-force in the circuit is, therefore, 125 volts, and as the total resistance is 5 ohms, the current is 25 amperes, giving an electromotive-force of 50 volts between the ends of the consumers' part of the circuit. Now, if the consumers' resistance increases to, say, 12 ohms, by some of the consumers ceasing to use their circuits, there is an instantaneous alteration of the electromotive-force produced in the dynamo to $12\frac{1}{2}$ volts, or $62\frac{1}{2}$ volts in the whole circuit, and $62\frac{1}{2}$ divided by 15 gives $4\frac{1}{8}$ amperes; and this means $4\frac{1}{8}$ multiplied by 12, or 50 volts as before, between the ends of the consumers' part of the circuit." (Here, again, the calculation is $\dfrac{x + 50}{15} = \dfrac{x}{3}$, whence $3x + 150 = 15x$, or $12\frac{1}{2} = x$.—S.P.T.)

[2] Some exception has been taken by Professors Ayrton and Perry to this figure on the ground that the magneto machine is drawn relatively too small. It was not intended that the figure should represent the sizes, but rather to indicate that the arrangement was essentially one of a *series* dynamo, plus an auxiliary independent excitement. So also with Fig. 80. The reader should refer to Professor Ayrton's lecture at the London Institution, Feb. 23, 1883, of which an abstract is given in *The Electrician*, March 10, 1883.

electromagnets in one and the same machine, is much older than the suggestions of either Deprez or Perry, having been described by Hjörth in 1854.

(3.) *Series and Shunt (Brush).*—A dynamo having its coils wound, as in Fig. 77, so that the field-magnets are excited

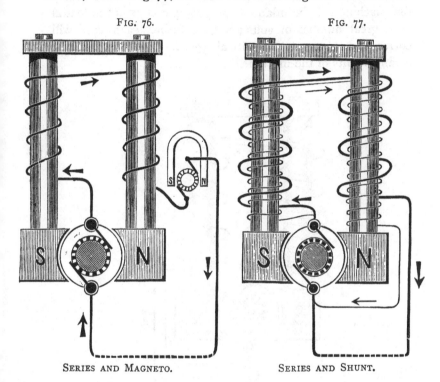

FIG. 76.   FIG. 77.

SERIES AND MAGNETO.   SERIES AND SHUNT.

partly by the main current, partly by a current shunted across the brushes of the machine, is no novelty, having been used in Brush dynamos[1] for some years past. The arrangement as used originally by Brush made the machine into one that was very nearly self-regulating, there being less than one volt of variation in the potential within a wide range of

[1] The shunt part of the circuit, originally called the "teazer," was adopted at first in machines for electroplating, with the view of preventing a reversal of the current by an inversion of the magnetisation of the field-magnets, but has been retained in some other patterns of machine on account of its usefulness in "steadying" the current.

current. If the shunt coils be comparatively few, and of high resistance, so that their magnetising power is small, the machine will give approximately a uniform potential of but few volts ; whereas, if the shunt be relatively a powerful magnetiser, as compared with the few coils of the main circuit, the machine will be adapted for giving a constant potential of a great number of volts ; but, as before, each case will correspond to a certain critical speed, depending on the arrangements of the machine.

Fig. 78.

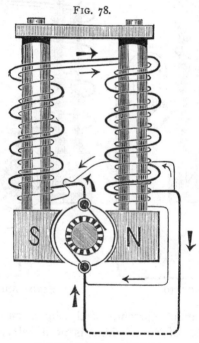

SERIES AND LONG SHUNT.

(4.) *Series and Long Shunt.*—In 1882 the author proposed to give this name to a combination closely resembling the preceding, which had not then, so far as he was aware, been actually tried for this purpose, though it had been, like the preceding, described by Brush. If, as in Fig. 78, the magnets are excited partly in series, but also partly by coils of finer wire, connected as a shunt *across the whole external circuit*, then the combination should be more applicable than the preceding to the case of a constant electromotive-force, since any variation in the resistance of the external circuit will produce a greater effect in the "long shunt" than would be produced if the resistance of the field-magnets were included in the part of the main circuit external to the shunt.

In 1882 it was the author's opinion that although the last two combinations were not such perfect solutions of the problem as those which precede, they were more likely to find an immediate application,[1] since they can be put into practice upon any ordinary machine, and do not require, as in the first two combinations, the use of separate exciters, or of independent magneto machines. This opinion has been fully justified in the great progress made since in the "compound" or "self-regulating" machines.

(5.) *Series and Separate-Coil.*—This method has not, apparently, been tried for direct-current dynamos. For alternate-current dynamos a modification of it has been tried by Zipernowsky with success, the "series" or main-circuit excitation being, in this case, replaced by an excitation derived from the main circuit by means of a small transformer. This system is explained in Chapter XX.

Combinations to give Constant Current.

(1.) *Shunt and Separate (Deprez).*—When it is desired, as in the case of a set of arc lamps in series, to maintain the current in the circuit at one constant strength, the previous arrangement of Deprez must be modified, as indi-

---

[1] The invention of the "series and shunt" winding is claimed for several rivals. Brush undoubtedly first used it, but whether with any knowledge of all its advantages is doubtful. It was, however, mentioned as having some advantages by Sir C. W. Siemens in *Philosophical Transactions*, March 1880. It is also claimed for Lauckert (see note by M. Boistel, p. 100 of his translation of first edition of this work); Paget Higgs (*Electrical Review*, vol. xi. p. 280, and *Electrician*, Dec. 23, 1882); J. W. Swan, see Bosanquet (*ib.*, Dec. 9, 1882); J. Swinburne (*ib.*, Dec. 23, 1882); S. Schuckert (*ib.*, Oct. 13, 1883); it is claimed in America by Edison; and it has been patented by Messrs. Crompton and Kapp (*ib.*, June 9, 1883). See also Hospitalier (*L'Électricien*, No. 20, 1882). Students should also consult a series of articles in *The Electrician*, vol. x., beginning Dec. 16, 1882, by Mr. Gisbert Kapp.

cated in Fig. 79, by combining a shunt-winding with coils for a separately-exciting current. This arrangement is, in fact, that of a shunt dynamo, with an initial magnetic field independent of the strength of the current in the circuit.

Seeing that the only object in providing the coils for separate excitation is to secure an initial and independent magnetic field, it is clear that other means may be employed to bring about a similar result.

FIG. 79.

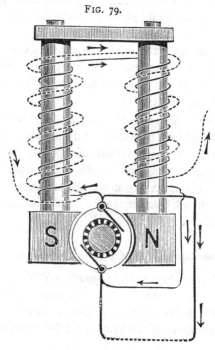

SHUNT AND SEPARATE.

(2.) *Shunt and Magneto* (*Perry*).—Perry's arrangement for constant current is given in diagram in Fig. 80, and consists in combining a shunt dynamo with a magneto machine of independent electromotive-force, this magneto machine being inserted either in the armature part or in the field-magnet part of the machine. As before, a certain critical speed must, on Professor Perry's plan, be found from experiment and calculation. In Chapter XX. on Constant-current Dynamos the theory of this method is given in the equations which show how the permanent initial excitation is to be proportioned to the permanent resistances in the circuit.

(3.) *Shunt and Series.*—There are, as mentioned above, a great many claimants to the discovery of the use of the " compound winding" for the purpose of obtaining a constant potential, but the use of the compound winding for the

Fig. 80.

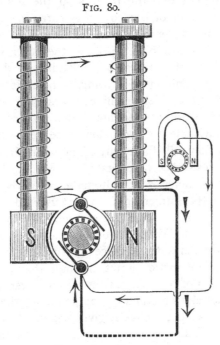

SHUNT AND MAGNETO.

purpose of obtaining a constant current was first described by the author of this book in his Cantor Lectures in December 1882, in the paragraph reprinted above, which states that for this purpose, the essential part of the magnetisation will be that due to the shunt coils. The theory of this winding in both its varieties was given for the first time in the first edition of this book, but the wide range of variations in the magnetism required by the method makes it useless.

Arrangements of Compound Winding.

Compound windings may be arranged in several different ways. If wound on the same core the shunt coils are some-times wound outside the series coils : less frequently the series coils are outside the shunt. In some of Siemens' dynamos they are wound on separate frames and slipped on side by side over the same core. In other cases, where (as in Siemens' usual patterns) the pole is at the middle of the magnet core, one end of the core may carry the shunt coils, the other the series; or both the coils on one of the cores may be series coils, and both those on the other, shunt coils. Mr. P. Higgs strongly advocates the winding of the two sets on separate cores. uniting at a common pole-piece. This is bad, for constant-potential machines. Practice seems to be drifting toward winding the shunt coils outside the series coils. It is advisable to keep down the resistance of the series coils, as they will form part of the main circuit; whilst the additional resistance necessitated by winding the wire in coils of larger diameter is not altogether a disadvantage in a shunt coil. In the former editions of this work the recom-mendation was made to wind the series coils nearer the pole than the shunt coils. With the better magnetic circuits now employed in dynamos this advice has ceased to have any weight. If the magnetic circuit through the iron-work be good, the position of the coils makes little difference.

Mr. Zipernowsky's method of compounding the alternate-current dynamo so as to make it give a constant (alternating) potential on the mains, consists in introducing into the main circuit that feeds the lamps the primary coil of a small trans-former, the secondary of which is connected through the commutator to the field-magnet circuit. This virtually makes the machine a series-excited machine combined with initial excitation derived by the separate-circuit method from one coil of the set in the armature.

The method of compound winding, though theoretically applicable to constant-current machines as well as to those for constant-potential, is until the present time a failure, not merely on account of the cost of the shunt winding, but because the systems appear to be unstable. For constant-current work, other methods of governing are therefore employed.

Other Methods of Automatic Regulation.

Brush's Automatic Regulator.—An ordinary series dynamo may be made to yield a constant current by introducing across the field-magnets a shunt of variable resistance, the resistance of the shunt being adjusted automatically by an electromagnet whose coils form part of the circuit. The system is shown in the accompanying diagram (Fig. 81).

FIG. 81.

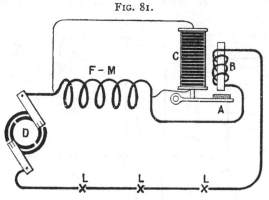

BRUSH'S AUTOMATIC REGULATOR.

The dynamo at D pours its current into the circuit, leaving the commutator (as drawn) by the upper brush, whence it flows through the field-magnets F M, and round the circuit of lamps L L, back to the negative terminal. Suppose now some of the lamps to be extinguished by switches which short-circuit them ; the resistance of the circuit being thus diminished there will be at once a tendency for the current to increase above its normal value unless the electromotive-force of the dynamo is at once correspondingly reduced. This is done by the solenoid B in the circuit. When traversed by the normal current it attracts its armature A with a certain force just sufficient to keep it in its neutral

position.   If the current increases, the armature is drawn
upward and causes a lever to compress a column of retort-
carbon plates C, which is connected as a shunt to the field-
magnets.   These plates when pressed together conduct well,
but when the pressure is diminished their imperfect contact
partially interrupts the shunt-circuit and increases its resistance.   When A rises and compresses C, the current is
diverted to a greater or lesser extent from the field-magnets
which are thus under control.

Edison's Regulator.—In Edison's system for supplying
mains at a constant potential a shunt dynamo is employed,
a variable resistance R being introduced into the shunt-
circuit (Fig. 82).   A lever moved by hand, whenever the

FIG. 82.

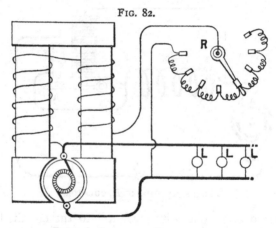

EDISON'S SYSTEM OF REGULATING.

potential rises or falls below its proper value, makes contact
on a number of studs connected with a set of resistances
and thus controls the degree of excitation of the field-
magnets.   A similar device has been used in several other
systems.   To make the arrangement perfect the variable
resistance should be automatically adjusted by an electromagnet the coils of which are an independent shunt across the
mains.   Edison has indeed used such a device.

Lane-Fox's Automatic Regulator—A relay, the coil of
which itself forms a shunt to the mains, is employed to

actuate a regulator to introduce more or less resistance into the shunt-circuit of the dynamo. A figure of this regulator was given at p. 145 of the second edition (1886) of this work.

Automatic Regulation by shifting the Brushes.—Several systems have been proposed for securing automatic regulation by shifting the brushes round the collector. Reference to the curve of potentials (Fig. 38), will show that if the brushes do not touch at the neutral points (marked + and −) the difference of potentials between them will be less than the maximum which the armature can give. Maxim prepared an automatic regulator based on this method. Similar adjustments have been used by Elihu Thomson and Hochhausen, and by Statter for constant-current machines. As applied to collectors of ordinary closed-coil dynamos the method is hardly successful. As applied to the commutator of the open-coil dynamo in the Thomson-Houston system (see Fig. 194) it answers satisfactorily.

Section Method of Regulation.—Another method has been suggested by Brush, who winds the field-magnets in a number of separate sections any number of which can be switched into circuit by a controlling electro-magnet. Very similar suggestions have been made by Cardew and by Deprez.

Constant-Current Governing.—The problem of getting a self-regulating constant-current machine has elicited several novel methods of governing. Messrs. Goolden and Trotter have ingeniously combined the section method of regulation with another principle, first suggested by Langley, namely that of applying a magnetic shunt to divert, more or less, the magnetic lines from passing through the iron core of the armature. Some account of this method is given in Chapter XXI.

Electric Governing of Engine.—Yet another way of accomplishing automatic regulation is possible in practice, and this without the condition of a constant speed of driving. Let the ordinary centrifugal governor of the steam-engine be abandoned, and let the supply of steam be regulated, not by the condition of the velocity of driving, but by means of an electric governor operated by the electric current itself.

K

Several of these governors are described in detail in Appendix VI. For many purposes they are more suitable and reliable than any of the arrangements which necessitate a constant speed of driving. One great advantage of the electric governor is that it cuts down the consumption of steam to the actual demands made upon the electric circuit, and prevents injury both to the dynamo and to the steam-engine.

Dynamometric Governing—One other method of govern·ing dynamos is too important to be omitted. Engineers are all aware that the horse-power transmitted along a shaft is the product of two factors, the speed and the moment of couple ; or, as it is now often termed, of speed and torque. If ω stands for the angular velocity and T for the torque (or turning moment) then

$$\omega\, T = \text{mechanical work per second, or "activity."}$$

But the "activity," or work per second, or horse-power, of a dynamo can be measured electrically, by the product of its electromotive-force into the current it drives through the circuit. If E stands for the electromotive-force, and i for the current, then

$$E\, i = \text{electric work per second.}$$

In a good dynamo the electric work, though not equal to the mechanical work, will exceed 90 per cent. of it. Now we know that, other things being equal, the electromotive-force E of a series dynamo or of a magneto machine is proportional to ω, the angular velocity or speed of driving. It follows at once that *the torque will be proportional to the current i.* This at once suggests that a series dynamo may be driven so as to give a constant current provided it be driven from a steam-engine governed *not by a centrifugal governor to maintain a constant speed,* but *by a dynamometric governor to maintain a constant torque or turning moment.* Some good transmission dynamometer, such as that of Morin, or one of the later varieties, such as those designed by Ayrton and Perry, or best of all that designed by the Rev. F. J. Smith,[1] and described

[1] See the excellent little book, published by Messrs. E. & F. N. Spon, from the pen of this able author, on ' Work-measuring Machines.'

in Chapter XXXII., may be adapted to work an equilibrium valve, and would fulfil the above condition of governing.

Prof. E. Thomson has suggested the use of a dynamo-metric apparatus to govern a constant-current dynamo by the method of shifting the brushes as explained at top of p. 71. A description of this governor was given in the second edition of this work.

Governing by Steam-Pressure.—It was remarked above that electric power and mechanical power are each a product of two factors. But in an ordinary steam-engine the work per second also consists of two factors, viz. speed of piston and steam-pressure; and the angular velocity of the shaft is proportional to the former, and its transmitted torque to the latter. Therefore the condition of maintaining a constant current ought to be fulfilled if the pressure is always constant. If the valves are such as to admit a fixed quantity of steam at each stroke, and if the boiler pressure is really kept up, then the average pressure behind the piston ought to be constant. In practice this is never attained on account of the friction of the steam against the steam-pipes and port-holes of the valves. The internal friction in the engine plays the same part in preventing absolutely true self-regulation, as does the internal electrical resistance in the dynamo. An approximation is all that is possible. In an experiment made by M. Pollard with a Gramme dynamo, the current gave deflections on a galvanometer, varying only from 59° to 54°, while additional resistances were introduced into the circuit, which caused the speed to run up from 436 to 726 revolutions per minute. Theoretically, therefore, a constant current ought to be one of the easiest things to maintain with a series dynamo. Have adequate boilers, keep the steam-pressure always at one point, abandon all governors, and admit equal quantities of steam at each stroke whatever the speed : the result *ought* to be a constant current. The condition of maintaining a constant potential cannot be similarly solved, except by employing a shunt dynamo under conditions that are both uneconomical and impracticable. But in the case of constant-current work it is possible to go further toward realising such results. The existing method of maintaining a constant steam-pressure is to put upon the boiler a pressure-gauge which indicates to the stoker when he is to add more fuel and when to damp down the fire. Let the pressure-gauge be abandoned, and instead, let there be provided at the side of the furnace an ampere-meter, and let the stoker feed or damp his furnace fires according to the requirements of the electric system of distribution. Is there any valid reason why such a method of government should not be efficient in practice, at least in the case of the series dynamo for constant currents?

Finally, to render the system truly automatic, it is conceivable that mechanical stoking appliances might be arranged, under the control of the ampere-meter or volt-meter, to supply the fuel in proportion to the number of lamps alight.

CHAPTER VII.

CONTINUOUS CURRENT MACHINES.

CLOSED-COIL ARMATURES.

Ring Armature Dynamos.

EXISTING types of dynamo machines may be broadly divided into those which furnish continuous, and those which furnish alternating currents; the latter type will be considered last. The continuous-current dynamos may again be classified under three heads: (1) closed-coil machines, (2) open-coil machines, (3) machines on the principle of the so-called unipolar induction. These last give really steady continuous currents, but are not yet commercially practical machines. The open-coil machines, so useful for arc-lighting, give currents which, though continuous in the sense of being without intermittence, nevertheless are always fluctuating with rapid periodical pulsations; whilst the closed-coil machines, so much used for incandescent lighting and for electroplating, give approximately non-pulsating continuous currents.

Accordingly we proceed to examine the actual dynamo machines in the following order:—

The method of closing the armature coils upon themselves, first invented by Pacinotti in the form of a ring, is adopted in the armatures of the Gramme, Siemens, and Edison dynamos, and in fact in the armatures of the majority of continuous current dynamos, which, though their armatures may have the form of a drum or a disk instead of a ring, are equally constructed with *closed coils;* that is to say, have the coils grouped in sections which communicate with successive bars of a collector, and which are connected continuously together into one closed circuit. The armatures of almost all the machines described in this and the next chapter are *closed-coil armatures.* There are other ways of winding a ring or a drum, in which the successive sections of the coil are not so connected together, and do not form a closed coil. These armatures may be called, for the sake of distinction, *open-coil armatures.* Dynamos having open-coil armatures are chiefly used for providing constant currents in arc-light circuits. Chiefest amongst them are the Brush dynamo and the Thomson-Houston dynamo. They are considered separately in Chapter X. on *Open-coil Dynamos.*

RING ARMATURES.

There are several ways of classifying ring-armatures, for example :—

(*a.*) Rings with teeth, between which the coils are wound (Pacinotti type).

(*b.*) Rings without teeth, having coils wound outside (Gramme type).

(*c.*) Rings having the coils or conductors threaded through perforations below the surface of the iron cores (Wenström type).

Again they may be classified as

(*a.*) Cylindrical rings, with the core elongated in the axial direction. In this case the polar surfaces are on the periphery.

(*b.*) Discoidal rings, with core extended in the radial direction. In this case the flat faces of the rings are usually used as the polar surfaces. There are, however, intermediate forms with elongation in neither direction, the cross-section of the ring-core being square or circular. There are others again in which, as in the European forms of Fein and of Fitzgerald, and in the American forms of Sperry and of Clark, an attempt is made to cause both internal and external periphery to act as a polar surface.

Again, armatures may be classified according to the method adopted for laminating the iron cores, the lamination being

(*a.*) Radial, if the core consists of thin iron washers ;

(*b.*) Tangential, if the core consists of ribbon ;

(*c.*) Filamentous, if the core consists of iron wires.

The essentials in the construction of good ring armatures are as follows :—

(*a.*) There must be sufficient cross-section in the iron core to carry the requisite magnetism. (Allow in continuous-current dynamo 1 square centimetre for each 16,000 magnetic lines.)

(*β.*) The copper conductors must have sufficient cross-section to carry the current, and should present sufficient surface to emit the heat developed in them. (Allow 1 square millimetre section for every 3 amperes ; and at least 5 square centimetres of surface for each watt lost in heat.)

(*γ.*) There must be a sufficient number of copper conductors in series to get the requisite electromotive-force. (See calculations in Chapter XV.)

(*δ.*) There must be a proper and sufficient means provided for transmitting the driving-power from the shaft to the conductors on the polar surface of the armature : for it is these, not the iron core, that require the power to drive them. (For calculations of the torque see Chapter XXX.)

(*ε.*) There must be proper insulation. No conductor should touch either against another conductor or against the iron core. The iron core disks or core wires should be properly insulated with mica, paper, or insulating paint. All supports and frames must be insulated so as to cut off eddy-currents.

(*ζ.*) There must be proper ventilation.

(*η.*) Means must be taken to prevent conductors from flying out centrifugally. In discoidal rings there is no need of special means : but with long cylindrical rings binding wires or thin binding bands must be used. These may consist of fine phosphor-bronze or tinned steel wires or of slips of thin tough sheet brass, soldered together at their ends, and protected by a layer of insulation from touching the con-ductors.

Methods of Winding Rings.

There are several methods of winding closed-coil ring armatures.[1]　The simplest and best-known is the method used by Pacinotti and by Gramme, in which there are as many "sections," or groups of coils, as there are "bars" in the collector, the end of each section being joined to the beginning of the next, and each bar of the collector being connected to such junction.　Simple modifications are possible—namely, such as providing twice as many sections as there are bars in the collector, each section being united either in series or in parallel with that diametrically opposite it, and the pair so united being treated as a single section in the coupling up of the ring.　For ring armatures that are to be used in multi-polar fields further modifications are possible, as will be shown at the end of this chapter.

CYLINDER-RING MACHINES.

This sub-class includes all machines having ring-armatures which are to be magnetised at their peripheral surfaces.

Pacinotti's Machine.—This machine, depicted in Fig. 83, was described in the Italian journal *Il Nuovo Cimento*, in 1864. The armature was an iron ring, shown separately in Fig. 84, having sixteen equal teeth supported by four brass arms B B. Between the teeth, on wooden frames, were wound sixteen coils, each of nine turns.　From the ends of each section, the wires were led downwards to the commutator.　This was a cylinder of wood having sixteen strips of brass let into grooves. Each strip of brass was soldered to the end of one section and to the beginning of the next.　Two metal brushes pressed against the commutator at points on a diameter at right angles to the line joining the poles of the vertical electro-mag-

---

[1] Some useful notes about right-handed and left-handed windings are given by von Waltenhofen in *Centralblatt für Elektrotechnik*, ix. 636, 1887.

nets, which were provided with wide pole-pieces. The ring was described by Signor Pacinotti as a "transversal electromagnet," and though the machine was denominated an electric

FIG. 83.

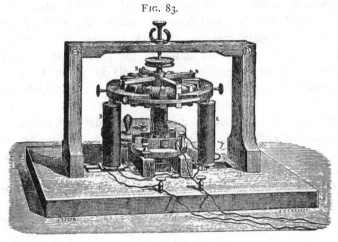

PACINOTTI'S DYNAMO.

motor, its function as a generator was announced in the most specific terms. "It seems to me," said Professor Pacinotti, "that that which augments the value of this model is the facility which it offers of being able to transform this electro-magnetic machine into a *magneto-electric machine with continuous currents.* If instead of the electromagnet there were a permanent magnet, and if the transversal electro-magnet [i. e. the ring] were set turning, one would have made it into a magneto-electric machine, which would give a continuous induced current directed always in the same sense." Professor Pacinotti also

FIG. 84.

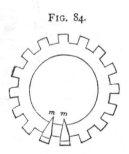

PACINOTTI'S RING.

separately excited the field magnets from a battery, obtaining on rotation a continuous current in the ring circuit. He also added the following significant remarks on the reversibility of his machine. "This model shows, moreover, how the electro-magnetic machine [motor] is the reciprocal of the magneto-electric machine [generator], since in the first the electric current which has been introduced by the rheophores [brushes], by circulating in the coils, enables one to obtain the movement of the wheel and its mechanical work ; whilst in the second, one employs mechanical work to cause the wheel to turn and to obtain, by the action of the permanent magnet, a current which circulates in the coils in order to pass to the rheophores, to be led thence to the body on which it is to act."

In spite of its extreme value, this model lay forgotten amongst the collection of instruments in the University of Pisa, until after all its fundamental points had been redis-covered and put into practice by other hands.

Gramme Dynamo.—The essential point of the Gramme machine is its ring. This is usually constructed as shown in Fig. 85. A quantity of soft iron wire is wound, upon a special

FIG. 85.

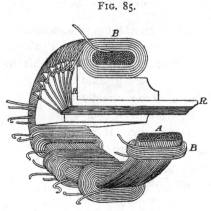

GRAMME'S RING.

frame or mould, into a ring to serve as the core. It is shown at A cut away to exhibit the internal structure. On this core the separate sections B B, of insulated copper wire, are wound. Each section is separately coiled by hand between temporary cheeks to ensure its being of exactly the right width ; the wire for each section being threaded in and out of the ring on a shuttle. The end of each coil and the beginning of the next are connected to one another, and to an insulated radial piece R, which forms one bar of the collector. When all the sections are wound, a wooden hub is driven in, and the ring mounted on a spindle. The power applied to the shaft is therefore transmitted to the coils by a purely frictional attachment, which, if the wooden hub shrinks, may fail altogether.

The specification of Gramme's British Patent (of 1870), states that the armature may be of the form of a solid or

Fig. 86.

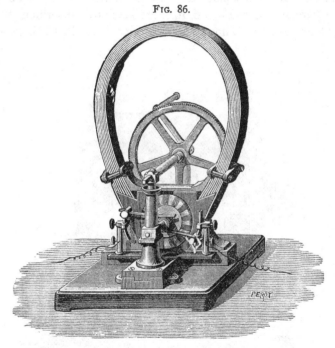

GRAMME MACHINE, LABORATORY PATTERN.

hollow ring cylinder or other suitable endless shape constructed either out of one piece of iron, or of a bundle of iron wires, and round the entire surface of which endless core is laid a series of coils of suitably isolated wire, of copper or other good conductor of electricity, in such a manner that the said coils or helices of wire may be considered as forming one

continuous series of small bobbins; each being connected
with the next so as to constitute one large endless bobbin.
As it rotates there will be set up in the core a continuous
displacement or advancing of the magnetism. The inventor
also states that the coils of wire may be replaced by coils
made of strips or ribbons of brass or other good conductor
suitably isolated.

Innumerable forms have been given to the Gramme
machine at different dates since its appearance in 1871, varying
from small laboratory machines with permanent steel magnets
such as are shown in Fig. 86, to large machines absorbing 30
or 40 horse-power. Those who desire more detailed informa-

FIG. 87.

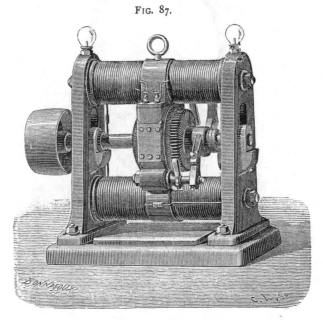

GRAMME DYNAMO, " A " PATTERN.

tion concerning the various patterns of Gramme dynamo,
should consult the treatise of the late M. A. Niaudet, entitled
*Machines electriques a courants continus, systemes Gramme
et congeneres* (1881). Fig. 87 shows the ordinary "A" Gramme,

the most frequent pattern in use. Many improvements in detail have been introduced into the Gramme dynamo, both in Europe and in the United States.

Fuller-Gramme Dynamo.—In the States the Fuller Company, which works the Gramme patents, has produced the machine depicted in Fig. 88, and several other forms in which mechanical skill of a high order is apparent. The

Fig. 88.

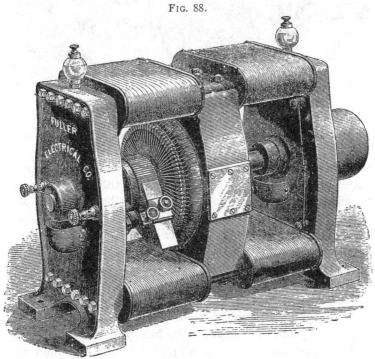

FULLER-GRAMME DYNAMO.

field-magnets, frames, and pole-pieces are very substantial. The ring, which is depicted separately in Fig. 89, is better built than the older European types, and is also connected to the shaft by an internal gun-metal spider, instead of being driven on to a wooden hub; and the collector-bars are prevented from flying to pieces by the addition of an insulated ring shrunk on over their ends. The cut also shows how the

coils are kept in their place by external bindings of fine wire
securely soldered together.

FIG. 89.

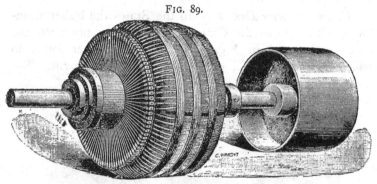

ARMATURE OF FULLER-GRAMME DYNAMO.

Deprez's Gramme Dynamo.—In France, too,* the machine
has received important modifications at the hands of
M. Marcel Deprez. M. Deprez's dynamo has two Gramme
rings upon one axle, which lies between the poles of two
opposing field-magnets, each of the two-branched, or so-
called horse-shoe form (see Fig. 34, p. 55). These are laid
horizontally, so that the N. pole of one is opposite the
S. pole of the other, and *vice versa;* the poles being provided
with curved pole-pieces between which the rings revolve.
In almost all the larger Gramme machines of ordinary
pattern, the pole-pieces are in the middle of long iron
cores, which are so wound as to give a consequent pole at
the central point. M. Deprez, who has given much attention
to the question how to design a machine which, with the least
expenditure of electric energy, gives the greatest actual couple
at the axle, is of opinion that the horse-shoe form of electromagnet is the most advantageous. The iron cores and yokes
of his field-magnets are very substantial; but the pole-pieces
are not very heavy. M. Deprez's machine has a very elaborate system of sectional windings of the field-magnets, and a

* Other modifications, chiefly of a mechanical nature, have been made by
M. Raffard, and by other French engineers; of these some account is given in
Industries for November 5th, 1886.

switch board enabling him to couple up the connections in
various ways. The circuits of the two rings are quite distinct, and each armature has its own collector and brushes.
M. Deprez has also constructed other Gramme machines,
with armatures of very fine wire, for his experiments on the
electric transmission of power.

Gramme's Later Dynamos.—Amongst the later forms given
in France to the Gramme machine, we find one of extremely

FIG. 90.

GRAMME DYNAMO, VERTICAL PATTERN (1881).

simple construction, represented in Fig. 90, having vertical
electro-magnets. These are made of cast iron put together in
four pieces. The structure is heavy, since, on account of the

lower permeability of cast iron, a much greater bulk than of
wrought iron must be employed. It will be observed that this
pattern differs from that of the better-known "A" Gramme
in using *salient poles*, instead of having the "consequent poles"
at the middle points of the electro-magnets. This is one of
the forms of machine employed by Deprez in the transmission

FIG. 91.

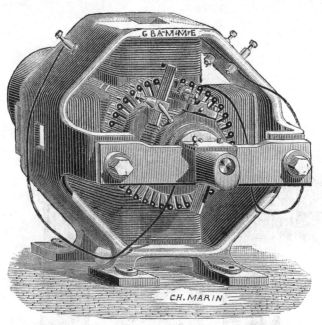

PORTABLE GRAMME DYNAMO (1885).

of power. In a still more recent form, four salient poles are
employed. The field-magnets of this type are still more
simple, for the four cores, the external octagonal frame, and
one of the two brackets which carry the shaft are all cast in
one piece (Fig. 91). The object of this is to secure a strong,
light, and portable machine, suitable for temporary lighting.
The coils are wound on a separate mould, and slipped on and
secured in their places. Some of these machines are con-

structed with two poles only. The weight of copper in them is relatively very small.

In yet another vertical pattern of machine designed by the firm of Breguet, the field-magnets are of cast iron, and are constructed so that the upper half can be removed from the lower.

FIG. 92.

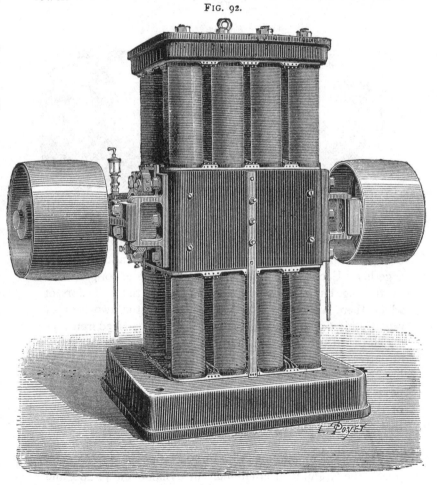

GRAMME DYNAMO FOR LARGE CURRENTS (1884).

For furnishing very powerful currents, M. Gramme has constructed machines of the type depicted in Fig. 92, which

L

somewhat resembles an older form, constructed in 1873, in having multiple cores united to a common pole-piece. As many as 14 columnar electro-magnets may be found in some machines. This is an entirely mistaken construction. The most interesting part of this machine is its armature, which is shown separately in Fig. 93. This consists of a hollow

FIG. 93.

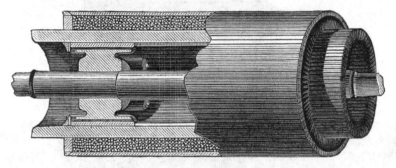

GRAMME ARMATURE FOR LARGE CURRENTS.

cylinder, built up of 100 wedge-shaped copper bars, each covered with a bitumenised paper wrapping, and then put together. Each bar has two radial projections of copper. The protruding ends of the copper rods form the collectors, of which there are two. The space between the two sets of radial projections is filled with windings of varnished iron wire which constitute the core, and finally 100 other bars of copper of flatter section are connected exteriorly from the projecting lug at one end of one bar to the lug at the other end of the next bar, so connecting the bars into a closed coil. Several of the inner bars are made thicker and of special form so that they may be keyed to spiders fixed upon the driving-shaft.

The Gramme dynamo for electro-plating is described on p. 281, the Gramme alternate-current dynamo on p. 260, and some Gramme motors in Chapter XXIX.

In 1885 a new form of field-magnets was adopted by M. Gramme, consisting of an inverted horse-shoe. In this machine, sketches of which are given in Figs. 94–97, the bed-plate, the magnet-cores, the pole-pieces, and the supports of the bearings are of cast iron all in one piece. The field-

FIG. 94.

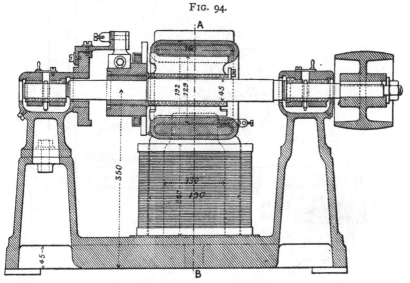

GRAMME DYNAMO (*Type Superieur*).

magnet coils are wound on separate mandrils and then slipped over the cores between cheeks of insulating substance. The ring-armature is driven upon a spider of bronze, keyed to the shaft, its outer edges passing between the insulated copper coils and supporting the core, which is of laminated iron. The collector bars are held together by an external ring of bronze, well insulated from them. The details are shown in the figures, and it will be noted that the field-magnet cores are for some reason cast hollow. Fig. 97 shows the arrangements of the brush-holders. One of these machines, intended for an output of 40 amperes with a pressure of 110 volts, at a speed of 1400 revolutions per minute, had the following dimensions :—

Internal radius of armature core, 6·5 cm.; external radius, 9·15 cm.; axial length of core, 16 cm.; total sectional area of armature core, 80 sq. cm.; distance from armature core to pole-

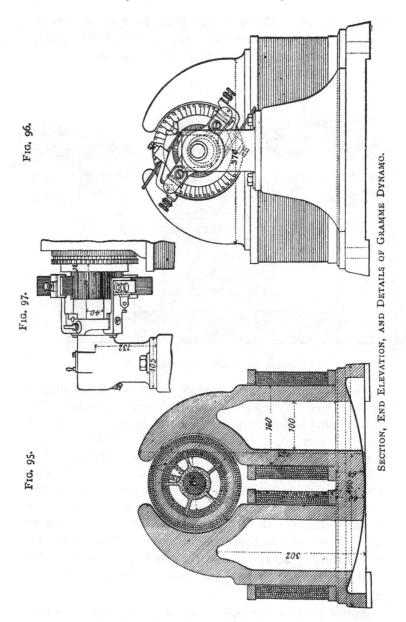

FIG. 96.

FIG. 97.

FIG. 95.

SECTION, END ELEVATION, AND DETAILS OF GRAMME DYNAMO.

piece, 0·85 cm.; estimated area of each polar surface, 366 sq. cm.; sectional area of each iron magnet core, 174 sq. cm.; estimated length of magnetic circuit within field-magnet, 81 cm.; commutator-bars, 60; total number of windings on the ring, 300; armature resistance, 0·174 ohm; shunt resistance, 46 ohm. The dimensions marked on the figure are in millimetres.

Maxim's Dynamo.—In the Maxim machine the ring is elongated in the axial direction so that it becomes a hollow cylinder, the wire being threaded through the interior and brought back over the exterior of the ring. The field-magnets used by Maxim are of the general form adopted in the vertical-pattern Siemens dynamo. This dynamo is also provided with an automatic device for regulating the conditions of supply of current, there being a small electro-magnetic motor mounted upon the machine to shift the brushes forward or backward, and by thus altering the lead to correct the variations of potential at the terminals.

Cabella's Armature.—A form of ring armature devised in 1883 by Signor B. Cabella closely resembles the form of Gramme armature shown in Fig. 93. According to Professor Ferrini, one of Cabella's armatures placed between the poles of a 60-light Edison (" Z," old pattern) instead of its ordinary armature, increased its powers so that it could be used for 150 lamps.

Jurgensen's Dynamo.—This machine, devised by Jürgensen and Lorenz, of Copenhagen, attracted considerable attention in 1882. Its armature is a hollow cylindrical ring placed between two salient poles of an arched electromagnet, the coils of which are wound most thickly close to the pole. There is also an electromagnet placed inside the ring to reinforce the polarity of the ring. This feature, indeed, gives its interest to the machine.[1]

Hochhausen's Dynamo.—In this dynamo, shown in Fig. 98, the armature is a cylindrical ring: but it is constructed in a novel fashion of four separate curved iron frames, upon which the previously wound coils are slipped, and which are then bolted together and secured to strong end plates. The

---

[1] For further description, see *Electrical Review*, Sept. 23, 1885.

field-magnets of this machine are disposed in a manner
similar to that of the Gramme machine shown in Fig. 90, the
ring being placed between two straight electromagnets placed
vertically over one another. The upper magnet is held in

FIG. 98.

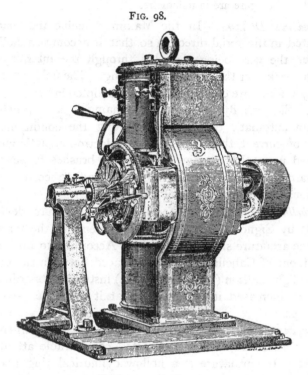

HOCHHAUSEN'S DYNAMO.

its place by curved flanking-pieces of iron, which run down
the two sides of the machine, and connect the topmost point
of the upper magnet with the lowest part of the lower.   This
arrangement strikes the eye as being both mechanically and
magnetically bad ; nevertheless the machine appears to be a
very good working machine. Like the Maxim dynamo it is
provided with a small motor arrangement to adjust the lead
of the brushes, and which is supposed to be automatic in its
operation. The collector segments, which are very massive,

have air gaps between them, and are bolted to a substantial disk of slate. This arrangement is exhibited in Fig. 99, and is very satisfactory. Another dynamo closely resembling

FIG. 99.

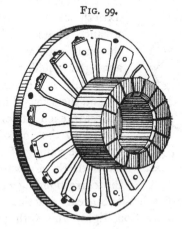

COLLECTOR OF HOCHHAUSEN DYNAMO.

this, and like it of American origin, is known from its inventor as the Van de Poele dynamo.

Burgin-Crompton Dynamo.—This dynamo is distinguished by its armature. The field-magnets are of a horizontal pattern, not unlike those of the horizontal Siemens machines, but of cast iron. The armature of the original Bürgin machine, as it came from Switzerland, consisted of several rings set side by side on one spindle, these rings being made of iron wire wound upon a square frame, and carrying each four coils. In this form it is described in Professor Adams' Cantor Lectures on Electric Lighting in 1881. But in the hands of Messrs. R. E. Crompton and Co., it underwent a remarkable course of development. Mr. Crompton changed the square form to a hexagon having six coils upon it (Fig. 100), and increased the number of rings to ten. It was thus described in 1882 :—" Each ring is made of a hexagonal coil of iron wire, mounted upon light metal spokes, which meet the corners of the hexagon. Over this hexagonal frame, six coils of covered copper wire are wound, being thickest at the six points intermediate between the spokes, thus making up the form of each ring to nearly a circle. Each of the six coils is separated from its neighbour, and each of the ten rings is fixed to the axis one-sixtieth of the circumference in advance of its neighbour, so that the sixty separate coils are in fact arranged equidistantly (and symmetrically as viewed from the end) around the axis. There is a 60-part collector, each bar of which is connected to the end of one coil and to the beginning of the coil that is one-sixtieth in advance ; that is, to the corresponding coil of the next

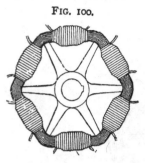

FIG. 100.

SINGLE RING FROM CROMP-TON-BÜRGIN ARMATURE.

ring. This armature has the great practical advantages of being easy in construction, light, and with plenty of ventilation."

This form, however, suffered from the harmful effects of induction between contiguous rings, and it was found advisable to alternate the positions of the rings, instead of placing them in a regular screw-order on the spindle as shown in most of the published drawings of this well-known machine. The next step was to increase the quantity of iron in the hexagonal cores, and to ascertain by experiment what was the best relative proportion of iron and copper to employ. At the same time, Mr. Crompton and Mr. Kapp introduced their system of "compounding" the windings of the field-magnets. Another change in the armature followed, the rings being made much broader and fewer in number, four massive hexagonal rings, united to a 24-part collector, replacing the ten slighter rings and their 60-part collector.

Crompton Dynamo.—This excellent machine shows what may be done in the way of improvement by careful attention to the best proportions of parts and quality of material. Its field-magnets are of the very softest Swedish wrought iron, compound-wound. The armature, as constructed in 1886, is a single ring of the elongated or cylindrical pattern, and its coils are

wound upon an iron core made up of disks of very thin soft iron
fixed upon a central spindle by means of short arms, which are
dovetailed into notches cut in the inner circumference of the
disks. At intervals gaps are left between the disks, for ventilation. The coils, ninety-six in number in some of these machines,
one hundred and twenty in others, are threaded through the
cylinder as in the Maxim ring, and kept in their places by small
boxwood wedges and by external bindings of thin brass wire.
Fig. 101 shows a sectional view of an armature as constructed

FIG. 101.

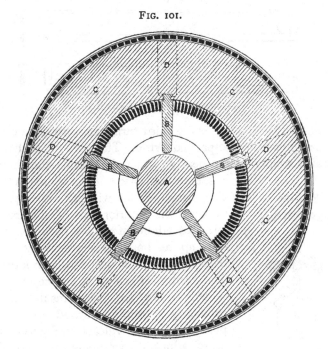

SECTION OF CROMPTON ARMATURE.

for these machines. This armature is 2 feet 4 inches in
length, and 12¾ inches in external diameter. The steel shaft
A is grooved with five deep slits to receive five flange-like
spokes B which dovetail into notches in the iron disks C.
Mr. Crompton's method of connecting with the driving
shaft by grooves in the latter is also shown in Fig. 102,

which illustrates a shaft with three grooves. At every
2 inches of the length there are inserted the pieces D,
which are ⅛ inch thick, to preserve ventilating gaps. There

FIG. 102.

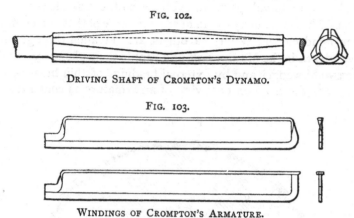

DRIVING SHAFT OF CROMPTON'S DYNAMO.

FIG. 103.

WINDINGS OF CROMPTON'S ARMATURE.

are 120 turns in the coil, every third turn being brought down
to one segment of a 40-part collector. The coils are built up
of drawn copper rod of nearly rectangular section and about
a square centimetre of sectional area.
The parts of the winding which pass
through the interior are of a narrow
form, to admit of closer packing, and
turn up at both ends somewhat like
the copper strips of the Cabella armature.
The forms of these conductors are
shown in Fig. 103. The insulation is
peculiarly carefully carried out with
pieces of flexible vulcanised fibre, cut so
as to admit of wrapping at intervals
round the copper conductors, but leaving ventilating spaces. A method of

FIG. 104.

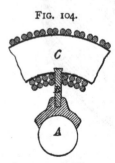

CROMPTON'S WIRE-
WOUND ARMATURE.

arranging the coils, in those cases where wires are used, is shown
in Fig. 104, one layer of coils only lying on the external surface.

Mr. Crompton has also designed armatures in which the
coils lie in deep grooves in the core, the principle of great
radial depth being still preserved.

As there were so few turns of wire on the armature, it was essential in this machine to employ a magnetic field of great power. Mr. Crompton's great aim was to have as complete a magnetic circuit as possible of iron of the best quality. He sought to increase the intensity of the field by having plenty of iron in the armature, and bringing that iron as closely as possible into proximity with the pole-pieces. The result was an extraordinary increase in the output of the machine.

In a machine of this pattern the armature was 12 inches in diameter, 28 inches long, and the radial depth of the core-plates 2½ inches. There were but 69 windings of copper ribbon upon it. At 440 revolutions per minute it gave 229 amperes of current, and a potential of 110 volts at the terminals. The cores of the field magnets were 3 feet 6 inches long, 24 inches broad, 4½ inches thick, and required about 24,000 ampere-turns, in total, to magnetise fully.

The quantity of copper in these machines was small as compared with the quantity of iron, especially in those machines that are designed to run at slow speeds. Two examples have been furnished by Mr. Crompton:—

(*i.*) *Fast-speed Machine* (1400 revolutions per minute), to feed 300 16-candle lamps :—

Weight of copper in armature coil .. 45 lbs.
Weight of iron in armature core .. .. 131 „
Copper : iron = 1 : 2·9.

(*ii.*) *Slow-speed Machine* (500 revolutions per minute), to feed 500 16-candle lamps :—

Weight of copper in armature coil .. 130 lbs.
Weight of iron in armature core .. .. 550 „
Copper : iron = 1 : 4·25.

The power of the field-magnets is such that at all speeds, and under all conditions of the external circuit, the intensity of the field is undisturbed by the magnetising action of the currents in the armature coils. There is, therefore, hardly any lead at all at the brushes, and what lead there is, is absolutely constant. There is no sparking, and it is impossible to tell by looking at the brushes whether the current is off or on. It was found that machines constructed with cast-iron field-magnets, instead of wrought-iron, give, at the same speed, an electromotive-force about 40 per cent. lower than those with wrought iron, all other things remaining the same.

During the progress of manufacturing and developing the Bürgin-Crompton dynamos, and those of the more recent type, Mr. Kapp was led to employ an empirical formula for calculating in a practical way the electromotive-force of dynamos of the cylinder-ring type. This formula, which was given in the former editions of this book, with some criticisms and explanations, has been abandoned in favour of the more complete formulæ, based on the principle of the magnetic circuit, of which an account is given in Chapter XIV., pp. 311 to 332.

Messrs. R. E. Crompton and Co. have more recently modified their machines in several details.

A general view of a compound-wound Crompton dynamo is given in Fig. 105, which shows vertical field-magnets with a double magnetic circuit.

The armature windings are formed of specially drawn copper wire of nearly rectangular section, cotton-covered, there being one turn of such wire to each segment of the commutator. The separate turns are bent to shape on proper formers before being placed on the ring, and are taped together in twos so that when put in place they lie with the longer dimension of their rectangular section radial, so getting great depth of copper radially on the armature. The two turns lie side by side on the external periphery of the ring, but radially above one another at the internal periphery. The armatures are usually provided with driving-projections of insulated metal, inserted at intervals between the conducting wires ; and after winding are strongly bound with binding-wires of tinned steel. In some of the most recent machines a single magnetic circuit only is employed. Another improvement, useful in machines for furnishing large currents, consists in dividing each conductor on the external periphery of the armature into two, which are crossed under one another at the middle, and united together at their ends. This construction greatly diminishes the eddy-currents which are set up in the conductors if they consisted of single rods or bars. Still more recently Mr. Crompton and Mr.

FIG. 105.

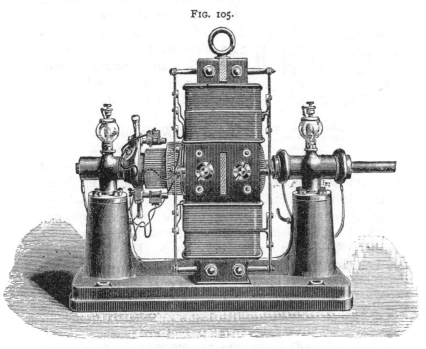

CROMPTON'S DYNAMO (1887).

Swinburne have designed a number of machines in which the armature conductors pass through holes punched in the core-disks near the edge (as in Wenstrom's armature), instead of being wound over the outside or between teeth. As a result, the external periphery of the armature runs much more true than when there are external windings, and the air-gap neces-sary for clearance between armature and pole-pieces may be reduced to an extremely narrow space. Machines made on this plan have magnetic circuits of extremely small resistance, and require a very few ampère-turns of excitation. The first design is sketched in Fig. 106. When a backward lead is given to the brushes of such dynamos, they become self-

exciting without any coils on the field-magnets ; but then a
small auxiliary pole-piece is required to reduce the sparking.

FIG. 106.

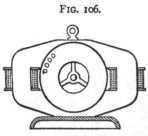

CROMPTON AND SWINBURNE'S
DESIGN FOR DYNAMO.

Special forms are given to the
polar surfaces of the magnets in
those machines, in which the circulation of current around the armature amounts to many ampere-
turns.   Still more recently Messrs.
Crompton and Swinburne have devised the drum armature described
on p. 208.

Kapp's Dynamos.—Mr. Kapp
has recently described[1] a form of
dynamo shown in Figs. 107 and 108, in which the two limbs of
the field-magnets are upright cylinders, the armature being
placed at the summit.   The armature-core consists of soft
charcoal iron wire coiled on a gun-metal supporting cylinder,
provided with end flanges, and subdivided lengthways by
pairs of flanges with air-ways between.   Driving pegs are
cast on the flanges, and are provided with fibre ferrules.   The
copper conductors are laid on in groups, with the insulated
driving pegs coming up between them to keep them apart
and admit of ventilation.   The supporting cylinder is mounted
on the shaft by two sets of gun-metal arms fitting into key-
ways at each end ; these arms are removed during winding.
Mr. Kapp has also designed dynamos of a kindred type
which work at very low speeds.   The field-magnets consist of
two semicircular wrought-iron cores, screwed to cast-iron pole-
pieces.   The coils are wound on six light cast-iron bobbins
which slide over the cores.   Fig. 109 is a transverse section of
this compact and powerful arrangement, which, though requiring a greater relative amount of copper, and more expensive to construct, is of much less weight for equal power.

Paterson and Cooper's Dynamo.—The " Phœnix " dynamo
constructed by Messrs. Paterson and Cooper has also a
modified cylindrical ring armature, built up of a number
of very thin rings of Swedish iron separated from one

[1] *Proc. Instit. Civil Engineers*, vol. lxxxiii., pt. 1, pp. 41 and 53.

another by paraffined paper and secured to two end frames
by three bolts passing through insulated brushes in the

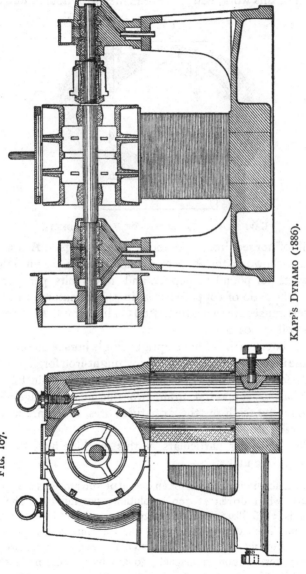

KAPP'S DYNAMO (1886).

FIG. 107.

plates. There are no air-spaces in the armature for ventilation, nor interior teeth to keep the coils apart ; there being

comparatively little heating in this armature. These machines, of which the first form was depicted on p. 137 of the second edition of this work, had a field-magnet with double magnetic

FIG. 109.

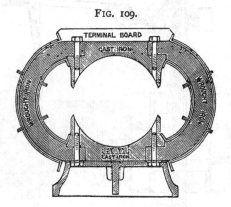

KAPP'S DYNAMO, SECTION OF FIELD-MAGNETS.

circuit of forged iron, and toothed core-rings. A Phœnix machine shown at the Antwerp Exhibition (of 65 units) gave 372 watts per pound of copper, and 1·31 volts per yard of coil. The ratio of copper to iron in the armature is 1 : 3·03. The resistances are : armature, 0·0055 ohm ; shunt, 10·00 ohms; series coil, 0·004 ohm.

The later machines have upright single horse-shoe magnets, in some instances made of a single wrought-iron forging slotted out to form the two limbs, and bored, as shown in Fig. 110. The shaft is supported from two gun-metal bridge-pieces. There are no teeth on the armature-cores, which are made of plain washers.

The following particulars of this machine have been furnished by the makers :—

The machine is designed for an output of 100 amperes at 250 volts pressure, when driven at 700 revolutions per minute. The total weight is 3136 lbs.; the useful output being 8 watts per pound of material. The external diameter of the armature core is 34·6 cm.; internal diameter of armature core, 20 cm.; length of armature core, 30 cm.; cross section of magnets, 30 cm. by 20 cm.; net sectional area of iron in armature, allowing for insulation, 193 sq. cm.; sectional area of iron in field, allowing for rounded corners, 612·7 sq. cm.; diameter of polar cavity, 37·5 cm. The armature is wound with 360 turns of 3·75 mm. square wire in two layers. The field is shunt wound, each link containing 3540 turns of 1·55 mm. wire, and the total resistance is 83 ohms. The radiating surface of both magnet limbs is 7740 sq. cm., and the energy used in exciting is 750 watts ; the ratio of the cooling surface being therefore 10·3 sq. cm. per watt.

From the above data we find that the total number of magnetic lines that cross the armature is about 6,200,000, and that the useful

FIG. 110.

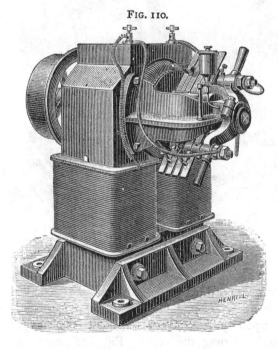

PHŒNIX DYNAMO (1887 Type).

induction through the armature is 16,000, and the induction through the field 10,100 lines per sq. cm. The peripheral speed of the armature core is 2530 feet per minute, and the total length of conductor wound on the armature is 11,200 inches, being at the rate of 45 inches per volt in the external circuit. At a peripheral speed of 3000 feet (which is a fair average in modern belt-driven dynamos), 38 inches of armature conductor would produce 1 volt.

Fig. 111 shows an alternative design, in which the field-magnets are cast in one piece. This is not so powerful a machine for equal weight, but can be made of equal power as

FIG. 111.

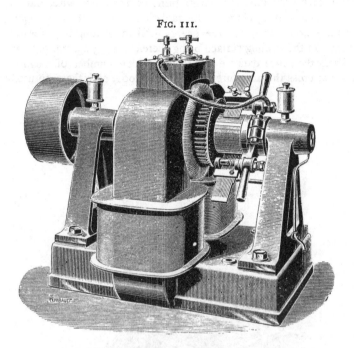

PHŒNIX DYNAMO (1887 Type).

the wrought-iron machine at lower cost. In both machines there is no joint in the magnetic circuit, and the magnet coils are wound upon special bobbins of sheet iron flanged with brass, slipped on over the cores. Fig. 112 shows the construction of the collector, and the manner in which the segments are insulated and secured from flying out of place.

The same makers have produced an arc-light dynamo to yield 10 ampères at 700 volts. The following are the data :—

Armature core, 32·5 cm. external diameter, 22·9 cm. internal; axial length, 15 cm.; wound with 1872 turns of wire 1·2 mm. in diameter, in 48 sections of 39 turns each in three layers. Armature resistance, 3·448 ohms. Field-magnet coils, 2, of 954 turns each, in series; their total resistance, 4·541 ohms. The maximum induction in armature is 19,080, in field magnet 10,800 lines per sq. cm.

FIG. 112.

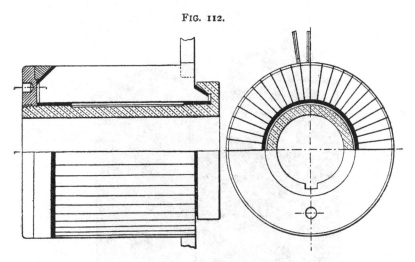

PATERSON AND COOPER'S COLLECTOR.

Dynamos of a pattern very similar to the last-mentioned machine have been constructed by Messrs. Hall and Co. and by Messrs. Elwell Parker and Co.

McTighe's Dynamo.—Fig. 113 depicts a machine which has been used with some success in the States, its armature being a modified Gramme ring. The interesting feature is, however, the field-magnet, which is of extremely simple form, there being two vertical cores on either side to receive the coils, united above and below by yokes of iron, which are specially formed so as to serve also as pole-pieces. Field magnets of an almost identical form have recently been used by Mr. Joel in his so-called "Engine Dynamo,"[1] and also by Messrs. Elwell and Parker.[2] In Joel's dynamo the pole-pieces are grooved on the polar surface. The armature is a modified Pacinotti ring built up in interlocking sections bolted together. Mr. W. A. Leipner has also designed a modified Pacinotti

[1] *Electrical Review,* xvi., p. 370, April 1885.
[2] *Ibid.,* p. 202, February 1885.

M 2

armature in which any single section of the coils can be removed and replaced in a few moments.

FIG. 113.

McTighe's Dynamo.

Mather and Hopkinson's Dynamo.—The field magnet of this well-designed machine, shown in Figs. 114 and 115, closely resembles the preceding; but the wrought-iron cores are cylindrical and the cast-iron yokes very massive. It was designed by Dr. Edward Hopkinson. The armature, designed by Dr. J. Hopkinson, F.R.S., and Dr. Edward Hopkinson, is a modified Gramme, with low resistance and careful ventilation. The collector is unusually substantial, and consists of 40 bars of toughened brass insulated with mica. It is usual in these machines so to shape the pole-pieces that there is a smaller

clearance opposite the highest and lowest points of the armature : this concentrates the magnetic field and helps to prevent
its distortion by the armature current.   In a 24-unit machine

FIG. 114.

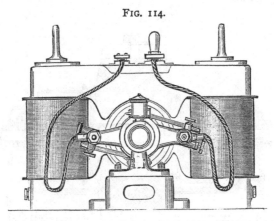

MATHER AND HOPKINSON'S DYNAMO (End Elevation).

(designed for 300 lamps) of this pattern the armature cores
are 12 inches long and 12 inches in diameter, with 120 turns

FIG. 115.

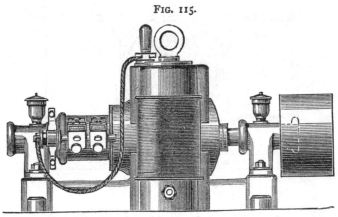

MATHER AND HOPKINSON'S DYNAMO (Front Elevation).

of wire.   The resistances are : armature, 0·023 ohm ; shunt,
19·36 ohms ; series coil, 0·012 ohm.   With a speed of 1050

revolutions per minute the current was 220 ampères, the machine being nearly self-regulating for 111 volts.   This machine is known as the " Manchester " dynamo; its efficiency is 90·9 per cent. [1]

FIGS. 116, 117.

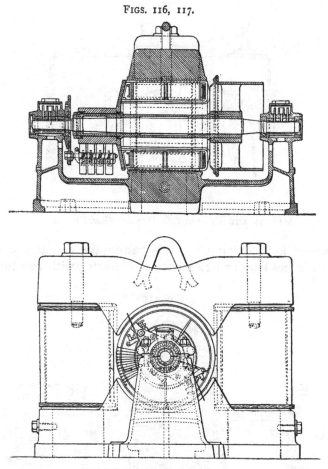

BROWN'S (OERLIKON) DYNAMO.

Foster and Andersen's Dynamos.—These dynamos, constructed by Messrs. Latimer Clarke, Muirhead, and Co., are

[1] One of these machines is very fully described in the paper by Drs. J. and E. Hopkinson in the *Philosophical Transactions* for 1886.

of two types. The "Regency" dynamo resembles Fig. 108, whilst the "Westminster" dynamo is not unlike Fig. 113. The armature is built up of sheet-iron washers supported on two strong gun-metal sleeves.

Brown's Dynamo.—The dynamos of O. E. Brown of Oerlikon also closely resemble the "Manchester" type, Fig. 114, but are even more massive. The form of the field magnet is given in Figs. 116 and 117, which represent a machine designed to give 450 amperes at 115 volts at 350 revolutions per minute; the dimensions of it are as follows :—

Diameter of ring, 61·4 cm.; length of ring, 55 cm.; length of shaft, 196 cm.; total height of machine, 120 cm.; wrought-iron cores, 60 cm. long, 40 cm. diameter; cast-iron yokes, 30 cm. thick, 44 cm. broad; diameter of pulley, 60 cm.; total weight, 6·5 tons; ampere-turns at full excitation, 35,000.

The armature of Brown's dynamo differs in one respect from those of the preceding machines. It is built up of thin iron disks, but these, instead of being toothed as in the Pacinotti forms, are perforated with a peripheral series of holes, as in Wenstrom's dynamo, to receive the "external" conductors, which lie thus about 1 millimetre below the periphery. This construction reduces the magnetic resistance of the air-gap to an exceedingly small quantity, and there is no tendency, as with most toothed armatures, to undue heating of the pole-pieces.

Holmes's Dynamo.—Messrs. J. H. Holmes and Co., of Gateshead, have introduced a compact and well-built form of Gramme machine, depicted in Fig. 118. It is known as the "Castle" dynamo; its yoke and pole-pieces are of cast iron, bored to receive the cylindrical upright wrought-iron cores. The ring armature core is made up of thin plates of charcoal iron.

Heinrich's Dynamo.—This machine has a ring armature with an iron-wire core of a U-shaped cross-section, enclosed almost entirely within the polar extensions of the field-magnets, which in some respects resemble those of the Siemens dynamo. The intent of this form of armature was to expose as much of the wire as possible to the action of the field. Some recently published tests of a series-wound Heinrich's dynamo, which, at a speed of 1680 revolutions per minute,

FIG. 118.

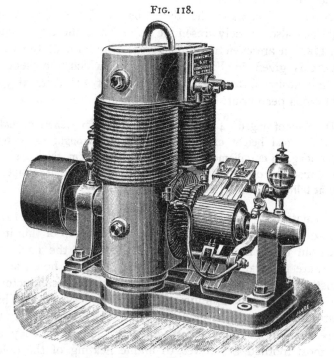

HOLMES'S DYNAMO.

gave 234 volts and 21·01 amperes, show a gross efficiency of 97 to 98·9 per cent., and a nett efficiency of about 83 per cent.

FLAT-RING MACHINES.

We next come to a sub-class of machines in which the armature is of the *flat-ring* type. The earliest of these was designed by Simon Schuckert, of Nuremberg.

Schuckert's Dynamo. — The armature of the Schuckert machine is a flat ring, the core of which is built up of a number of thin iron disks, insulated from one another. The winding is identical with that of a Gramme or Pacinotti machine, and the field-magnets resemble, in general, those of the typical Gramme. But the ring is almost entirely enclosed between wide pole-pieces, each of which covers nearly half the ring. The ordinary pattern of machine is shown in Fig. 119,

FIG. 119.

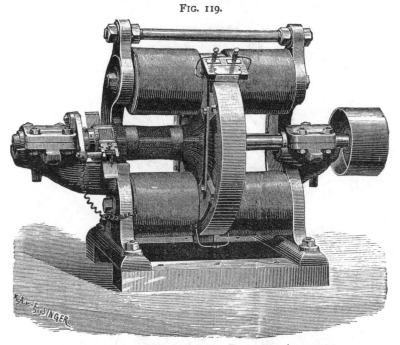

SCHUCKERT'S DYNAMO, WITH FLAT-RING ARMATURE.

which shows the device for removing the armature. The flat ring was designed by Herr Schuckert, to give better ventilation and employ less idle wire than the cylindrical pattern of ring. There is also another type of machine, having four poles. which was illustrated in the former editions of this work. This machine, which is designed to supply 350 incandescent lamps, is compound-wound. Its resistances are 0·01 ohm in the armature, 0·015 ohm in the series coil, and 32 ohms in the shunt coil of the field-magnets. A still newer type of 4-pole machine, by Schuckert, having only two field-magnet coils, is described in the *Elektrotechnische Zeitschrift*, Aug. 1887, p. 353.

MULTIPOLAR RING MACHINES.

When an ordinary ring armature is placed in a multipolar field, having, say, 4, 6, or 8 poles of alternate polarity, there

FIG. 120.

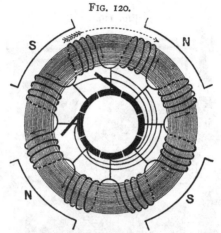

MORDEY'S METHOD OF MULTIPOLAR CONNECTING.

FIG. 121.

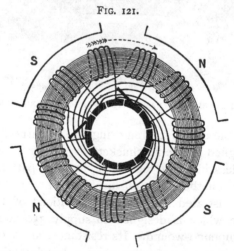

PERRY'S METHOD OF MULTIPOLAR CONNECTING.

will be 4, 6, or 8 neutral points : hence, unless some alteration is made in the winding, it will be necessary to put 4, 6, or 8

brushes at symmetrical points of the collector. To obviate
this, several special methods of connecting up the windings of
the armature have been devised. The best known of these
is that introduced by Mr. Mordey, of which mention is made
below. It consists in adding to the usual Gramme winding a
system of cross-connexions between those portions of the
armature-circuit which arrive simultaneously at equal potentials. This may be done by cross-connecting either the bars
of the collector or the wires of the winding. In 4-pole
machines each bar must communicate with that situated at
180° from it; in 6-pole machines, with those situated at 120°
from it. Mordey's method, as applied to a 4-pole machine, is
sketched in Fig. 120, which shows connexions of a simple
8-part ring. It will be noted that now only two brushes—
and these at 90° apart—are required to collect the currents.

Another method, suggested by Professor Perry, in 1882, is
applicable only to armatures wound with an odd number of
sections. The sketch in Fig. 121 relates to an 11-part armature
in a 4-pole machine. In this method the successive sections of
the coil are not connected together as in Gramme's winding,
but each coil is connected across to that coil which lies nearest
the diametrically opposite point, or, if there are c sections,
each section is connected to the section $\frac{1}{2}c - 1$ beyond.
The coils still form a closed circuit, but the total electromotive-force from brush to brush is the sum of the electromotive-forces in half the coils, whilst in Mordey's method it is
but one quarter. Mordey's method has the contrary advantage of reducing the resistance to one quarter, and is preferable
for low-potential machines. Methods somewhat similar to the
above have been suggested by Wodica[1] and by Muller.[2]

The Schuckert-Mordey Dynamo.—The Anglo-American
Brush Electric Light Corporation has produced a dynamo of
the flat-ring type, under the patents of Schuckert, Mordey,
Wynne, and Sellon, to which the not very apt name of the
"Victoria" dynamo has been given. There are two types of

[1] *La Lumiere Electrique*, xxv. 44, 1887. See also Rechniewsky, in *ib.*, xxiv. 514, 1887.
[2] See Bradley, in *Electrical World*, x. 80, 1887.

this dynamo, one having four, the other six poles arranged round the ring. The development of the Victoria machine from the original Schuckert machine commenced, under the auspices of the electricians of the Brush Corporation, with the discovery by Mr. Mordey, by the aid of his method of examining the distribution of potentials round collectors, that there was in the machine a point at a considerable distance in front of each brush, having the same potential as the brush, and that the whole of the portions of the armature between these equipotential points were useless, or worse than useless, as they were occupied only in producing an opposing electromotive-force. In some of the Schuckert dynamos these useless portions occupied more than half of the armature. By reducing the size of the pole-pieces, space was found for a 4-pole field, the effect of the change being that, from the same ring as employed by Schuckert with a 2-pole field, the electrical output was doubled, without increase of speed. Great attention has been given to the form of the pole-pieces. These pole-pieces in the earlier Schuckert machines, consisted, as mentioned above, of hollow iron shoes or cases which occupied a large angular breadth along the circumference of the ring. Similar hollow polar extensions were long used in the Gülcher machines. By long extended experiments Mr. Mordey arrived at a narrower form of pole-piece, not covering more than 30° of angular breadth of the circumference of the armature, which completely obviated these effects. As will be seen from Fig. 122 which represents the 4-pole Victoria dynamo as now constructed, the pole-pieces, though they embrace the ring through its whole depth, from external to internal periphery, are not quite so narrow. This has been rendered possible by changes in the proportions of the dynamos, which have reduced the inductive effect of the armature. The pole-pieces are of cast iron shrunk upon the cylindrical cores of soft wrought iron which receive the coils. The armature of the Victoria dynamo has several times been modified, and its core is now made of almost square section. It is built up of charcoal iron tape, coiled upon a strong foundation ring, contact between successive layers being prevented by coiling paper between. Special

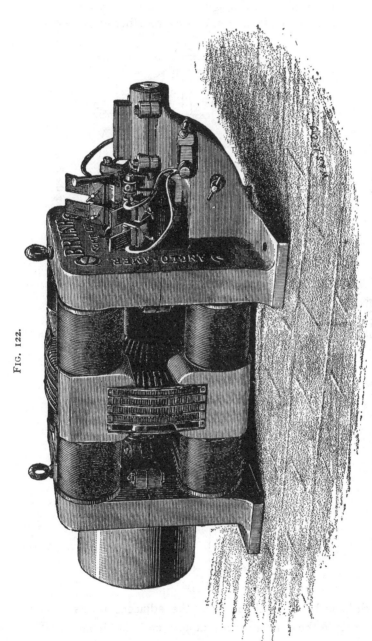

FIG. 122.

VICTORIA (SCHUCKERT-MORDEY) DYNAMO OF ANGLO-AMERICAN (BRUSH) CORPORATION.

pains have been taken throughout to ensure that there are no
electric circuits made in the bolting together of these cores ;

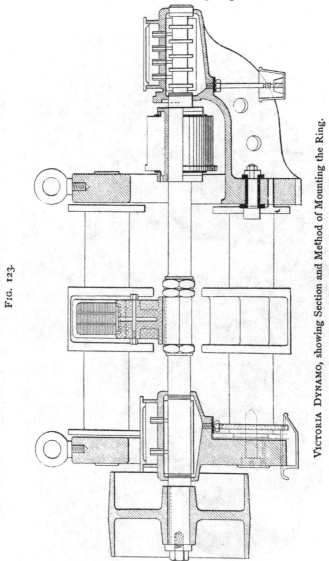

FIG. 123.

VICTORIA DYNAMO, showing Section and Method of Mounting the Ring.

each layer being insulated from the adjacent layers. Eddy
currents in the core are thus almost entirely obviated. The

foundation ring and some of the inner convolutions of tape are
slotted out to receive the gun-metal arms, of which there are
two sets clamped together, one on either side. Fig 123 shows
this construction, and the method of securing the ring to the
shaft by lock-nuts. Square wire is used for winding the

FIG. 124.

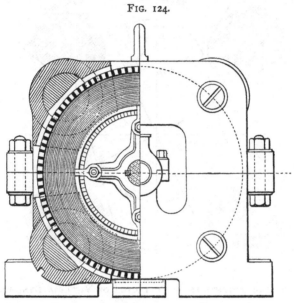

VICTORIA (SCHUCKERT-MORDEY) DYNAMO: End View and Transverse Section.

armature coils, and as they do not cover the entire external
periphery of the armature core there is ample ventilation.
End-play is prevented by the use at one end of a deeply
grooved Babbitt-metal thrust-bearing. Formerly, in a 4-pole
machine, four brushes were necessary—as in the Gulcher
dynamo and the 4-pole Gramme. Mr. Mordey, as mentioned
above, reduced the number to two, by the device of cross-
connecting. Two brushes only are then necessary, and these
are 90° apart. Fig. 125 gives the actual diagram of the
potentials at the collector. Suppose that there are sixty
sections in the ring, then there will be fifteen segments of
the collector from the negative brush to the positive. The

potential rises steadily from the negative brush, and becomes a maximum at the positive brush at 90°, whence it again diminishes to zero at 180°. The bars of the collector

FIG. 125.

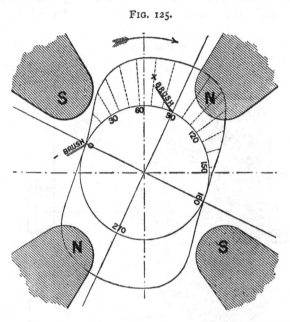

DIAGRAM OF POTENTIALS AT COLLECTOR OF VICTORIA DYNAMO.

being connected, it will be remembered, to those diametrically opposite to them, it follows that the potential will rise from 180° to 270°, precisely as it rose from 0° to 90°, and will again fall to zero in passing from 270° to 0°. If the curve from 0° to 180° were plotted again horizontally, we should clearly see how nearly regular the rise and fall is. If from this curve we were to construct another one, in which the heights of the ordinates should correspond to the tangent of the angle of slope of this potential curve—in other words, if we were to differentiate the curve—we should obtain a second curve—the curve of induction. It would show a positive maximum at about 30°, and a negative maximum at about 120°, where the slope up and slope down are steepest in the potential curve. These maxima of induction are situated very nearly opposite the edges of the pole-pieces, on the side toward which the armature is rotating. Apparently the lines of force of the field are the densest here. In this displacement of the maximum of induction, we have, probably, the explanation of the inferiority of some of the earlier machines with broad polar expansions and weakly excited magnets. In those machines the maximum position of induction was displaced to the very edge of the broad pole-piece, and therefore the induction was sudden and irregular. It is a singular result that while in those machines in which the ring armature is extended cylinderwise, there must be wide-embracing pole-pieces, in those in which the ring is flattened into a disk shape the pole-pieces must be narrow unless the field magnetism is very intense.

The Victoria dynamo is now usually compound-wound, having all the eight magnet-cores wound with main-circuit coils inside and shunt coils outside. The external "characteristic" of this machine is remarkably straight (see Fig. 299). In a "D^2" machine, wound for a potential of 60 volts, the following values were obtained:—Open circuit, 58 volts; 10 amperes, 58·5 volts; 20 ampères, 59 volts; 60 ampères, 59·7 volts; 90 ampères, 59·9 volts; 120 ampères, 60 volts. It will be seen that for small loads the potential drops a little; but it is under these circumstances that the engine speed usually rises slightly in practice, so that the constancy of the potential between the mains is somewhat better than the figures would show. In actual practice, the regulation is perfect. The author has opened the circuit of a Victoria dynamo which at the time was feeding 101 lamps, 100 being at a distance, 1 lamp attached to the terminals of the machine. On detaching the main wire from the terminal, the 100 lamps were suddenly extinguished. The solitary lamp on the machine did not even wink, and there was no flash at the brushes. The sparking was so slight it was impossible to tell whether the machine was an open circuit or whether it was doing full work. The lead was the same under all loads. Some of these machines are wound for slow speeds for ship lighting.

N

These machines have an electrical activity slightly higher, and an efficiency slightly lower, than the high-speed machines. They also have field-magnet cores slightly heavier, requiring, therefore, the expenditure of rather more electrical energy in maintaining the field. These remarks refer, of course, to a comparison between machines wound to light an equal number of lamps, and to work at an equal electromotive-force. The machine depicted in Fig. 122 weighs 1904 pounds, and has an output of 18,000 watts at 1000 revolutions per minute. Such machines are now guaranteed under tender to run at a commercial efficiency of 92 per cent., as tested by Hopkinson's method.

A larger type of Victoria machine, having six poles alternately N. and S. set round the ring, has also been constructed by the Anglo-American Corporation. As each segment of the collector is connected with those situated at 120° and 240° distance round the set, only two brushes are required. A figure of this machine was given in the previous edition of this work.

The present drift toward multipolar dynamos of this type, with a discoidal ring armature, is very significant. There is little difference save in detail between the Victoria dynamos and the 4-pole machines of Gülcher. The advantage originally claimed for this construction, namely, that it allows less of the total length of wire to remain "idle" on the inner side of the ring, is rather imaginary than real, for the total resistance of the armature is but a small fraction of the whole resistance of the circuit; and it is possible to spread the field so as to make all parts of the wire active without any gain whatever, if by this spreading there is no increase on the whole in the total number of lines of force in the field. The real reasons in favour of multipolar flat-ring armatures appear to be the following :—First, their excellent ventilation ; second, their freedom from liability to be injured by the flying out of the coils at high speeds ; third, their low resistance, due to the fact that the separate sections are cross-connected, either at the brushes or in the ring itself, in parallel arc. To these may be added that, with an equal peripheral speed, the armature rotating between four poles undergoes twice as much induction as when rotating between two poles; since it cuts the lines of force twice as many times in the former case as in the latter.

Gulcher's Dynamo.—In this machine as originally brought out, the ring core consisted of a number of flat washers of sheet-iron clamped together, rotating in a 4-pole field, and furnished with four collecting brushes at the commutator. There were four hollow box-like pole-pieces of iron, cast upon wrought-iron cores, embracing a considerable angular breadth. In recent years this machine has been greatly improved by the engineers of the Gülcher Electric Lighting Company. They have followed the same line of progress as that pursued in the design of the Victoria dynamo. The lamination of the cores has been tangential instead of radial ; the field-magnets have been made more powerful; the pole-pieces have been reduced in breadth ; and the number of the brushes has been reduced by cross-connections in the armature to two. Fig. 126 shows an 8-pole type of machine which has been largely used by this Company both for lighting and for electro-

FIG. 126.

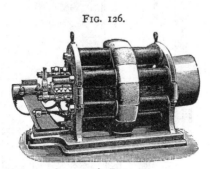

GULCHER'S DYNAMO.

deposition. In this cut the armature is covered from sight by guards of perforated metal. These machines have been made in all sizes up to 32 units.

The latest type of Gulcher dynamo is depicted in Figs. 127 and 128. In Fig. 127 the lower half shows a section in the vertical plane along the line A C, whilst the upper half

N 2

represents a section in the oblique plane of the line A B so as to show the true section of the polar portions. The drawings relate to a 4-pole 10-unit machine, designed to give

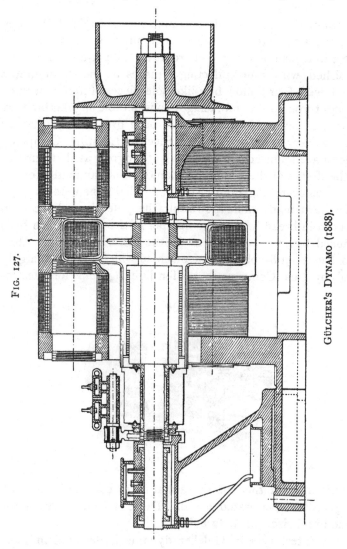

FIG. 127.

GÜLCHER'S DYNAMO (1888).

185 amperes, at 65 volts, at a speed of 900 revolutions per minute. The dimensions and details of construction are as

FIG. 128.

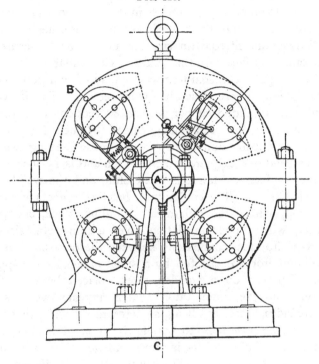

GÜLCHER'S DYNAMO (End View).

FIG. 129.

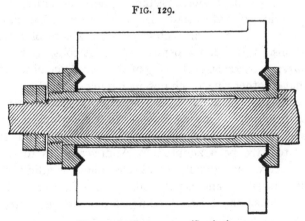

GÜLCHER'S COLLECTOR (Section).

follows : The end-yoke frames and the pole-shoes are of cast iron.   The magnet-cores, 4 inches in diameter and five inches long, are screwed into the pole-shoes, and their other ends are turned down and fitted with circular nuts so as to secure a large contact-surface with the yokes.   The armature is built up of rectangular charcoal iron wire wound upon a gun-metal pulley, forming a core of nearly square section.   This is over-wound with insulated copper wire in 80 sections of 4 turns each, the wire being rectangular, ·12 × ·85 inch, lying in a single layer outside and in two layers inside the ring, which when finished has a square 3-inch section and is 15 inches in diameter.   The resistance is ·025 ohm, and the weight of copper is 18½ lbs.   The commutator has 80 segments of hard-drawn copper insulated with mica ·75 millim. thick, clamped together, with proper insulation, over a gun-metal sleeve. The requisite cross-connexions between the opposite bars of the commutator are effected by a series of insulated copper rings, supported on a wooden sleeve placed between the armature and the commutator.   Each ring has two lugs at opposite sides, these lugs being soldered to the wires that run from the ring windings to oppositely situated bars of the commutator.   This machine is compound-wound, the windings being as follows : Shunt coils of wire ·075 inch diameter, 8 layers of 44 turns each in each coil ; total weight of the 8 shunt coils, 77 lbs. ; total resistance joined in series, 7·805 ohms.   Series coils of rectangular wire, ·13 × ·26 inch, in 2 layers of 14 turns each on each core ; total weight of 8 series coils, 38 lbs. ; total joint resistance of series coils, ·005 ohms.

Andrews' Dynamo.—The ring of Andrews' machine differs from that of other ring machines in having the coils that lie at opposite ends of a diameter connected together in series with one another.   This method of winding, which was first suggested for multipolar machines by Prof. Perry in 1882, appears to have been adopted by Andrews so as to enable him to secure the advantages of compound winding whilst keeping the shunt and series coils of the field-magnets on separate cores ; the series windings are upon one core, the shunt windings upon the three other cores.

Rankin Kennedy's Dynamo.—To this dynamo, which is shown in Fig. 130, its inventor has given the name of the *iron-clad dynamo*, on account of the construction of the field-

FIG. 130.

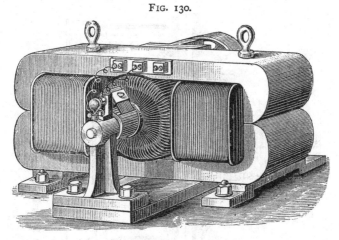

KENNEDY'S IRON-CLAD DYNAMO.

magnets, which enclose the exciting coils.[1] This form of field-magnet, which is discussed in Chapter XIV., p. 328, has 4 poles, two of them being salient poles protruding directly through the exciting coils on either side of the armature, the other two poles being consequent poles formed at the middle-points of the upper and lower portions of the iron framework which here are bored away to form polar surfaces. These machines are supplied chiefly for ship-lighting at slow speeds ; a machine weighing 1½ ton, having an output of 10,000 watts at 350 revolutions per minute. The field-magnets, which, by their design, have no joints, are forged out of special faggoted iron. Swedish iron bars cut up and laid together in faggots, and welded under a very high temperature by a very heavy hammer, make a material which the inventor prefers to hammered scrap. For in the faggoted iron there is a slight grain which runs always in the direction of the lines of magnetisation. The

[1] An earlier type of "iron-clad" dynamo by the same inventor is described in *Industries*, vol. i. p. 137, 1886.

magnets have a cross section of 120 square inches.   The ring
armature is mounted on brass spokes, well-ventilated and

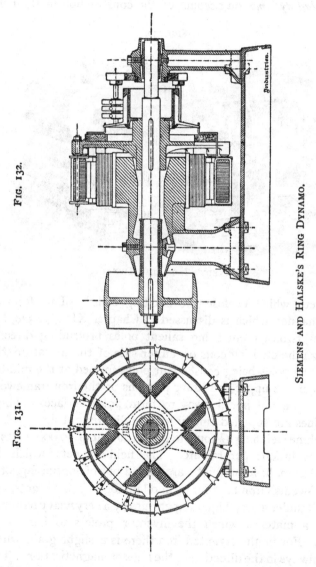

FIG. 132.

FIG. 131.

SIEMENS AND HALSKE'S RING DYNAMO.

perfectly balanced.   Its external diameter is 13½ inches,
internal, 8½ inches ; external axial length, 15 inches.   The

wire is rectangular, 2·5 × 4 millims. cotton-covered, wound in 60 sections laid on edge, and cross-connected so as to require only two brushes. Still more recently Mr. Kennedy has constructed dynamos with a single exciting coil of a form resembling Fig. 244, No. 31, p. 330.

Siemens and Halske's Ring Dynamo.

Toward the end of 1886 a form of multipolar ring machine, with ring external to the field-magnets, was brought out almost simultaneously by Messrs. Ganz of Buda-Pesth, Messrs. Fein of Stuttgart, and by Messrs. Siemens and Halske of Berlin.[1] It will be sufficient to describe the machine of the latter firm.

The field-magnet, as shown in Fig. 131, is stationary and internal to the ring. It consists of a substantial cross-shaped mass of cast iron, through the centre of which passes the driving shaft. The four poles, after receiving the exciting coils, are furnished with polar expansions which approach close to the inside of the ring. The ring itself is made up of thin iron washers bolted together, and supported on one side by a brass spider keyed to the shaft. A machine of this type, weighing 2660 lbs., with an output of 25,000 watts at 480 revolutions per minute, had a ring 20 centimetres broad and of 64 centimetres internal diameter. The advantages of this type are the ease of repair, the immense cooling surface of the armature, and the non-necessity of applying binding-wires. In some of these machines the commutator is unusually large, consisting of stout bars of iron with an air-insulation.

Other multipolar ring-dynamos have been constructed by Gramme, Wenström, Einstein, Paris and Scott, Elwell Parker and Co., and other makers.

Other Ring-armature Machines.

Lahmeyer's Dynamo.[2] —This machine, the magnets of which belong to the iron-clad type, is depicted in Fig. 133.

[1] For further information about the various machines of this type see *Elektrotechnische Zeitschrift* for April and May, 1887; *La Lumiere Electrique,* vol. xxiv. p. 182, 1887; *Centralblatt für Elektrotechnik,* vol. ix. pp. 186, 410, and 581, 1887.

[2] See *Centralblatt für Elektrotechnik,* vol. ix. pp. 71 and 411, 1887; also *Elektrotechnische Zeitschrift,* vol. ix. p. 89, 1888.

It has been submitted to elaborate tests by Professor W. Kohlrausch. The machine, weighing 597 kilogrammes, gave 60 ampères at 65 volts with a speed of 1250 revolutions per

FIG. 133.

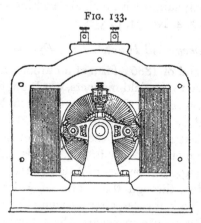

LAHMEYER'S DYNAMO.

minute. The current density in the armature was 8·3 amperes per square millimetre of copper in section. The commercial efficiency was over 80 per cent. Though the field-magnets of cast iron were magnetised up to only 36 per cent. of their maximum, the armature current showed very little demagnetising tendency. According to the inventor the waste magnetic field is less than 8 per cent. of the total field generated. The armature is wound on a plan suggested by Arnold of Riga, and independently suggested by Crompton, namely, the conductors are wound between teeth in the periphery of the core, after which the whole exterior of the armature is served with a thin layer of insulating material, and over this a layer of iron wires is wound.

Other machines with ring armatures have been devised by Lever,[1] who grooves the polar faces to admit the iron teeth protruding from the armature, and by Swan,[2] who also uses a toothed armature-core. A machine was shown in 1881 at the Paris Exposition by Messrs. Siemens, in which the field-

[1] *Electrical Review*, xxi. 406, 411, 1887.
[2] *Electrician*, xviii. p. 370, 1887.

magnet, of simple shuttle-wound form, rotated within a fixed ring armature. The brushes of this so-called " Topf-Maschine " rotated against a fixed collector. Information respecting several of these dynamos, and of others of less note, may be found in Dredge's *Electric Illumination,* in the official report of the Munich Exhibition of 1882, in Schellen's *Magnet- und Dynamo-Elektrischen Maschinen,* and in the periodical journals devoted to electricity.

CHAPTER VIII.

CONTINUOUS-CURRENT MACHINES (CLOSED COIL).

Drum-armature Dynamos.

THE drum armatures of all types may all be regarded as modifications of Siemens' well-known longitudinal shuttle-form armature of 1856, a multiplicity of sections of the coils being employed to afford practical continuity in the currents. The drum pattern was invented in 1872 by Von Hefner Alteneck, of the firm of Siemens and Halske of Berlin.

The advantages of the drum form of armature appear to be (1) that they require somewhat less wire than the ring armature of equal size ; (2) are free from liability to false inductions (p. 75) and therefore more independent of the form of the pole-pieces ; (3) have smaller cross-magnetising tendency than ring armatures. Their disadvantages hitherto have been (1) greater difficulty of construction ; (2) greater difficulty of securing proper insulation on account of overwrapping of conductors ; (3) greater difficulty of executing repairs.

Siemens' Dynamo.—In some of the earlier patterns of Siemens' machines the cores of the drum were of wood, over-spun with iron wire circumferentially before receiving the longitudinal windings. In another of their machines there was a stationary iron core, outside which the hollow drum revolved ; in other machines, again, there was no iron in the armature beyond the driving-spindle. In all the Siemens armatures the individual coils occupy a diametral position with respect to the cylindrical core, but the mode of connecting up the separate diametral sections is not the same in all. In the older of the Alteneck-Siemens windings the sections were not connected together symmetrically, the connexions (for an 8-part collector) being as in Fig. 134. In the more recent machines a symmetrical plan has been adhered to, as shown in Fig. 135. In this system, as in the Gramme ring, the successive sections of coils ranged round the armature are connected together continuously, the end of one section and the beginning of the next being both united to one segment or bar of the collector. A symmetrical arrangement is, of course, preferable, not only for ease of construction, but because it is important that there should never be any great difference of potential between one

FIG. 134.

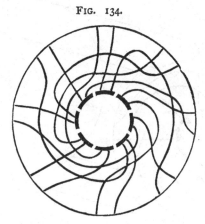

DIAGRAM OF CONNEXIONS OF SIEMENS WINDING (OLD).

segment of the collector and its next neighbour; otherwise there will be increased liability to spark and form arcs across the intervening gap.

In Fig. 135 the eight sections are shown for simplicity as if each consisted of one turn only. For example, beginning at the segment of the collector marked 1, the wire ascends to 1', thence passes along the drum, descends along a diameter, returns longitudinally along the lower side of the drum to the point marked 1", whence it is brought up to the second segment of the collector; from the second segment the second section of the winding proceeds in like manner. In ordinary Siemens armatures there are in reality eight

turns in each section : the wire after ascending to 1, wrapping eight times round the cylinder and finally turning down
to join the second segment. The process of constructing the
armature employed down to the year 1885 is shown in Fig. 136.
Upon the axle are secured by pins two stout cheeks of gun-
metal to form the ends of the drum. Between these, and
resting on an inner projecting rim, is wrapped a thin sheet of
iron, and over this a quantity of soft iron wire is wound to

FIG. 135.

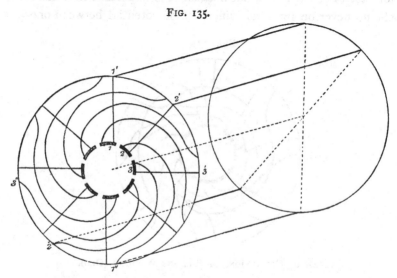

DIAGRAM OF CONNEXIONS OF SIEMENS WINDING (NEW).

form a core, as in the Gramme armature. A number of cuts
equal to the proposed number of segments, are sawn radially
in the end faces of the gun-metal cheeks, and in these cuts
small boxwood wedges are inserted to facilitate the winding.
The coiling of the sections is done in the following manner:—
The wire is carried along the drum, as shown, four of the
strands passing to the left, four to the right of the axle at
either end, and is then ready to be turned over to meet the
connecting piece of the second segment of the collector.
From this point the next section will start in a like manner ;
but before the second section is wound the drum is turned

completely over, and the section diametrically opposite to section No. 1 is wound on the top of the eight strands already wound. Thus, in a 16-part armature, section No. 9 lies on the top of section No. 1, No. 10 on the top of No. 2, and so on, there being two layers of coils all over the drum. The object of this arrangement is to ensure that parts which are at very different potentials shall never come in contact with one another. Although, for the sake of rendering the

FIG. 136.

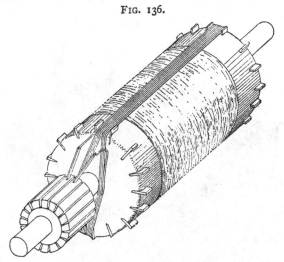

METHOD OF WINDING SIEMENS ARMATURE.

connexions more intelligible, the collector is shown in Fig. 136 in its place on the axle, it is not, as a matter of fact, put into its place until after all the sections have been wound, the ends of the wires being temporarily twisted together until all can be soldered to the connecting strips of copper. In some of the armatures intended for electro-plating machines there are four layers of wire, and the wires are connected together four or eight in parallel to reduce the resistance.

The field-magnets in the earlier patterns of Siemens dynamo were horizontally placed (Fig. 137) and consisted of a row of wrought-iron cores, arched where they passed near the armature, and bolted together at their outer extremities : the

coils, which are wound on flat brass frames, being slipped
on before the cores are bolted down. In some more recent
machines a vertical position is preferred, as in Fig. 138. In
all these forms the arched cores are removable, and in the
larger sizes the top half of the machine can be unbolted
and removed from the lower half. In some machines
there is a provision for ventilating the cores of the magnets,
an air space being left in the brass frames of the coils on the
outer side. Some account of the compound-wound machines

Fig. 137.

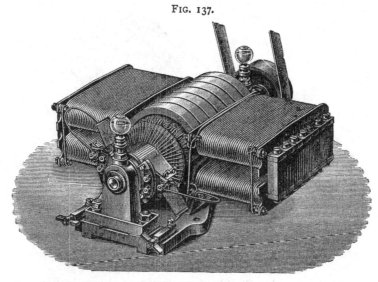

SIEMENS DYNAMO, HORIZONTAL PATTERN.

made by the firm has been given by Herr E. Richter in the
Elektrotechnische Zeitschrift. It appears that three methods
of combination have been tried. The shunt and series coils
have been wound on different arms of the magnets; they
have been wound on separate short frames, and slipped on
to the cores side by side, and they have been also wound
over one another. The series coils are generally wound
outside the shunt windings. The regulation, judging by the
curves given by Herr Richter, is not perfect. The best
regulation was from a " *g* D 17 " machine, of which two of

the magnet limbs were wound with shunt coils of twenty-
nine layers of a 1-millimetre wire, and the other two with two
layers of a 3·5-millimetre wire. The potential varied from
64 to 69 volts when the number of lamps was reduced from
twenty to nine (see curve, Fig. 299).

Since then the firm has constructed dynamos in which
the potential does not vary 1 per cent. between the limits of

FIG. 138.

SIEMENS DYNAMO, VERTICAL PATTERN.

working.[1] Some very large machines of the vertical pattern
but with many improvements in detail, including three 112-unit

---

[1] See series of papers in the *Elektrotechnische Zeitschrift*, March—June 1885,
by Dr. O. Frölich.

compound-wound dynamos, "B 13" pattern, were used at the Inventions Exhibition of 1885. Each of these is capable of yielding 450 ampères at a potential of 250 volts, making an output of 112,500 watts, when running at 300 revolutions per minute only. The armatures are wound with flat strip-copper.

In 1886 Messrs. Siemens and Halske, after trying some intermediate forms (depicted on pp. 159 and 160 of the second edition of this book), adopted the type depicted in Fig. 139. The field magnet of this machine consists of a single very massive casting, over which the exciting coils are slipped. The collectors of these machines are more massive, and

FIG. 139.

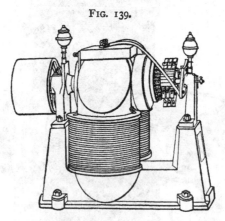

SIEMENS MACHINE (Type of 1886).

insulated by air-gaps. The brushes are adjustable; and they are trimmed with oblique ends so as to touch two collector-bars at once.

The Siemens ring-machine with internal field-magnet, described on p. 184, is still more recent.

The Siemens alternate-current dynamo (Fig. 217) is described on p. 267; the Siemens unipolar dynamo (Fig. 203), on p. 253; and the Siemens electro-plating machine (Fig. 228), on p. 283.

Edison's Dynamo.—There are several forms of Edison dynamo having drum armatures rotating between heavy iron

Fig. 140.

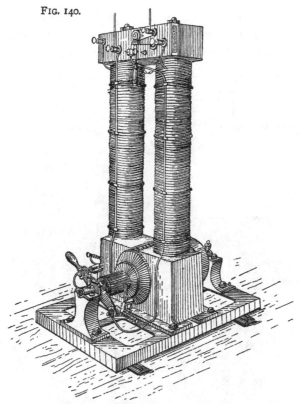

Edison Dynamo, "Z" Pattern.

Fig. 141.

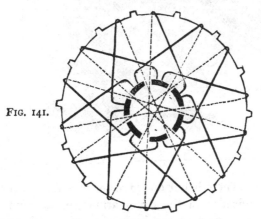

Diagram of Edison Armature.

O 2

pole-pieces. The "Z" machine, capable of working sixty
Edison lamps, is shown in Fig. 140. The field-magnets are
of extraordinary length, and of circular section, with a heavy

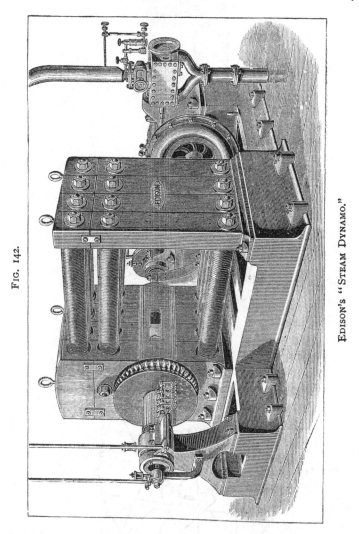

FIG. 142.

EDISON'S "STEAM DYNAMO."

yoke at the top. There is but one layer of wire on the cores.
In Edison's modification of the drum armature, the winding,
though symmetrical in one sense, is singular, inasmuch as the

number of sections is an odd number. In the first machines there were seven paths, as shown in Fig. 141, taken from Edison's British Patent Specification. In his giant "steam-dynamo" (Fig. 142) the number of sections is forty-nine. One consequence of this peculiarity of structure is, that if the brushes are set diametrically opposite to one another, one will touch the middle of a bar of the collector at the instant when the other slides from bar to bar. In Edison's larger dynamos the armature is not wound up with wire, but, like some of Siemens' electro-plating dynamos, is constructed of solid bars of copper, arranged around the periphery of a drum. Fig. 143 shows the armature removed from the machine.

FIG. 143.

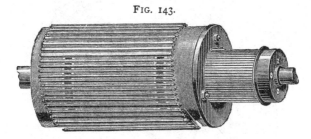

ARMATURE OF EDISON DYNAMO.

The ends of the bars are connected across by washers or disks of copper, insulated from each other, and having projecting lugs, to which the copper bars are attached. Such disks present much less resistance than mere strips would do. To make the mode of connexion plainer the dia-grammatic sketch of Fig. 144 is given. The connexions are in the following order:—Each of the forty-nine seg-ments of the collector is connected to a corresponding one of the forty-nine disks at the anterior end of the drum ; and this disk is connected, by a lug-piece on one side, to one of the ninety-eight copper bars. The current generated in this bar—say, for example, the highest of the three bars shown in Fig. 144—runs to the further end of the machine, enters a disk at that end, crosses the disk, and returns along a bar diametrically opposite that along which it started.

The anterior end of this bar is attached to a lug-piece of the
next disk but one to that from which we began to trace the
connexions: it crosses this disk to the bar next but one to
that first considered, and so round again.   The two lug-pieces
of the individual disks at the anterior end are, therefore, not
exactly opposite each other diametrically, as the connexions
advance through $\frac{1}{49}$ of the circumference at each of the forty-
nine paths.   To simplify matters, in the drawing the alternate
disks and bars are only indicated in dotted lines.   Just as the
two bars shown at the bottom are the returns for the currents

FIG. 144.

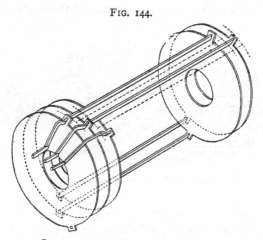

CONNEXIONS OF EDISON ARMATURE.

in the top bars, so there must be top bars provided as returns
for the currents in those bars (not shown) which start from
the segments of the lower half of the collector.   The dotted
lines show the position of these return bars.   It will be
noticed in Fig. 142,[1] that the collector is very substantially built
and that a screen is fixed between the collector and the rest
of the armature, to prevent any copper-dust from flying back
or clogging the insulation between the bars or disks.   There
are no fewer than five pairs of brushes, the tendency to

[1] Taken, by permission of Messrs. Macmillan and Co., from the author's
Elementary Lessons in Electricity and Magnetism.

sparking being thereby greatly reduced. The figure does not show the structure of the armature itself, nor indicate the means taken to suppress eddy currents. The core of the armature is made of very thin disks of iron, separated by mica or paper from each other, and clamped together. Some exception may be taken to the use of such stout copper bars, as being more likely to heat from local currents than would be the case if bundles of straps, or laminæ of copper were substituted; and, indeed, the presence of the 4 horse-power fan to cool the armature, is suggestive that continuous running is liable to heat the armature.

Dr. J. Hopkinson's efforts to improve this machine resulted, as detailed below, in a better design.

The field-magnets of all the larger machines turned out by Edison prior to 1884 had a number of long iron columns as cores to receive the coils. Since that date the more compact arrangement of a single magnetic circuit with short stout magnets has been adopted by the Edison companies on both sides of the Atlantic. The present form (1888) of Edison dynamo, as used in the States, is depicted in Fig. 145. The field-magnets are of cast iron, with a massive yoke, and stand upon a high footstep of zinc to diminish short-circuiting through the bed-plate. These machines are shunt-wound, and are intended for incandescent lighting work. The bearings are longer and the mechanical arrangements in every way superior to those of the older machines.

Edison-Hopkinson Dynamo.—The Edison dynamo has received very material improvements at the hands of Dr. John Hopkinson. Some of these improvements relate to the field magnet; others to the armature. Dr. Hopkinson, in the first place, abolished the use of the multiple-field magnets, which in Edison's original "L," "K," and "E" machines were united to common pole-pieces; and instead of using two, three, or more round pillars of iron, each separately wound, he puts an equal mass of iron into one single solid piece of much greater area of cross-section and somewhat shorter length. One such iron mass, usually oblong in cross-section, is attached solidly to each pole-piece, and the two are united at the top by

a still heavier yoke of iron. The machines have, consequently, a more squat and compact appearance than before
(Figs. 146 and 147). Dr. Hopkinson also introduced the
improvement of winding the magnets with a copper wire of

FIG. 145.

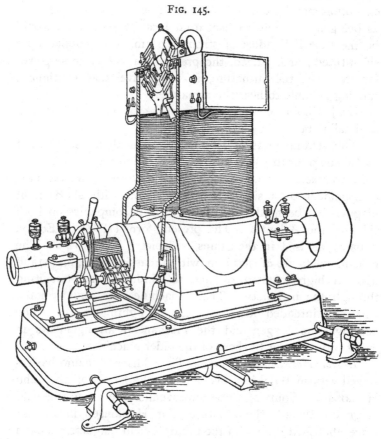

EDISON DYNAMO (1888).

square section, wrapped in insulating tape. This wire packs
more closely round the iron cores than an ordinary round
wire. In the armature the following change has been made.
The iron core in the older Edison machines was made of
thin iron disks, separated by paper, slipped on over a sleeve

of *lignum vitæ*, and held together by six longitudinal bolts passing through holes in the core plates, and secured by nuts to end plates. These bolts are now removed, and the iron plates are held together by great washers, running upon screws cut on the axle of the armature. The size of the central hole in the plates has been diminished, thus getting into the interior more iron, and providing a greater cross-section for the magnetic induction. By these improvements, a machine occupying the same ground space, and of about the same weight as one of the older "L" 150-light machines, was made capable of supplying 250 lights, the economic coefficient being at the same time higher. In the new 250-light machine, the diameter of the armature is 10 inches; its resistance, cold, is 0·02 ohm; that of the magnets is 17 ohms. The magnetisation curve of the machine shows that even when doing full duty, the field magnets are far from being saturated. It will be remarked that, in the older construction, the bolts and their attached end-plates furnished a circuit in which idle currents were constantly running wastefully round, with consequent heating and loss. An Edison 60-light "Z" machine of the older pattern, tested by the Committee of the Munich Exhibition, was found to give an efficiency which, if measured by the ratio of external electric work to total electric work, exceeded 87 per cent.; but its commercial efficiency—the ratio of external electric work to mechanical energy imparted at the belt—was only, at the most, 58·7 per cent.

A remarkably complete account of the construction and performance of one of these Edison-Hopkinson dynamos was published in 1886.[1] As this machine is often referred to in the theoretical chapters of this book, a detailed account of it is important. Its design may be gathered from Figs. 146, 147, and 148.

The machine described is intended for a normal output of 320

[1] See paper on *Dynamo-electric Machinery*, by Drs. J. and E. Hopkinson, in the *Philosophical Transactions* for 1886, Part I. This most valuable paper was reprinted, but without the plates, in the *Electrical Review*, vol. xviii., 1886. It was also reprinted in the *Electrician*, vol. xviii., pp. 39, 63, 86, and 175, in issues of Nov. 19th and 26th, and Dec. 3rd and 31st, 1886, where the figures of the plates are printed in the text.

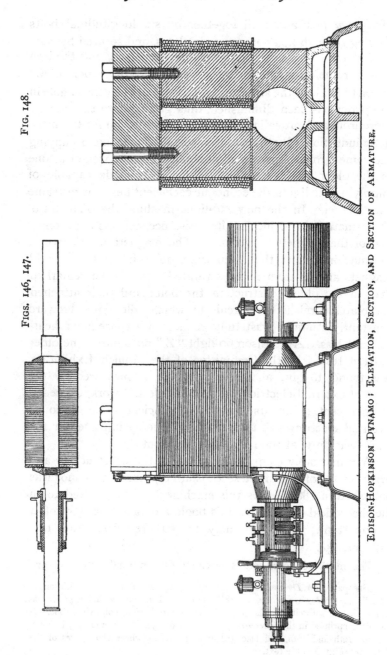

FIG. 148.

FIGS. 146, 147.

EDISON-HOPKINSON DYNAMO: ELEVATION, SECTION, AND SECTION OF ARMATURE.

amperes at a pressure of 105 volts, running at 750 revolutions per
minute. The field magnet consists of two limbs connected by a yoke
of rectangular section. Each limb, together with its pole-piece, is
formed of a single forging. The wrought iron used for these and the
yoke is of annealed hammered scrap; the magnetic properties being
those described in Chapter XIV. The section of the limbs is nearly
rectangular, with rounded corners. The yoke is bolted to the limbs,
the joints being well surfaced. The bed-plate is of iron, a zinc base
12·7 cm. high being interposed. The armature core is built up of
about 1000 thin plates of soft wrought iron, insulated from the shaft,
and separated by paper from one another. They are held between
two end plates, one of which is secured by a washer shrunk on the
shaft, and the other by a screw-nut and lock-nut.

The following are the dimensions of the iron parts :—Diameter of armature
core, 25·4 cm. ; of internal hole, 7·62 cm. ; of shaft, 6·98 cm. ; length of core,
50·8 cm. Length of field-magnet limb, 45·7 cm. ; breadth, 22·1 cm. ; width
(parallel to shaft), 44·45 cm. Length of yoke, 61·6 cm. ; width, 48·3 cm. ;
depth, 23·2 cm. Diameter of bore of field magnets, 27·5 cm. ; depth of pole-
piece, 25·4 cm. ; width (parallel to shaft), 48·3 cm. ; width between pole-pieces,
12·7 cm. Area of section of iron in armature core, 810 sq. cm. Angle sub-
tended by bored face of pole-pieces, 129°. Actual area of pole-piece, 1513
sq. cm. ; effective area, 1600 sq. cm. Thickness of gap space, 1·5 cm. Area of
section of limbs, 980 sq. cm. ; ditto of yoke, 1120 sq. cm.

The windings are as follows :—Magnetising coils, 11 layers on
each limb of copper wire, 2·413 mm. diameter. Total convolutions,
3260; total length, 4570 metres. Armature, 40 convolutions in two
layers of 20 convolutions of stranded copper wire, consisting of
16 strands of wire 1·753 mm. diameter. Resistance (at 13·5° C.) :
field magnet, 16·93 ohms; armature, 0·009947 ohm. Normal
magnetising current, 6 ampères. Commutator, 40 copper bars insulated with mica.

Recent tests with Edison-Hopkinson dynamos constructed
by Messrs. Mather and Platt, of Manchester, show that they
have an economic coefficient of over 95 per cent., and an
actual commercial efficiency of over 93 per cent. These
machines have usually from two to five separate brushes at
either side, capable of separate removal, so that they may be
trimmed without stopping the machine. In order to bring
the neutral points of the commutator to convenient positions
right and left, the connecting pieces which join the commutator bars to the armature windings are carried spirally

through about 90°. The makers of these machines have lately modified in detail the winding of the armatures,[1] enabling them to use copper bars instead of stranded wire. They shape the pole-pieces to diminish distortion of field, and connect the armature bars across the ends of the armature by

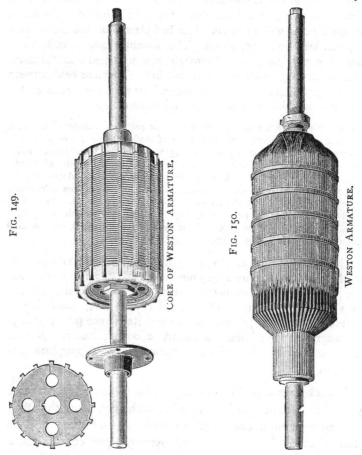

FIG. 149.

CORE OF WESTON ARMATURE.

FIG. 150.

WESTON ARMATURE.

spiral connectors closely resembling those used in Siemens's electro-plating dynamos, Fig. 229, p. 284.

Weston's Dynamo.—A good many successive forms of dynamo have been designed by Mr. Weston; one of the

[1] See *Industries*, ii. 549, 1887; and Specification of Patent 4884 of 1886.

earliest being a small electro-plating machine with a pole-armature. In Weston's electric light dynamos, the characteristic feature is the armature, a drum, the core of which is built up of thin iron disks or washers having projecting teeth between which the wire coils are laid. The skeleton of the armature is depicted in Fig. 149, together with one of the disks. The completed armature is shown in Fig. 150. Weston devised a method of winding the armature with two circuits, so that the accidental short-circuiting of any two adjacent segments of the collector shall not cause the armature to break down · by the over-heating of the short-circuited section.

The recent Weston machines show substantial design, and many improvements in detail upon the older forms. In this machine, as is also evident from Fig. 151, the pole-pieces

FIG. 151.

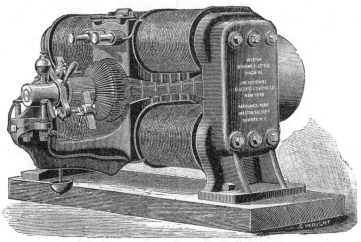

WESTON DYNAMO.

are laminated, to obviate eddy currents and heating. The magnets are shunt-wound; and Mr. Weston succeeded in working up the intensity of the field to such a degree that the number of turns of wire on the armature could be reduced in a manner previously unattained, though since surpassed.

His armature has a 24-part collector, and there are twenty-four sections each of one turn only and of very thick wire. The resistance of the armature is therefore practically negligible. As a consequence of the very small number of coils, the armature current produces very little shifting of the field; and in consequence of the very small internal resistance, the difference of the potential at the terminals is, with a constant speed, practically constant. With this machine, which is not compound-wound, but which is self-regulating simply because its internal resistance is *nil*, it is possible to turn off 99 out of 100 lamps at once, leaving one lamp burning. The armature resistance is inappreciable in comparison with that of the 100 lamps in parallel.

Wenström's Dynamo.—This machine, which made its first appearance in Sweden in 1882, has several remarkable features.

FIG. 152.

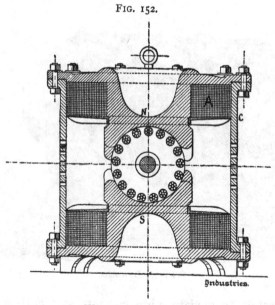

ꝋnbustries.

WENSTRÖM'S DYNAMO.

The field-magnet is of the "iron-clad" type, with two salient pole-pieces upon which the coils are borne; an external cylinder of iron, which encloses the machine, serving as the magnetic yoke. Fig. 152 shows the main features of construction. The armature has the peculiarity that the coils are wound in circular holes cut in the periphery of the armature core. The inventor has also devised a six-pole machine,[1] also of iron-clad type.

Crompton and Swinburne's Drum Armatures.—Two improvements in the construction of drum armatures have recently been made. In order to lessen the tendency of the armature currents to produce cross-magnetising and demagnetising actions, Mr. Swinburne so couples the peripheral conductors that the armature currents flow in alternate directions in adjacent conductors in those parts of the periphery which for the time being lie in the regions where they are not cutting the lines of the magnetic field. This is accomplished by carrying the end connexions not across a diameter of the circumference, but across a chord subtending an arc equal to half the angular breadth of the polar surface plus the angular breadth of one of the gaps between the poles.

Mr. Crompton and Mr. Swinburne have devised a method of connecting the conductors of a drum armature which enables the core to be ventilated. The fundamental point in this construction is illustrated by the simple form sketched in Figs. 153 and 154. There is only one layer of conductors outside the core, and these consist of copper wire of narrow rectangular section set edgeways. Every alternate conductor of the set is bent radially inwards for a certain distance, and then recurved outwards again parallel to the axis of the drum. The conductors that lie between them are prolonged straight outwards to a certain length. The result of this arrangement is that there are two concentric sets of projecting ends. Connection is made by a spirally bent strap of copper from the ends of one of the outer straight set to one of the inner recurved set; for example, to the one that lies next to that at the diametrically opposite point of the periphery. In Figs. 153 and 154, which depict the connexions at one end of a simple six-part armature (the six-part commutator at the

---

[1] See *La Lumière Électrique*, xx. 20, 1886.

other end is not shown), conductor 1 is connected in this manner to conductor 1. At the other end of conductor 1 (not shown) there would be a similar cross-connexion to the

FIG. 153. FIG. 154.

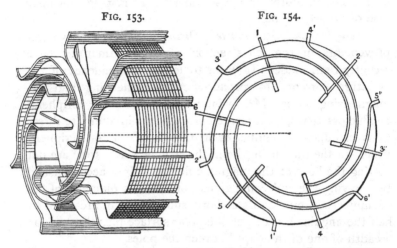

DIAGRAM OF CROMPTON AND SWINBURNE'S METHOD OF DRUM-WINDING.

far end of conductor 2 : and the near end of 2 is joined to 2′ ; and so forth. Or the same method of connecting by spiral straps may be applied to winding across a shorter chord, as on Swinburne's plan mentioned above. The method of connecting by spirals of copper differs from that adopted in the Siemens electro-plating dynamos, Fig. 229, p. 284, in the use of the cranked pieces. In the actual construction of large drum armatures the Crompton and Swinburne construction is more complex, as it is impossible with a very large number of conductors to find room for all the spiral connectors in one whirl. Accordingly, there are two separate layers of spiral conductors, the projecting extremities of the conductors at the ends of the drum being themselves made alternately long and short, the short projections being connected by one layer of spirals, and the long projections being connected by another layer of spirals. The armature as so constructed is depicted in Fig. 155.

Still more recently Messrs. Crompton and Swinburne have

proposed to construct a dynamo with revolving field-magnet inside an external drum armature. The idea is illustrated in

FIG. 155.

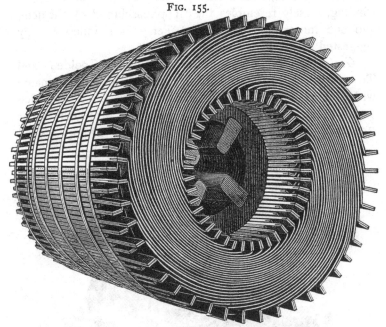

CROMPTON AND SWINBURNE'S DRUM-ARMATURE.

Fig. 156, in which it will be noted that the plan of carrying the armature conductors through holes in the periphery of

FIG. 156.

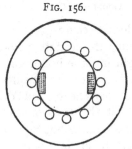

CROMPTON AND SWINBURNE'S DESIGN FOR DRUM MACHINE
WITH INTERNAL FIELD MAGNET.

the core is intended to be adopted. With this arrangement the actual length of the magnetic circuit is so short that

P

very few ampère-turns are required upon the field-magnet, which is a simple shuttle-wound grooved iron cylinder. This form of machine, which, save in the adoption of a drum-winding, closely resembles the "Topfmaschine" of Siemens, and the motors of Ayrton and Perry, was independently suggested by M. Cabanellas.

Goolden's Dynamos. — Formerly Messrs. Goolden and Trotter constructed a modified type of Gramme dynamo with ring armature. More recently Messrs. W. T. Goolden and Co. have introduced, for working at pressures under 250 volts, machines with drum armatures having several

FIG. 157.

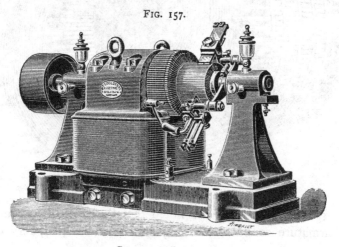

GOOLDEN'S DYNAMO.

peculiarities.[1] In order to secure good ventilation with compact form, they construct the armature core hollow, with the bearing projecting right into the hollow of the armature. The windings are of round copper wire insulated with silk, lying close together at the periphery, and driven by means of driving-horns inserted at the ends of the drum. The field-magnets are usually of the horseshoe type, with poles upwards; but for coupling direct to engines with a low shaft they are inverted, and a third coil is then placed on the yoke-piece. The form of the over-type machine is shown in Fig. 157.

[1] Specifications of Patent 5303 and 5505 of 1887.

MULTIPOLAR DRUM MACHINES.

The Elphinstone-Vincent Dynamo.—The earliest of multi-polar drum machines was the now obsolete Elphinstone-Vincent Dynamo. In this dynamo the drum armature is of a distinct order, the separate coils being laid upon the sides of a hollow *papier-mâché* drum in an overlapping manner, and curved to fit it. The-field magnet arrangement was a complex one, with six external and six internal poles, as shown in Fig. 158. The sections of the armature of this

FIG. 158.

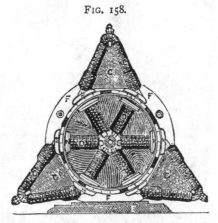

ELPHINSTONE-VINCENT DYNAMO (Section).

machine were wound separately upon parallelogram formers, and the separate sections were then fixed upon the periphery of the *papier-mâché* cylinder, which was mounted so as to rotate between the external and internal field-magnets whose poles reinforce one another. The idea of the inventors was to use these internal magnets to complete the magnetic circuits ; but the mechanical difficulties involved in the construction of the machine prevented its commercial success.

A very similar machine, having a hollow drum armature, was designed by Zipernowsky.[1]

Thury's Dynamo.—A very similar machine, having a drum armature built upon a hollow core of laminated iron,·

[1] *Electric Illumination*, p. 310.

was designed by Thury.[1]   Fig. 159 shows a six-pole machine
based on this principle.   The field-magnets, with slab cores,

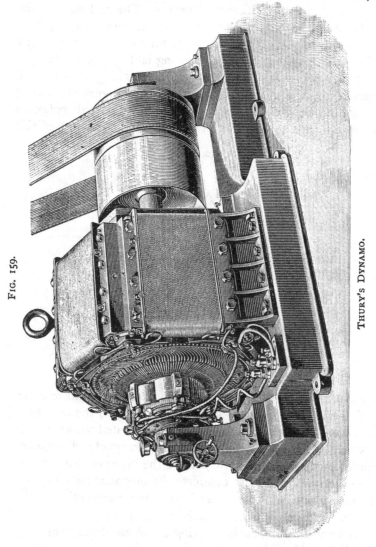

FIG. 159.

THURY'S DYNAMO.

are bolted together into a strong hexagonal shape, with six
inward-facing pole-pieces of alternate polarity.   The windings

[1] See also *Electrician*, xvii. 493, 1886.

on the drum must embrace chords of 60°, and six brushes must be used unless the armature or collector bars are respectively cross-connected at 120° all round. Machines of this type are made from 1300 kilogrammes' weight, giving 22,000 watts at 600 revolutions per minute, to 4500 kilogrammes, giving 66,000 watts at 350 revolutions per minute.

Other Drum Machines.—Messrs. Chamberlain and Hookham[1] have constructed drum machines having the coils wound between projecting ridges of teeth ; and the brushes are pressed longitudinally through slide-holders. Messrs. Paris and Scott[2] wind the conductors of their armature in very deep narrow slots in the core-plates, which are driven direct by being threaded upon a hexagonal shaft. A dynamo of the " iron-clad " type, but having a drum armature, has been devised by Eickemeyer.[3] The Stanley dynamo, which has a drum armature, belongs to the open-coil type described in Chapter X. The Schuyler dynamo, also an open-coil machine, has a drum armature.

For discussions upon the various methods of winding drum armatures, the reader is referred to the following authorities:—Dr. O. Frölich : *Elektrotechnische Zeitschrift*, various papers ; M. Aug. Guérout : *La Lumière Électrique*, xii. 212, xiii. 81, 1884 (the latter paper discussing the difference between various drum-windings) ; Prof. Carl Hering : *Electrician and Electrical Engineer*, iv. 423, 1885, and v. 84, 1886 ; Prof. von Waltenhofen : *Centralblatt für Elektrotechnik*, ix. 658, 1887 ; Fritsche: *ib.*, ix. 649, 1887.

[1] *Electrician*, xvii. 426, 1886.
[2] *Manufacturer and Inventor*, January 1888.
[3] *Electrical World*, ix. 41, 182, 1887 ; and *Electrician*, xviii. 437, 465, and 485, 1887.

CHAPTER IX.

CONTINUOUS-CURRENT MACHINES.

Disk-armature Dynamos.

IN the dynamos of this class the coils are carried round to different parts of a magnetic field such that either the intensity differs in different regions, or more generally the lines of force run in opposite directions in different parts of the field. Fig. 9 (p. 31) illustrates this principle ; and we shall now consider how it is carried out in practice. In the early machines of Saxton, Clarke, and Stöhrer, single pairs of coils were mounted so as to pass in this fashion through parts of the field where the magnetic induction was oppositely directed. Such a machine will, therefore, give alternate currents, unless a commutator be affixed to the rotating axis. Niaudet's dynamo, which may be regarded as a compound Saxton machine, having the separate armature coils united as those of Gramme and Siemens into one continuous circuit, was furnished with a radial collector instead of a cylindrical one. In the Wallace-Farmer dynamo was very nearly realised the condition of field of Fig. 9, there :being, as shown in Fig. 160, a pair of poles at the top arranged so that the N. faces the S. pole, and another pair at the bottom where the S. faces the N. pole. The coils are carried round, their axis being always parallel to the axis of rotation, upon a disk ; there being two sets of coils on opposite faces of two disks of iron set back to back. They are united (see Fig. 161) precisely as in Niaudet's dynamo, and each disk has its own collector. Each bar of the collector is, moreover, connected, as in the dynamos of Pacinotti, Gramme, Siemens, &c., with the end of one coil and the beginning of the next. In fact, the Wallace-Farmer machine is merely a double Niaudet dynamo with cylindrical collectors. There is a serious objection to the employment of solid iron disks such as these. In a very short time they grow hot from the internal eddy currents engendered in them as they rotate. This waste reduces the efficiency of the dynamo. Disk-armatures, composed of a number of separate small bobbins, with or without iron cores, set side by side in the cirumference of a disk, are, as will be seen in Chapter XII., of very general use in *alternate-current machines.* There seems, however, to be some difficulty in constructing continuous-current dynamos with armatures of this kind; possibly on account of the fact that when two adjacent sections are having similar polarity induced in them, the currents in the

FIG. 160.   FIG. 161.

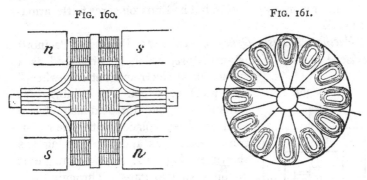

WALLACE-FARMER DYNAMO.   WALLACE-FARMER DYNAMO.

wires of the two sections are flowing in opposite directions in the space between the adjacent cores ; possibly also because the reversals of magnetism in the separate cores are sudden, and cause deleterious internal inductions. Several inventors, including Lane-Fox, Brockie, and Leipner, have essayed dynamos on this type. Von Hefner Alteneck, however, succeeded in converting into a continuous-current machine the alternate-current Siemens dynamo described in Chapter XII. This he accomplished by the device of employing a disk-armature, in which the number of coils differed by two, or some other even number, from those of the field, and by the employment of a multiple bar collector with complicated cross connexions.

In the dynamo of Hopkinson and Muirhead, shown at the Paris Exhibition of 1881, the disk-armature took a more reasonable construction. Instead of a solid plate of iron to support the coils, there is a disk built up of a thin iron strip wound spirally round a wooden centre. The coils, of approximately quadrangular shape, and flat form, were wound upon the sides of this compound disk. The separate coils were connected to one another, and to a collector of ordinary type. Hopkinson and Muirhead's dynamo was wound with strip copper instead of ordinary wire. The coils wound upon the front of the disk were alternate in position with those wound on the back of the disk.

The dynamo of Ball (the so-called "Arago-disk" machine) was similar in many respects, but had no iron cores to the armature coils.

Pacinotti's Disk Dynamo.—In 1875 Professor Pacinotti devised[1] a form of disk-armature, which he described as a "transversal electromagnetic fly-wheel." The machine, which was exhibited at Paris in 1881, had for field-magnet two electromagnets placed with their contrary poles juxtaposed, forming, as shown in Fig. 162, a single magnetic circuit with two gaps. Through these two gaps passed a disk-armature, constructed of radial conductors arranged to cut the intense magnetic fields. The electromotive-forces induced in these conductors would on the one side be directed radially inwards, on the other radially outwards. The method devised by Pacinotti for connecting the radial conductors into a single closed coil is shown in Fig. 163. It will be noted that the outer end of each radial conductor is carried round by a peripheral connecting piece to join the end of another radial conductor, which for a two-pole machine would be the one

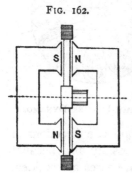

FIG. 162.

FIELD MAGNETS OF PACI-
NOTTI'S DISK DYNAMO.

---

[1] *Nuovo Cimento* [3], x., Sept. 1881.

lying next but one to that which is diametrically opposite. The schematic figure relates to a 10-part armature, made up of 20 radial conductors. They are numbered so that the order of connexions may be traced. The diameter of commutation being *d d*, the currents flow radially inwards in one half and radially outwards in the other half of this disk. The

FIG. 163.

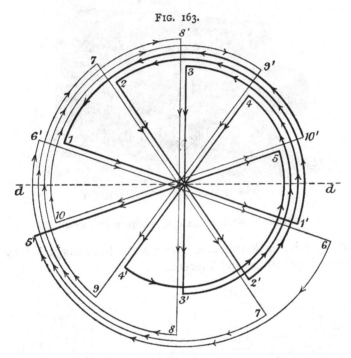

PACINOTTI'S DISK ARMATURE.

construction of Pacinotti's experimental machines is described in his original paper.

Ayrton and Perry's Oblique-coiled Dynamo.—This dynamo is a 4-pole machine, the ring passing successively through four regions, in two of which the lines of force pass from right to left, and in the alternate two from left to right. The ring is of wood, having iron pieces thrust through it, as shown in section in Figs. 165 and 166, and the wire is so coiled that each turn passes radially up one face of the ring, thence obliquely along its periphery for about 90°, then radially inwards along the other face, and back obliquely. The separate sections overlap. They are connected to one another and to the collector in the usual way, but there are four brushes.

FIG. 164.

AYRTON AND PERRY'S OBLIQUE-COILED DYNAMO.

A very similar disk-armature dynamo has been suggested by Messrs. Elphinstone and Vincent.

FIG. 165.   FIG. 166.

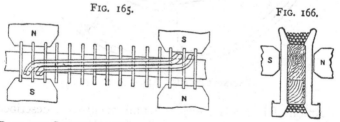

PLAN AND SECTION OF WINDING OF AYRTON AND PERRY'S ARMATURE.

Edison's Disk Dynamo.—Another machine of this category, having a disk-armature, is due to the indefatigable Mr. Edison (Fig. 167), who built up an armature of radial bars connected at the outer ends by concentric hoops, and at the inner by plates or washers. The general arrangement of the disk is indicated in Fig. 168. Each radial bar communicates through one of the hoops with the bar opposite to it; and the disk thus built up is rotated between the cheeks or pole-pieces A¹,

FIG. 167.

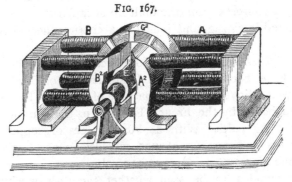

EDISON'S DISK DYNAMO.

A², and B¹, B², of powerful field-magnets. I cannot hear of any of these dynamos having yet come into practical use.

FIG. 168.

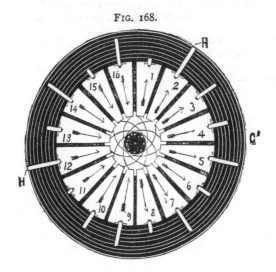

ARMATURE OF EDISON DISK DYNAMO.

Sir W. Thomson's Wheel Dynamo.—Another type of disk armature has been invented by Sir W. Thomson. It belongs, however, to the open-coil class of armatures, and is briefly described on p. 248.

Bollmann's Disk Dynamo.[1]—This is a multipolar machine, having a complex armature built up of radial strips of copper connected in zigzag, and joined to a cross-connected commutator.

Recent Disk Dynamos.—More recently machines of this class have been devised by Desroziers,[2] Robin,[3] Jehl and Rupp,[4] and Sayers.[5] In the machine of Jehl and Rupp, which is brought out by the French Edison Co., there are four pairs of opposing poles. With 24 kilogrammes of copper in the armature, and running at 735 revolutions per minute, this machine gives 350 ampères at 110 volts.

It remains to be seen whether the disk machine will establish itself as a permanent type, or whether the mechanical difficulties of construction will be too great to permit it to compete with the ring and drum types of machine.

---

[1] For detailed drawings and description, see *Centralblatt für Elektrotechnik*, ix. 7, 1887. A very similar machine is described by Matthews in Specification of Patent 3334 of 1882.

[2] See *La Lumière Électrique*, xxiv. 293, 294, and 517, 1887.

[3] Ibid., xxiv. 544, 1887.

[4] Ibid., xxiv. 343, 1887. See also detailed illustrations and description in xxv. 368, 1887 ; and in *Electrician*, xix. 94, 1887.

[5] Specification of Patent 717 of 1887.

CHAPTER X.

CONTINUOUS-CURRENT MACHINES.

Open-coil Dynamos.

As explained on p. 28, it is possible to construct armatures in which the separate coils or sections of the windings are not united together into one closed circuit. An example is given in Fig. 169. This diagram (which should be compared

FIG. 169.

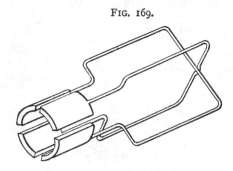

SIMPLE OPEN-COIL ARMATURE.

with Fig. 16, p. 39) shows an armature consisting of two separate loops, set in planes at right angles to one another, so that when one is passing through the inactive region the other is in the position of maximum action. There is no reason why these two loops should not have each a separate 2-part commutator like that of Fig. 16; and one pair of brushes might press on both commutators. It is, however, obviously more convenient to unite these two commutators into a single one of four parts, as in Fig. 170; and then it will at once be seen that as this rotates between its pair

of brushes one loop only will be in action at once, the other
loop being cut out of circuit for the time being.   It would
clearly be possible to arrange any number of loops or coils
in this way so that only that loop or coil which was passing
through the position of maximum action should be feeding
the brushes, all the rest being meantime open-circuited.   A
ring armature wound in sections might of course be similarly
arranged, so that the pairs of sections had each a separate
commutator ; and Fig. 170 (which should be compared with

FIG. 170.

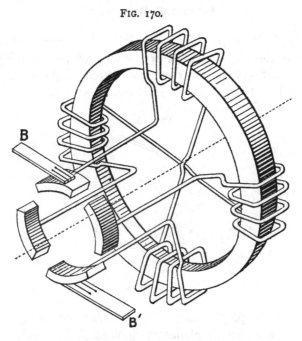

FOUR-PART OPEN-COIL RING ARMATURE.

Fig. 23, p. 41) shows such a ring, but with the two commutators cut down and formed into a 4-part collector.

It will be noticed that each coil is joined at the back to
the one diametrically opposite to it, and that the front ends
of the coils pass to the commutator.   As a matter of fact, it
would make no difference in either of these armatures were

the wires which cross at the back all united where they meet. In order further to follow the action we will refer to Fig. 171.

FIG. 171.

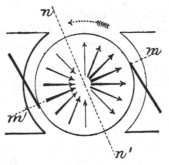

INDUCTION OF CURRENTS IN ARMATURE.

This diagram represents by means of radial arrows the electromotive-forces induced in a loop or loops rotating (*left-handedly*, as it happens, in this figure) in a magnetic field. The action is a maximum along the line of the resultant magnetic field *m m'* (which would be horizontal were it not for the reactions explained in Chapter V.), and is a minimum along the line *n n'*. The reader will remember that line *n n'* is the neutral line which lies at right angles to the line of maximum magnetisation; and that, for those armatures in which all the coils are joined together in a closed circuit, it is at *n n'* that the brushes have to be placed. But when each coil is independent of the others it is no use putting the brushes at *n n'*; they must be put at *m m'*; the line of maximum action coinciding in this case with the diameter of commutation. But those coils which lie very near the line *m m'* are undergoing induction very nearly as strongly as the actual pair that lie in that line: it would therefore naturally occur that the current might be simultaneously collected from more than one coil at once, either (1) by making the pieces of the commutators overlap, or (2) by connecting to the brushes that touch on the line *m m'* another pair having either a forward or a backward lead. If we now consider Fig. 172 we shall see this a little more

clearly. This figure is a diagram of such an armature, the
coils or loops being here represented merely by wavy lines.

The wavy line A C may represent either a pair of coils
such as there are in Fig. 172 on the ring, or may represent
a single loop or group of windings round a drum. There
is a pair of commutator-plates for A C, and another at right
angles for B D. Coils A and C are just coming into the position of best action ; they are delivering a current to the
brushes P P, and this current will accordingly increase a little,
and then decrease again. Meantime coils B and D are idle.

FIG. 172.

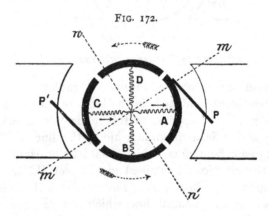

If the four parts of the compound commutator each occupy
just a quarter of the circumference, it is clear that when A
comes into action its plane makes an angle of 45° with *m m'*,
and that just as it leaves contact with the brush makes again
an angle of 45° on the other side, being in contact in all
intermediate positions ; and so with each coil as it passes
the brushes. There will be a momentary break of current
and a spark as the two successive segments pass under the
brush, unless the brush touches both at once. Remembering that Fig. 21, p. 40, represents the alternating currents
from a single loop or pair of coils, and that Fig. 22, p. 41,
represents the same currents rectified by the use of a simple
2-part commutator, we shall be able to represent the effect
of our new arrangement by some such diagram as Fig. 173.

The angles marked below are reckoned from the neutral line *n n'*. When coil A has gone round 90° from this position it is in the position of maximum induction : but because the segment A of the commutator is itself 90° in breadth, the current will be collected from 45° to 135°. The shaded portions of the curve show the discontinuous effect due to the coils A and C coming into circuit during two quarters of the rotation. The coils B and D come in

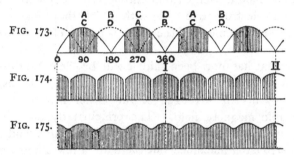

CURVES ILLUSTRATING THE PRODUCTION OF CURRENTS BY USING AN OPEN-COIL FOUR-PART ARMATURE.

in the intervals, as indicated by the dotted lines. The induced currents will therefore present an approximate continuity depending on the arrangements of the commutator and the brushes. Fig. 174 represents the effect when there are gaps between the segments of the commutator; and it will be noticed that the currents, though all of the same sign, are discontinuous. If the brushes thus left contact with one segment of the commutator before the next came into contact there would inevitably be a considerable amount of sparking. Fig. 175 shows the result of making contact with one set before the other set is cut out ; the induced current being now continuous, but with undulating fluctuations of strength. During the time when both sets of coils are in contact with the brushes, they are, of course, in parallel with one another. During this stage of the action the resistance of the armature is half as great as when one of the coils is cut out ; but it is necessary to cut out the idle

Q

coil, otherwise some of the current from the active coil would flow back uselessly through the idle coil that was in parallel with it. During the time when the two sets of coils are in parallel they are not equally active. The induced electro-motive-force is increasing in one and diminishing in the other; there is but a moment when they are equally active— when they make equal angles with *m m'*. At all other moments the higher electromotive-force of the more active coil tends to send a back-current through the less active coil : and the net electromotive-force with which they act on the brushes will be the mean of their two separate electromotive-forces.

From what has now been said it will be clear that open-coil armatures may be constructed either as rings, drums, or disks. They may be arranged to run either in a simple or in a multiple magnetic field. The principal dynamos constructed upon this plan are the Brush machine and the Thomson-Houston machine : but there are a few others which also come within the category of open-coil dynamos.

Brush's Dynamo.—One of the best-known and least under-stood of these machines is the Brush dynamo. Its general form and the disposition of the field-magnets may be gathered from Fig. 176. The field-magnets are very substantially built. The magnet heads are insulated with sheets of the so-called vulcanised fibre, thoroughly varnished. The field-magnet cores are, however, first surrounded with a thin sheet of copper soldered together at the edges so as to form a continuous tube or envelope. The object of this copper coating is to absorb the induced extra currents which otherwise would be set circulating in the core whenever a variation of the magnetism occurred. Over the copper envelope are wound four or five thicknesses of very heavy paper saturated with shellac varnish to insulate the wire from the iron. In some of the Brush dynamos, there is a double winding, a shunt or "teazer" circuit being added to maintain the magnetism of the field magnets when the main circuit is opened. The automatic regulator, described briefly on p. 127, is used with arc-lighting Brush machines.

The armature—a ring in form, not entirely over-wound with coils, but having projecting teeth between the coils, like the Pacinotti ring—is a unique feature. Though it so far resembles Pacinotti's ring, it differs more from the Pacinotti armature than that armature differs from those of Siemens, Gramme, Edison, Bürgin, &c. ; for in all those the successive sections are united in series all the way round, and constitute, in one sense, one continuous bobbin. But in the Brush armature there is no such continuity. The ring itself was formerly made of malleable cast iron. The wire-spaces are planed or milled out, and all angles and corners are carefully rounded. All iron parts which are to adjoin the wire of the "bobbins" are covered first with a layer of strong heavy canvas saturated with shellac varnish, and in the case of the armatures of the larger machines there are additional layers of tough paper saturated with shellac varnish. A sheet of strong cotton-cloth inserted occasionally separates contiguous layers of wire from each other both in the armature bobbins and in the coils of the field-magnets. All the bobbins are wound by hand in the same direction, and the inner ends of diametrically opposite bobbins are soldered together and carefully insulated from all other wires and adjacent metal. The free outer ends of each pair of bobbins are separately carried along the shaft, through the journal, and connected to diametrically opposite segments of the commutator. In the "sixteen-light" machine the ring is 20 inches in diameter. In each of the eight coils there are about 900 feet of wire of ·083 inch gauge. Thus connected, the machine is adapted to deliver a current of ten amperes. By connecting the two bobbins of each pair in parallel, instead of in series with one another, the machine may be used to deliver a 20-ampère current. For electro-plating, much stouter wires are used.

For each pair of coils there is a separate commutator. In the No. 7 size of machine, which is depicted in Fig. 176, there are eight coils on the armature, four commutators grouped in two pairs, and two sets of brushes. This size is commonly known as a "sixteen-light" machine, though it will, as now improved, supply from 24 to 25 arc lights.

FIG. 176.

BRUSH DYNAMO. No. 7 (TWENTY-FIVE LIGHT).

The larger Brush dynamos (" sixty-light machines ") have twelve coils upon the ring, connected as six pairs. There are three pairs of brushes, and three pairs of commutators, each pair being set one-twelfth of a rotation (30°) in advance of the next pair. These machines have the enormous electro-motive-force of nearly 3000 volts ; and, with the recent improvements, will supply sixty or more arc lights in a single circuit. They formerly supplied forty lights only.

The brushes are arranged so as to touch at the same time the commutators of two pairs of coils, but never of two adjacent pairs ; the adjacent commutators being always connected to

FIG. 177.

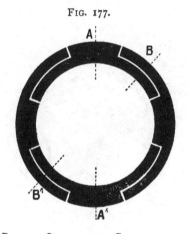

PAIR OF OVERLAPPING COMMUTATORS.

two pairs of coils that lie at right angles to one another in the ring. Continuity is obtained in the currents by making the two parts of the commutator of each pair of coils overlap those of the commutator belonging to the pair of coils that is at right angles, one pair of brushes resting on both commutators. Fig. 177 is a diagram illustrating this device. Each pair of segments overlaps the other to the extent of 45°. Each of the two pairs of coils is thus cut out twice during a revolution : it is twice in circuit alone, as when the brushes are at A A¹, and four times in circuit along with

the pair that are at right angles, when the brushes are at
B B¹.   Fig. 178 shows the way in which the commutator
is arranged in all 8-coil Brush armatures—that is, in machines
for supplying from one to twenty-five arc lights.   There
are really four commutators here, corresponding to the
four pairs of coils, grouped in pairs ; one pair of commutators being set one-eighth of a rotation (45°) in advance
of the other.   It will be seen from this figure that while the
brushes A A' (shown in dotted lines) are receiving current

Fig. 178.

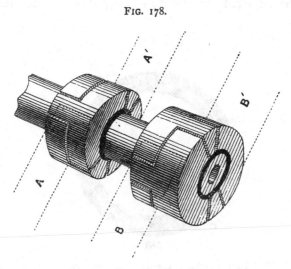

BRUSH COMMUTATOR FOR ARMATURE WOUND WITH FOUR PAIRS
OF COILS.

from *one pair* of coils only, the brushes B B' are at the same
instant receiving the current from *two pairs* of coils which
are joined in parallel with one another in consequence of
both of their commutators touching the same pair of brushes.
The arrangement may be still further studied by the aid of
Fig. 179, which also illustrates the way of connecting the
brushes with the circuit.   In this figure the eight coils
are numbered as four pairs, and each pair has its own

commutator, to which pass the outer ends of the wire of each coil, the inner ends of the two coils being united across to each other (not shown in the diagram). In the actual

FIG. 179.

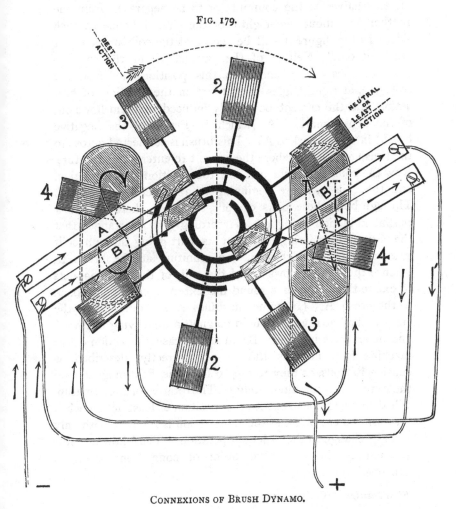

CONNEXIONS OF BRUSH DYNAMO.

machine, each pair of coils, as it passes through the position of least action (i.e. when its plane is at right angles to the direction of the lines of force in the field, and when the number of lines of force passing through it is a *maximum*, and the rate of change of these lines of force a *minimum*) is cut out of connexion. This is accomplished by causing the two halves of the commutator to be separated from one another by about one-eighth of the circumference at each side. In the figure it will be seen that the coils marked 1, 1 are "cut out." Neither of the two halves of the commutator touches the brushes. In this position, however, the coils 3, 3, at right angles to 1, 1, are in the position of best action, and the current powerfully induced in them flows out of the brush marked A (which is, therefore, the negative brush) into that marked A'. This brush is connected across to the brush marked B, where the current re-enters the armature. Now the coils 2, 2 have just left the position of best action, and the coils 4, 4 are beginning to approach that position. Through both these pairs of coils, therefore, there will be a partial induction going on. Accordingly, it is arranged that the current, on passing into B, splits, part going through coils 2, 2 and part through 4, 4, and reuniting at the brush B', whence the current flows round the coils of the field-magnets to excite them, and then round the external circuit, and back to the brush A. (In some machines it is arranged that the current shall go round the field-magnets after leaving brush A', and before entering brush B ; in which case the action of the machine is sometimes, though not correctly, described as causing its coils, as they rotate, to feed the field-magnets and the external circuit alternately.) The rotation of the armature will then bring coils 2, 2 into the position of least action, when they will be cut out, and the same action is renewed with only a slight change in the order of operation. The following table summarises the successive order of connexions during a half-revolution :—

First position. (Coils 1, 1 cut out.)

A–3–A' ; $B\overset{4}{\underset{2}{<\ >}}B'$; Field-magnets – External circuit – A.

Second position. (Coils 2, 2 cut out.)

$A\overset{1}{\underset{3}{<\ >}}A'$; B–4–B' ; Field-magnets – External circuit – A.

Third position. (Coils 3, 3 cut out.)

A – 1 – A′; B $<{}^2_4>$ B′; Field-magnets – External circuit – A.

Fourth position. (Coils 4, 4 cut out.)

A $<{}^3_1>$ A′; B – 2 – B′; Field-magnets – External circuit – A.

From this it will be seen that whichever pair of coils is in the position of best action, is delivering its current direct into the circuit; whilst the two pairs of coils which occupy the secondary positions are always joined in parallel, the same pair of brushes touching the respective commutators of both; and the remaining pair of coils being cut out.

One consequence of the peculiar arrangement thus adopted is, that measuring the potentials round one of the commutators with a voltmeter gives a wholly different result from that obtained with other machines. For one-eighth of the circumference on either side of the positive brush, there is no sensible difference of potential. There then comes a region in which the potential appears to fall off, but the falling off is here partly due to the shorter time during which the adjustable brush connected with the voltmeter and the fixed positive brush are both in contact with the same part of the commutator. Further on there is a region in which the voltmeter gives no indications, corresponding to the cut-out position; and again, on each side of the negative brush, there is a region where the polarity is the same as that of the negative brush.

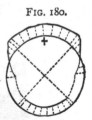

FIG. 180.

POTENTIALS AT COMMUTATOR OF BRUSH DYNAMO.

Fig. 180 is a diagram of a 6-light Brush taken at one commutator, the main + brush being, however, allowed to rest (as in its usual position) in contact with both this commutator and the adjacent one.

From the foregoing considerations, it will be clear that the four pairs of coils of the Brush machine really constitute four separate machines, each delivering alternate currents to a commutator, which commutes them to intermittent unidirectional currents in the brushes; and that these indepen-

dent machines are ingeniously united in pairs by the device
of letting one pair of brushes press against the commutators
of two pairs of coils: Further, that these paired machines are
then connected in series, by bringing a connexion round from
brush A′ to brush B.

The core of the Brush ring, as formerly made, was of
malleable cast iron channelled at the sides, so as in some
degree to obviate eddy currents. The coils are wound in
deep radial recesses. The form of the ring, with its coils,
may be discerned in Fig. 176, and the channelled cast-iron
ring itself is shown in Fig. 181. The solid masses of iron in

FIG. 181.

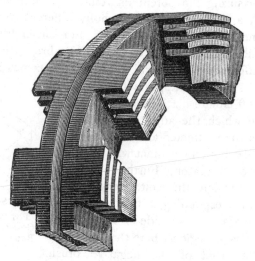

CORE OF BRUSH RING (Old pattern).

this now obsolete form of armature gave rise to wasteful
eddy currents, the production of which heated the ring and
absorbed considerable power. In the newer Brush machines
a ring is employed which is built up of a thin iron ribbon
1·5 millimetres thick. Figs. 182 to 184 show its construction, though in reality a larger number of pieces of
thinner iron than is shown are used. The ribbon is wound

upon a circular foundation ring A', projecting cross-pieces
of the same thickness and of the form shown in Fig. 184
(and also marked H in Figs. 182 and 183), being inserted

FIG. 182.   FIG. 183.

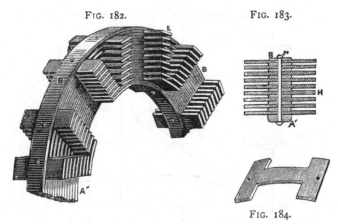

FIG. 184.

CORE OF BRUSH RING (New pattern).

at intervals to separate the convolutions, admit of ventilation, and form suitable projections between which to wind
the coils.   In the larger armatures there are 45 turns of
the ribbon.   It is secured by well-insulated radial bolts.
The gain in coolness is great; and the old machine, which
formerly supplied 40 arc lamps, when provided with the
new ring can supply 65 lamps at the same speed as before.
The smaller 16-light machine with an armature of the new
type lights 25 lamps, showing that the deleterious reactions
have been fairly eliminated ; and it may even be run at higher
speeds with perfect safety.
    The most striking way of realising the great improvement
which has thus been made is to compare the speeds required
to develop equal electromotive-forces in the two machines.
The experiments were made with identical machines.   Both
armatures were of the same size, with same length and
weight of wire ;   and the field-magnets were identically
excited with a 10-ampère current.   The results are shown
in the two curves of Fig. 185.   At 800 revolutions the

old cast-iron armature gave about 730 volts: the new laminated armature gave over 1000 volts.[1]

FIG. 185.

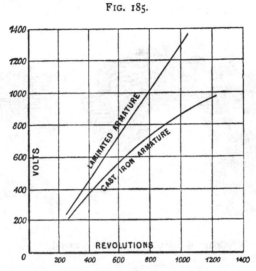

CURVES OF OLD AND NEW PATTERNS OF BRUSH MACHINE.

Some large shunt-wound Brush machines have been recently constructed for electro-metallurgical purposes. The largest of these,[2] having an armature 42 inches in diameter, running at 405 revolutions per minute, gave 3200 ampères at 83 volts. This colossal machine weighs nearly 10 tons, the weight of copper on the field-magnets being 5424 lbs., that on the armature 1600 lbs. This is at the rate of 1 lb. of copper on the armature for every 160 watts of output, or 43 watts for every 1 lb. of copper on the machine.

Thomson-Houston Dynamo.—This machine, which is equally remarkable, was designed by Professors Elihu

---

[1] More surprising still, the power absorbed in driving the old armature giving 730 volts was 17 horse-power, whilst that absorbed in driving the new, giving 1020 volts was only 16 horse-power. The old armature could not be run above 800 revolutions per minute without dangerous heating. The new armature may be safely run at a much higher speed, without risk of heating or other danger.

[2] See paper by Prof. Thurston, in 'Journal of Franklin Institute,' Sept. 1886; or see *Electrician*, xvii. 469, 1886.

Thomson and Edwin J. Houston of Philadelphia. Its spherical armature is unique among armatures; its cup-shaped field-magnets are unique amongst field-magnets; its three-part commutator is unique among commutators. A general view of the machine is given in Fig. 186, and a

FIG. 186.

THOMSON-HOUSTON DYNAMO.

sectional one in Fig. 187. As will be seen from these cuts, the field-magnet core consists of two flanged iron tubes furnished at their inner ends with hollow cups cast in one with the tubes, and accurately turned to receive the armature. Upon the tubes are wound the coils C C', and afterwards the two parts are united by means of a number of wrought-iron bars *b b* which constitute the yoke of the magnet and at the same time protect the coils. The magnets are carried on a framework, which also supports the bearings for the armature shaft X. The armature, which is spheroidal rather than spherical, is constructed as follows. Upon the shaft are keyed

Fig. 187.

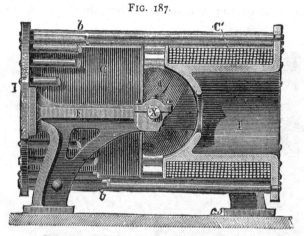

THOMSON-HOUSTON DYNAMO (part Section).

two concave iron disks S, S (Fig. 188), the space between them being bridged by light ribs of wrought iron, *d, d*. Wooden pins J, J are driven at intervals into appropriate holes drilled in the iron shell to facilitate the winding of the coils. The winding itself is very remarkable. There are but three coils. The inner ends of these are united together (at *h*, Fig. 189), *and not connected to any other conductor*. The three wires are then wound over the shell (which is covered with varnished paper) in three sets of windings making 120° with one another, and arranged to be at equal average distances from the core by the following device. Beginning at the junction at *h*, half the No. 1 coil is wound. Then the armature is turned through 120° and the No. 2 coil is wound on to half its length. Then coil No. 3 is wound, starting from *h*, and finishes off at 3. Then the second half of No. 2 is completed ; lastly, the second half of coil No. 1. The three coils are therefore, on the average, equidistant from the iron core: their overlapping makes the external form nearly spherical. They are held

in place by the binding wires *g g*. When this armature is
rotated within the cavity between the cup-shaped poles
alternate currents are generated in each separate coil in turn,

FIG. 188.   FIG. 189.

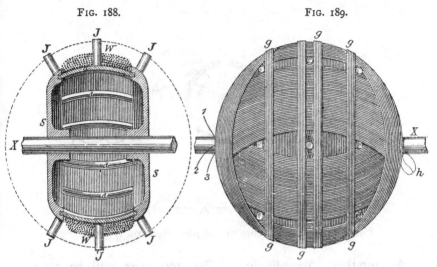

THOMSON-HOUSTON ARMATURE (Section).   THOMSON-HOUSTON ARMATURE.

and it now remains to consider how these alternate inductions
are rectified and combined by the commutator. In the
diagrams which follow, the rotation is represented as left-
handed, as viewed from the commutator-end of the shaft, as it
is in practice. Fig. 190 represents the arrangement in diagram.
The three coils represented diagrammatically by the three
lines A B C, are united at their inner extremities, each
outer end being led to one segment of a 3-part commutator.
There are two positive brushes P and F, and two negative
brushes P′ and F′. The current delivered to P and F first
flows round one of the field-magnets, thence goes to the outer
circuit of lamps, returning through the other field-magnet to
P′ and F′. The reader should now compare this diagram with
Fig. 171, p. 223, and observe that in that figure the neutral line
n n′ divides the rotation obliquely into two halves, the induced
currents flowing outwardly from centre to commutator in all

coils that are rising through the right-hand half of this
obliquely divided circle ; and inwardly from commutator to
centre in all coils descending through the left-hand half of

FIG. 190.

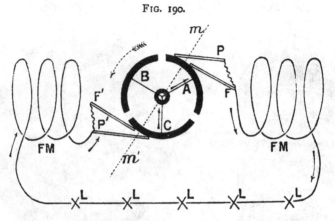

COMMUTATOR AND CIRCUIT OF THOMSON-HOUSTON DYNAMO.

the rotation. Accordingly in Fig. 190 there will be an
outward current in A and an inward one in C ; B being for
the moment cut out of circuit as it passes through the neutral
position. Continuity is obtained by the device mentioned on
p. 223, of having the second pair of brushes F F′ following the
pair P P′. In this position of the armature A and C make
about equal angles with the line of maximum action *m m′*,
hence the two electromotive-forces in these coils are for the
moment about equal, but that in A is increasing, that in C
decreasing. As these coils are now in series, their separate
electromotive-forces are of course added together. A moment
later we shall have arrived at the state of things represented
in Fig. 191, which is a twelfth of a turn advanced. A is now
in the position of maximum induction ; C is rapidly approaching the neutral position but is not yet cut out ; B has again
begun to have electromotive-force induced in it, and has just
come into circuit. B and C are in parallel with one another and
in series with A. The next twelfth of a turn brings us to the
stage shown in Fig. 192. C is now at the neutral position

and is out of circuit, A has passed the maximum on one side and B is approaching the maximum on the other ; and they

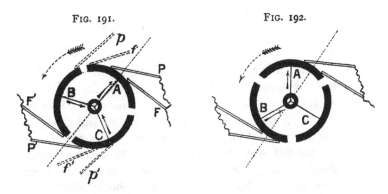

FIG. 191.   FIG. 192.

are in series. Another twelfth of a turn and we arrive at Fig. 193. A is fast approaching the neutral position; B is at its maximum ; C has passed the neutral stage, and has just come into circuit again by touching the positive following brush. Another instant and C will occupy the place where A was in Fig. 190, a whole third of a revolution having been completed, and the actions recommence, A occupying the place of B, and B that of C.

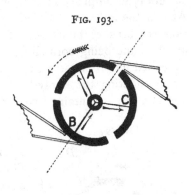

FIG. 193.

The following table exhibits the round of changes during a third of a revolution :—

From external circuit ──⟨ F' / P' ⟩ C ─ A ⟨ P / F ⟩── to external circuit,

 ,,   ,,   ,, ──⟨ F'─B / P'─C ⟩ A ⟨ P / F ⟩── ,,   ,,   ,,

 ,,   ,,   ,, ──⟨ F' / P' ⟩ B ─ A ⟨ P / F ⟩── ,,   ,,   ,,

R

From external circuit ——$\left\langle\begin{matrix} F' \\ P' \end{matrix}\right\rangle B \left\langle\begin{matrix} A-P \\ C-F \end{matrix}\right\rangle$—— to external circuit,

„   „   „   ——$\left\langle\begin{matrix} F' \\ P' \end{matrix}\right\rangle B-C \left\langle\begin{matrix} P \\ F \end{matrix}\right\rangle$—— „   „   „

and so forth.

If the width of the gaps between the segments of the commutator be equal to the width between the adjacent brushes, each coil will be cut out of circuit whenever it is more than 60° from the position of maximum action, and the time during which any two coils are in parallel will be practically nil. But if the following brushes F F' are at a considerable angle—about 60° in practice—behind the brushes P P', there will be a considerable duration of the stage during which two coils are in parallel.

The regulation of this machine to maintain a constant current is accomplished by an automatic shifting of the brushes. At first the inventor adopted a method of "forward" regulation, the brushes being set at only 35° in advance of one another, and being both shifted forward whenever the current exceeded its proper strength. The effect of this will be seen by considering the dotted lines $p\,p'$ and $f\,f'$ in Fig. 191. Clearly, if p maintained contact with B *after* it had passed the neutral line, its electromotive-force would tend to diminish that of A which is in contact with f. This method was found in practice unsatisfactory, and has been abandoned. The actual method now used is termed "backward" regulation. In this system the following pair of brushes F F' is shifted backward, to $f\,f'$, as shown in Fig. 194, whilst at the same time the leading brushes P P' are shifted forward through an angle one-third as great towards $p\,p'$.

FIG. 194.

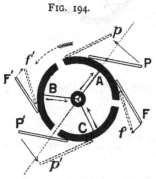

If, as stated above, the brushes are 60° apart under normal conditions, there will be exactly 120° on either side between the positive brushes P F and the negative brushes P′ F′ ; and, as 120° is the exact length of each segment of the commutator, no coil will be cut out, and parallelism will subsist between two coils through angles of 60°: that is to say, there will always be two of the three coils in parallel with one another and in series with the third coil. The six stages of change will be :—

From external circuit ──⟨ F′−B / P′−C ⟩A⟨ P / F ⟩── to external circuit,

,, ,, ,, ──⟨ F′ / P′ ⟩B⟨ A−P / C−F ⟩── ,, ,, ,,

,, ,, ,, ──⟨ F′−A / P′−B ⟩C⟨ P / F ⟩── ,, ,, ,,

,, ,, ,, ──⟨ F′ / P′ ⟩A⟨ C−P / B−F ⟩── ,, ,, ,,

,, ,, ,, ──⟨ F′−C / P′−A ⟩B⟨ P / F ⟩── ,, ,, ,,

,, ,, ,, ──⟨ F′ / P′ ⟩C⟨ B−P / A−F ⟩── &c.

Now suppose the current to become too strong owing to reduction of number of lamps in circuit, the following brushes are made to recede. This will shorten the time during which any single coil in passing through the maximum position is throwing its whole electromotive-force into the circuit, and will hasten the moment when it is put in parallel with a comparatively idle coil. During such movements of regulation the whole machine is momentarily short-circuited six times during each revolution by F receding so far towards P′ and F′ receding so far towards P as that both touch the same segment of the commutator at one instant. The action is assisted by the slight advance of P and P′, but the main object of this advance is to lessen the sparking. If the current is too weak, then the pairs of brushes must be made to close up, thereby reducing the time during which the

most active coils are in parallel with those that are less
active. This motion of advance and retreat is accomplished
by the simple mechanism shown in Fig. 195. The brushes are
fixed to levers Y Y and $Y_2 Y_2$ united by a third lever *l*. The
automatic movement is imparted by the regulating electromagnet R, whose pole M, of paraboloidal form, attracts its
armature N according to the current flowing round it. A
dash-pot J, attached to the arm A, prevents too sudden

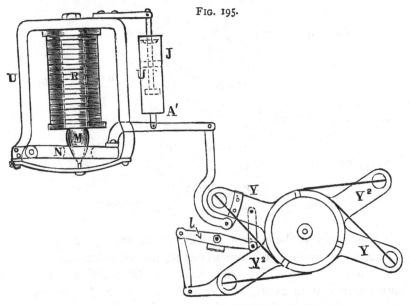

FIG. 195.

REGULATING MAGNET AND GEAR OF THOMSON-HOUSTON SYSTEM.

motions. The circuits which operate this mechanism are
further shown in Fig. 196. Normally the electromagnet
R is short-circuited through a by-pass circuit, and only acts
when this circuit is opened. At some convenient point of the
main circuit two solenoids are introduced, their cores being
supported by a spring; and the yoke of the cores operates
the contact lever S. If the current becomes too strong this
contact is opened, and the regulating magnet R raises the
arm A. During running the lever S is continually vibrating

up and down, and so altering the brushes to the requirements of the circuit. A carbon shunt of high resistance *r* is
added to minimize the destructive spark at S.

It might be expected that with only three parts to the
commutator, the sparks occurring as the segments pass under

FIG. 196.

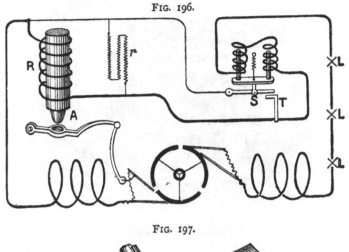

FIG. 197.

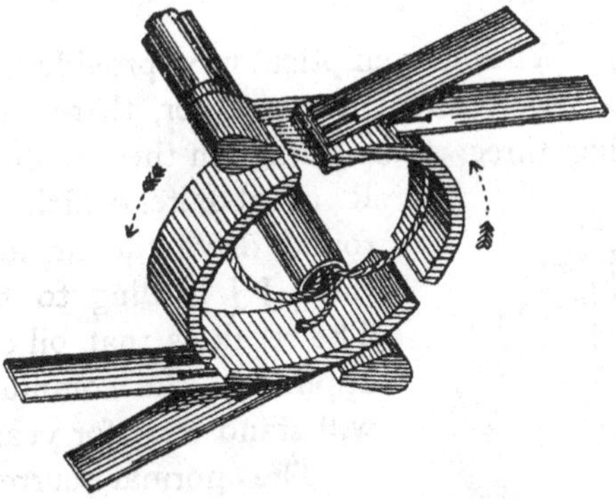

COMMUTATOR, BRUSHES, AND BLAST-NOZZLES.

the brushes would speedily destroy the surface. This difficulty has been met by Prof. Thomson in the boldest manner.
He blows out the sparks by an air blast delivered exactly at
the right place and time. The three segments of the commutator are separated by gaps; and in front of each of the

leading brushes (as shown in Fig. 197) there projects a nozzle
which discharges a blast (alternately) three times in each
revolution. The blast is itself supplied by a very small,
simple, and ingenious blower, fixed upon the shaft immediately behind the commutator. The blower is shown in Figs.

FIG. 198.

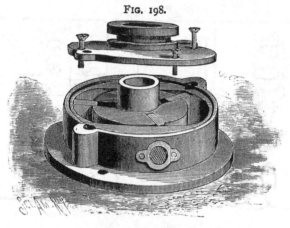

THOMSON'S BLOWER.

198 and 199. Within an elliptical box, provided at the sides
with perforations I I where air can enter, there rotates a steel
disk H having three radial slots. In these slide three wings

FIG. 199.

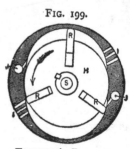

THOMSON'S BLOWER
(Section).

R of ebonite, which as they fly
round drive the air into the aper-
tures J J leading to the nozzles.
The result is that oil can be freely
applied to the commutator, which
will stand wear for years.

The normal current for this
dynamo is 9·6 ampères, though
they are also wound for stronger
currents. One of the larger machines will maintain 63 arc lights in
a single circuit, with an electromotive-force of 3000 volts. The
ordinary machine supplying 34 arc lamps at 45 to 46 volts
each with a current of 9·6 ampères, has an internal resistance

of 10·5 ohms in the armature and 10·5 ohms in the field-magnets. The armature is 23⅛ inches in external diameter; its wire being of 0·081 inch gauge. That of the field-magnets is 0·128 inch gauge. The usual speed is 850 revolutions per minute; but the peculiar arrangements for automatic regulation render the machine nearly independent of irregularities in the speed. There is in the armature about 1 yard of wire for every 0·53 volt of the induced electromotive-force. The ratio of weight of copper to iron in these dynamos is very high compared with that of many machines, being 1 of copper to 2·25 of iron.

Professor E. Thomson has more recently designed, for supplying large currents at low potential, a form of dynamo which is outwardly of similar structure to the machine described above. But in this new machine, which is intended for incandescent lamps in parallel, the spherical armature is wound as a closed-coil in sixteen sections. The field-magnets are much more massive, and the output of the machine more than five times as great in proportion to its size. It is compound-wound; the series coils being nearest to the armature, with the shunt coils behind them.

Newton's Dynamo.—This machine has an armature somewhat resembling that of Weston, being built up of cast-iron cogged disks. The winding is, however, entirely different. There are only eight circuits, each of fifty-six turns of wire united in parallel, and the eight circuits are coupled in four twos. These are independently connected to two commutators, as in the smaller Brush machines, but the commutators differ in being obliquely cut. One pair of brushes presses on both commutators.

Schuyler's Dynamo.—This also has a drum armature wound in eight parts, and coupled up so that the coils are successively cut out at the neutral points.

Sir W. Thomson's Mousemill Dynamo.—Several points in this machine—the form of its field-magnets and their coils, and the internal electromagnets, have been noticed in other chapters. The armature is a hollow cylinder made up of parallel copper bars, arranged like the bars of a mousemill (whence the name of the machine). These bars are insulated from each other, but are connected all together at one end. At the other, they serve as collector bars and deliver up the current generated in them to the "brushes," which in this case are rotating disks of springy copper. All the bars are therefore cut out of circuit except the two which are at the moment in contact with the brushes. As the armature is a hollow barrel, with fixed electromagnets within, it cannot be rotated on a spindle, but runs on friction rollers. In spite of its great originality, this form of machine has not shown itself to be a practical one.

Sir W. Thomson's Wheel Dynamo.—This machine also is an experimental form, necessitating too high a speed to be practical. The armature is a flat wheel, very like a flattened bicycle wheel. The radial arms or spokes, in which the currents are induced, are all connected at their outer ends to a copper rim ; but at their inner ends they are carefully insulated and connected each to a segment of an ordinary collector. This wheel is made to rotate between two closely approximated field-magnets. All the spokes are out of circuit except the pair which are passing the brushes. The complete machine was figured in the first edition of this work.

Other Open-coil Dynamos.—Other forms of open-coil dynamo have been proposed by Bain and by Dr. Hammerl. The armature of the latter consists of a Gramme ring wound in sections, but having the inner ends of all the sections united together at a common junction, and their outer ends brought each to a separate bar of a collector.

Advantages of Open-coil Dynamos.—The two great typical open-coil dynamos—those of Brush and of Thomson-Houston —appear to have certain qualities which render them specially applicable as constant-current dynamos for arc lighting. Three quarters of all the arc lights in the world are run by one or other of these machines. It would seem that the closed-coil dynamos, whether of the ring or of the drum type are not so well adapted for furnishing the very high electro-motive-forces needed for this work. The collector (which is the indispensable adjunct of the closed-coil armature), with its many parallel bars insulated with mica or asbestos, rapidly deteriorates when exposed to the inevitable sparking and wide alterations of lead which are inseparable from the constant-current method of working (see p. 97). For this method of distribution of electric energy, nothing will stand wear and tear so well as the simple air-insulated commutators described in this chapter. As a partial set off against these advantages may be reckoned the fluctuations in the currents which arise from the employment of so few coils or groups of coils.

Fluctuations of Current in Open - coil Armatures.—The calculations of the fluctuation, given in Chapter XIII., for closed-coil armatures are also applicable to open-coil armatures, provided there is no cutting out of coils. When idle coils are cut out, the fluctuations are less marked, but the calculations are less simple. That such fluctuations exist is abundantly verified by the interference of arc-light circuits with neighbouring telephone lines in which they induce buzzing noises. Rapidly fluctuating currents are affected by the self-induction of any electromagnets in the circuit, which always tends to tone down the rapid fluctuations and steadies the current.

CHAPTER XI.

CONTINUOUS-CURRENT MACHINES.

Unipolar Dynamos.

THOSE dynamos in which rotation of a conductor effects a continuous increase in the number of lines of force cut, by the device of arranging one part of the conductor to slide on or round the magnet, are known as *unipolar* machines.

The earliest machine which has any right to be called a dynamo was, in fact, of this class. Barlow and Sturgeon had shown that a copper disk, placed between the poles of a magnet (Fig. 200) rotates in the magnetic field when traversed by an electric current from its axis to its periphery, where

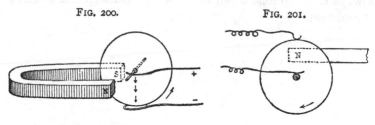

FIG. 200.   FIG. 201.

STURGEON'S WHEEL.   FARADAY'S DISK DYNAMO.

there is a sliding contact. Faraday, in 1831, showed that by rotating a similar disk mechanically between the poles of a magnet, continuous currents were obtained. These he drew off by collecting springs of copper or lead, one of which touched the axis (see Fig. 201), whilst the other pressed against the amalgamated periphery. He was thus "able to construct a new electrical machine."[1] " Here, therefore, was

---

[1] *Experimental Researches*, § 83.

demonstrated the production of a permanent [i. e. continuous] current of electricity from ordinary magnets." But Faraday did not stop short with ordinary magnets ; he went on to employ the principle of separate excitement of his field-magnets. " These effects were also obtained from *electromagnetic poles*, resulting from the use of copper helices or spirals, either alone or with iron cores. The directions of the motions were precisely the same ; but the action was much greater when the iron cores were used, than without." [1] The invention of the dynamo dates, therefore, from 1831, and Faraday was its inventor, though he left to others to reap the fruits of his splendid discovery.[2] Such a machine, however, is impracticable, for several reasons ; the peripheral friction is inadmissible on any but a small scale ; moreover, the disposition of the field magnets necessarily evokes wasteful eddy currents in the disk, which, even if slit radially, would not be an appropriate form of armature for such a limited magnetic field.

Another method of obtaining a continuous cutting of the lines of force is indicated in Fig. 202, where a sliding conductor travels round the pole of a magnet. Faraday also generated continuous currents by rotating a magnet with a sliding connexion at its centre, from which a conductor ran round outside, and made contact with the end-pivots which supported the magnet.

A similar arrangement was devised by Mr. S. Alfred Varley about the year 1862. He rotated an iron magnet in a vertical frame having a mercurial connexion at the centre. The current which flowed from both ends of the magnet toward the centre was, in this machine, made to return to the machine, and to pass through coils surrounding the poles of the rotating magnet, thus anticipating the self-exciting principle of later

---

[1] *Experimental Researches*, § 111.

[2] *Ibid.*, § 158 :—" I have rather, however, been desirous of discovering new facts and new relations dependent on magneto-electric induction, than of exalting the force of those already obtained, being assured that the latter would find their full development hereafter." Can any passage be found in the whole range of science more profoundly prophetic, or more characteristically philosophic, than these words, with which Faraday closed this section of his researches?

date. Mr. Varley also proposed to use an external electro-magnet to increase the action.

More recently, the same fundamental idea has been worked upon by Messrs. Siemens and Halske, who have produced a

FIG. 202.

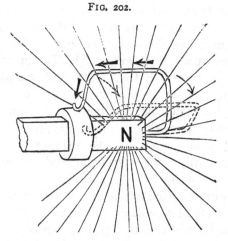

CONDUCTOR SLIDING ROUND POLE OF MAGNET AND CUTTING LINES OF FORCE.

so-called " unipolar " machine,[1] depicted in Fig. 203. In this remarkable dynamo there are two cylinders of copper, both slit longitudinally to obviate eddy currents, each of which rotates round one pole of a U-shaped electro-magnet. A second electro-magnet, placed between the rotating cylinders, has protruding pole-pieces of arching form, which embrace the cylinders above and below. Each cylinder, therefore, rotates between an internal and an external pole of opposite polarity, and consequently cuts the lines of force continuously by sliding upon the internal pole. The currents from this machine are of very great strength, but of only a few volts of electromotive-force. To keep down the resistance, many collecting brushes

---

[1] This sounds like a *lucus a non lucendo*, for the machine has two poles. But the name is derived from the term "unipolar induction," which Continental electricians give to the induction of currents by the process of "continuous cutting" which we are now dealing with. I do not adopt the term, as it is needlessly mystifying.

press on each end of each cylinder. This dynamo has been used at Oker for electro-plating. Other unipolar machines have been designed by Delafield, Rapieff, Voice, Ferraris, and Garrett. No details of their performances are known.

FIG. 203.

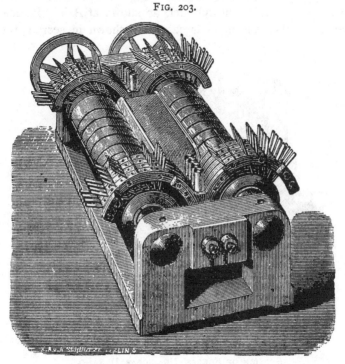

SIEMENS' "UNIPOLAR" DYNAMO.

Much attention has been paid of late years to machines of this type, and the author himself designed such a machine, in which two Faraday disks, coupled at their peripheries outside an internal stationary pole-piece, rotate in a symmetrically-uniform magnetic field. Mr. S. A. Varley has also worked in the same direction. Mr. Willoughby Smith has shown that if an iron disk be used instead of a copper disk in the Faraday apparatus, a much more powerful effect is obtained, and the electromotive-force is more nearly proportional to the speed than is the case with a copper disk.

Forbes' Dynamo.—Prof. George Forbes has constructed a machine of this class. Originally he began by employing an iron disk which rotated between two cheeks of opposite polarity, the current being drawn from its periphery. He then doubled the parts. The next stage was to unite the two disks into one common cylinder, as shown at A in Fig. 204. Here the coils lying in their cases are shown in section, the

FIG. 204.

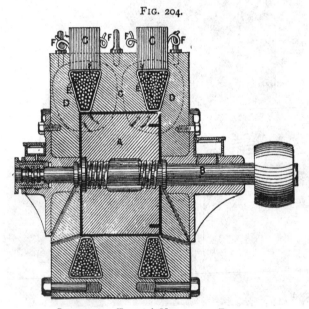

SECTION OF FORBES' NON-POLAR DYNAMO.

dotted lines showing the direction of the lines of magnetic force induced in the iron. These are practically closed on themselves so that there is no external field at all. For this reason the inventor prefers to call this type of dynamo "non-polar." A rubbing contact—for which purpose Prof. Forbes at one time used carbon "brushes," and at another a number of springy strips of metal foil—is maintained at the two extremities of the periphery. One of the earlier forms of machine, with a single disk 18 inches in diameter, was stated to give 3117 amperes at a potential of 5·8 volts when running at

1500 revolutions per minute. One of the later machines, in which the "armature" is a cylinder of iron 9 inches in diameter and 8 inches long, is designed to give, at 1000 revolutions per minute, a current of 10,000 ampères at a potential of one volt. The electromotive-force of such machines increases as the square of the diameter.[1]

The theory of the unipolar disk machine has been given by Sir W. Thomson,[2] who has shown that such a machine is not self-exciting except above a certain critical speed, dependent on the resistance of the circuit.

[1] For some reason or other, probably the inherent difficulties of peripheral collection of currents, these dynamos have not proved successful. For further information respecting the "unipolar" type of dynamo, see Ferraris in *Zeitschrift des elektrotechnischen Vereins in Wien*, vol. i., October 1883; de Tromelin in *l'Électricité*, December 15, 1883, p. 594; *La Lumière Électrique*, May 1884, p. 307, and September 1884, p. 309; *Centralblatt für Elektrotechnik*, Bd. vii., 1884, p. 327 and p. 402. See also much recent correspondence between Prof. Edlund, Herr Hummel, and others respecting "unipolar" induction.

[2] On a uniform electric current accumulator.—*Philos. Magazine*, January 1868, and *Reprint of Papers*, p. 325.

CHAPTER XII.

ALTERNATE-CURRENT MACHINES.

IN these machines electric currents are induced in the armature coils by causing the amount of magnetic induction through them to alternately increase and decrease. Most frequently there is not simply an alternate increase and decrease, but an actual rapid reversal in the direction of the magnetic induction. In some of these machines, as in the majority of continuous-current dynamos, the armature part rotates while the field-magnet part of the machine stands still. In others, however, the armature part—that is to say, the part comprising the coils through which the magnetic induction is to be rapidly reversed or varied—is a fixture, whilst the field-magnets are made to rotate. In a third class of machines both armature part and field-magnet part are fixed, the amount of magnetic induction passing from the latter through the former being caused to vary or alternate in direction by the revolution of appropriate pieces of iron.

In the older machines the field-magnets were either of steel permanently magnetised, or else electromagnets separately excited. About 1869 began the practice of making these machines self-exciting by the method of diverting a small current from one or more of the armature coils which were for this purpose separated from the rest, this current being passed through a commutator which rectified the alternations and made it suitable for magnetising the field-magnets. Such commutators were used by Wilde and by Holmes, and have in general the form depicted in Fig. 205, consisting of two metal cylinders cut like crown wheels, having the teeth of one projecting between the teeth of the other. They are insulated from one another, one being connected to one end of the wire of the armature coils that are to be used for exciting, whilst the other is connected to the other end of that wire. Two brushes are set so that one presses against a tooth of one, whilst the other presses against a tooth of the other part. Such commutators had previously been used in small motors.[1] Holmes used a system of parallel bars (like the Gramme collector) connected together alternately into two sets. If the field-magnets are wound with fine wire, such a commutator may be used to rectify a fraction of the current from the whole of the armature coils, thus making the machine virtually a self-exciting shunt machine.

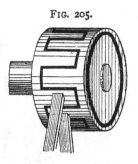

FIG. 205.

COMMUTATOR FOR SELF-EXCITING ALTERNATE-CURRENT DYNAMOS.

For collecting the alternating main current of the dynamo, extremely simple means are required. In those machines in which the armature part is fixed, mere terminals are required. In machines with rotating armatures simple sliding connections are required.

The usual method of collecting is shown in Fig. 206. Two undivided insulated metal rings, forming the terminals of the armature coil, slide each under a single collecting-brush.

As it is requisite in alternate-current working to have many alternations in every second, and as mechanical considerations forbid very high speeds, it is the general practice to make this class of machines multipolar, with a considerable number of poles of alternate polarity arranged symmetrically around a common centre. The number of complete alternations per minute in machines of different systems varies from 2500 to 12,000, or more, whilst the number of symmetrical poles varies from 12 to 32.

There are two ways of coupling up the coils of alternate-current dynamos. For lighting incandescent lamps from parallel mains, it is usual to unite the coils in parallel, as shown

---

[1] See Joule, in Sturgeon's *Annals of Electricity*, ii. 122, 1838.

in Fig. 207, so as to reduce the internal resistance. For arc-lighting, and for supplying distant transformers, in both of

FIG. 206.

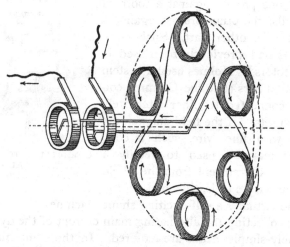

COLLECTING-RINGS OF ALTERNATE-CURRENT DYNAMO.

FIGS. 207, 208.

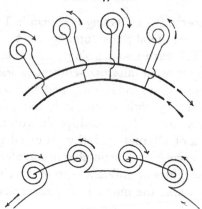

DIFFERENT MODES OF COUPLING UP ARMATURE-COILS OF ALTERNATE-CURRENT DYNAMO.

which cases high electromotive-force is required, the more usual mode of connecting is to join the several coils in series, as in Figs. 206 and 208.

Classification of Alternate-current Armatures.

Alternate-current dynamos may be classified according as their armatures belong to the ring, drum, pole, or disk type.

Ring-armature Alternate-current Dynamos.

This type was invented in 1878 almost simultaneously by Gramme[1] and by Wilde,[2] the main difference between them being that, whilst Gramme rotates his field magnet within a large stationary ring, Wilde rotates his ring armature within an external system of inwardly pointing field-magnet.poles (see Fig. 241, No. 28). When ring armatures are used in this type of dynamo, they must not be wound in the same manner

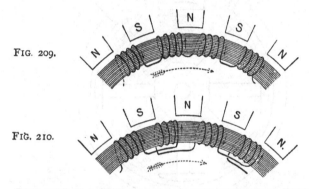

FIG. 209.

FIG. 210.

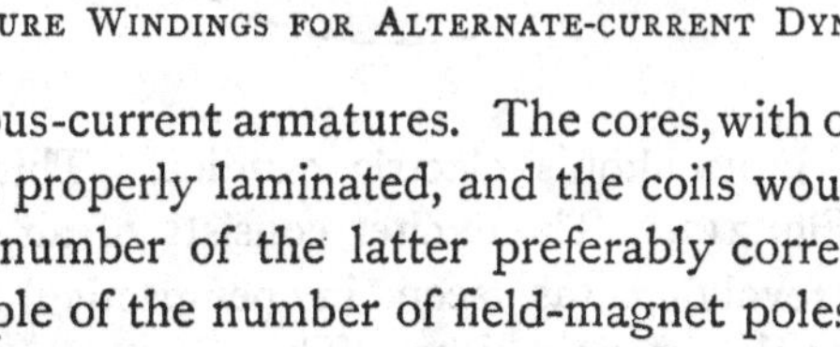

RING-ARMATURE WINDINGS FOR ALTERNATE-CURRENT DYNAMO.

as for continuous-current armatures. The cores, with or without teeth, must be properly laminated, and the coils wound on in sections; the number of the latter preferably corresponding to some multiple of the number of field-magnet poles. If the successive sections are to be connected up consecutively, then they must be wound, as shown in Fig. 209, alternately with right-handed and left-handed windings. If all the sections are coiled right-handedly, then they must be connected, as shown in Fig. 210; for the electromotive-force induced in a coil as

[1] Specification of Patent 953 of 1878.
[2] Specification of Patent 1228 of 1878.

it passes under a north pole will circulate around the armature core in an opposite direction to that induced in the neighbouring coil that is passing under a south pole.

Gramme's Dynamo.—A diagram of the Gramme alternate-current machine is shown in Fig. 211. The sections of this machine were four times as numerous as the poles, and might be coupled to feed four separate circuits.

A convenient machine in which the alternate-current dynamo and its exciter are combined in one, has been devised by Gramme for the purpose of supplying the currents re-

FIG. 211.

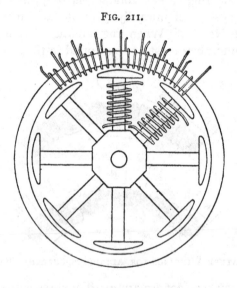

quired by Jablochkoff's electric candles. This machine is shown in Fig. 212. The exciter consists of an ordinary ring armature revolving (as seen in perspective) between an upper and a lower pole-piece. This ring armature supplies current to the set of field-magnets which rotates on the same shaft at its further end within an external cylindrical ring.

Messrs. Elwell and Parker employ a ring armature with 22 sections surrounded by 22 external poles. The ring is 36 inches in diameter and 30 inches in length. On it are 660 turns of wire in one layer only. At 600 revolutions per minute this machine gives 30 ampères at a pressure of 2000 volts. The armature resistance is 2·4 ohms ; that of the

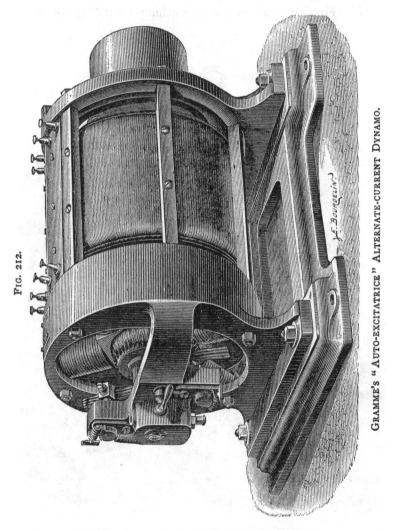

FIG. 212.

GRAMME'S "AUTO-EXCITATRICE" ALTERNATE-CURRENT DYNAMO.

field magnets 19 ohms. The machine, which weighs 4 tons, is separately excited by a small exciter giving 10 ampères.

M. De Meritens has also devised machines having a number of steel horseshoe magnets set with poles, alternately presented to the periphery of a ring, the latter being wound like Fig. 210.

Mr. Mordey has devised a machine consisting of two

FIG. 213.

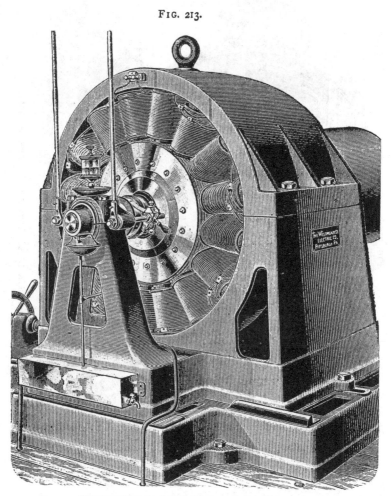

WESTINGHOUSE ALTERNATE-CURRENT DYNAMO.

Pacinotti rings, one laminated as armature, one not laminated as field magnet, and both wound as indicated in Fig. 209.

Mr. R. Kennedy employs as armature a flat ring, the core

of which is of hoop iron. Mr. Kapp has designed a very similar machine. In both these, the rings are magnetised from the sides, as in the Victoria and Gülcher dynamos. In Kennedy's magnet all the N-poles are on one side of the ring, alternating between the S-poles on the other.

Drum-armature Alternate-current Machines.

Stanley's Dynamo —A drum armature is employed in the alternate-current machine[1] of the Westinghouse Company (Stanley's), depicted in Fig. 213. In this machine there are 16 radial poles pointing inwards and fixed externally to a common cylindrical iron yoke. The armature core, which is 23 inches in diameter and 12 inches in length, is built up of thin iron disks, with holes perforated for ventilation. The winding consists of 16 coils wound in flat hanks on special formers, and then laid side by side upon the periphery of the core, not overlapping, and secured by well-insulated binding wires ; the ends of the hanks being bent over and secured against the flanks of the drum. This machine weighs 3 tons, and gives 1100 volts at 1080 revolutions per minute.

Pole-armature Alternate-current Dynamos.

Lontin's Dynamo.—This type was introduced by Lontin. A skeleton diagram is given in Fig. 214. In this machine the field - magnet (separately excited with a continuous current) consisted of a set of radiating poles, and it rotated within an outer set of coils which served as a fixed armature. These coils were wound upon short cores of solid iron. This machine, which has long since been superseded, had many defects, not the least of which was the great mass of iron in which so many internal eddy currents were induced that the machine was very prone to become overheated. Indeed, more power was required to drive it on open circuit than when the machine was supplying its maximum number of lamps.

A modification of this machine, having laminated iron cores in the armature part, has been used by Kennedy.

---

[1] See Specification of Patents Nos. 9725, 9727, of 1887.

A very large alternate-current machine, having many points in common with that of Lontin, was shown at Vienna, by Messrs. Ganz, of Buda-Pesth, in 1883. It was capable of furnishing light for 1200 Swan lamps (20 candle-power each). This dynamo was constructed according to the Mechwart-Zipernowsky system. The thirty-six bobbins of the field magnet were set concentrically on an iron frame, and rotated within an outer circle of thirty-six armature bobbins. The field-magnet coils were, in fact, the fly-wheel of the high-pressure compound engine which drove the dynamo and its

FIG. 214.

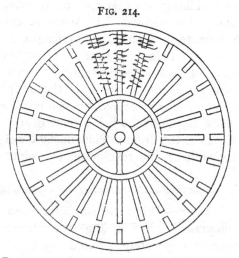

DIAGRAM OF PRINCIPLE OF LONTIN'S MACHINE.

exciter, The diameter of the rotating part was 2½ metres. A salient feature of this machine was the fact that any one of the coils, either of armature or field-magnets, could be removed from the side of the machine, in case such were needed. The whole fly-wheel could, in this way, be taken down by one man in a few minutes. The armature coils were attached flat against the inner periphery of a large ring of iron wire.

Messrs. Siemens and Halske[1] also exhibited at Vienna a machine closely resembling the Lontin type. Quite recently

[1] See *Elektrotechnische Zeitschrift*, viii. 232, 1887.

Drs. J. and E. Hopkinson have designed machines on this same type of construction.

Disk-armature Alternate-current Dynamos.

By far the most numerous of alternate-current machines are those in which the armature coils are arranged around the

Fig. 215.

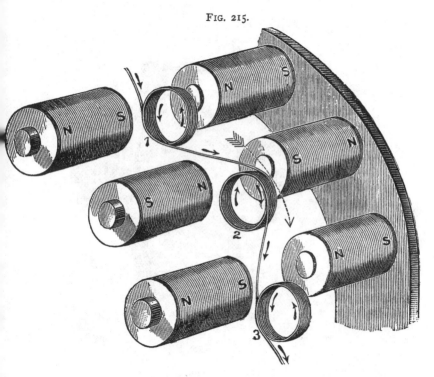

periphery of a disk. The old machines of Nollet and Holmes, and the so-called "Alliance" machine [1] (all of which had permanent magnets of steel), belonged to this class. The more modern type, with electromagnets, was created by Wilde, in 1867. The field-magnets consist of two crowns of fixed coils, with iron cores arranged so that their free poles

---

[1] The theory of the " Alliance " machine was treated by Le Roux in *Ann. Chim. Phys.* [3], l. 463, 1857; and later by Jamin and Richard, *ib.* [4], xvii. 276, 1869.

are opposite to one another, with a space between them
sufficiently wide to admit the armature, Fig. 216.   The
poles taken in order round each crown are alternately of
N. and S. polarity ; and opposite a N. pole of one crown faces
a S. pole of the other crown.   This description will apply to
the magnets of the alternate-current machines of Wilde
and Siemens, to the Ferranti machine, and in part to those of

FIG. 216.

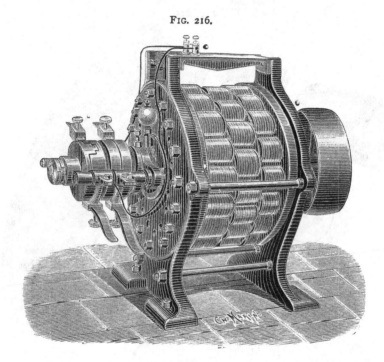

WILDE'S DYNAMO.

Lachaussée and of Gordon.   The principle will be best understood by reference to Fig. 215, which gives a general view of
the arrangement.   Since the lines of force run in opposite
directions between the fixed coils, which are alternately S—N,
N—S, as described above, the moving coils will necessarily
be traversed by alternating currents ; and as the alternate
coils of the armature will be traversed by currents in oppo-

site senses, it is needful to connect them up, as shown
in Figs. 207 or 208, so that they shall not oppose one
another's action.

Wilde's Dynamo.—In Wilde's dynamo, the armature coils

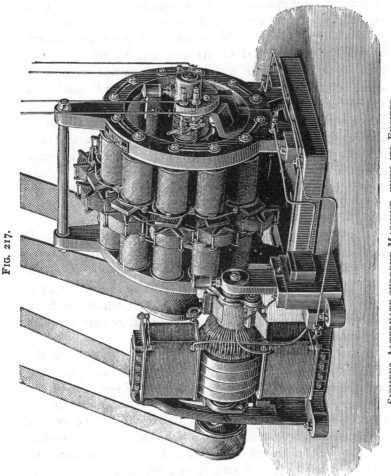

FIG. 217.

SIEMENS ALTERNATE-CURRENT MACHINE, WITH ITS EXCITER.

have iron cores, and the machine is provided with a commutator on the principle indicated in Fig. 205, p. 257. This
commutator Wilde usually applied to a few, or only one, of
the rotating coils, and utilised the current thus obtained to

magnetise the field magnets. The two simple rings for collecting the main current are also shown in the figure.

Siemens's Dynamo.—Siemens prefers to ·use a separate continuous-current machine to excite the field-magnets of alternate-current dynamos. In the armature of the latter (see Fig. 217), the coils are wound usually without iron, upon wooden cores. Copper ribbons insulated from one another by strips of vulcanised fibre are also used for the coils; the connexions being made by soldering the strips with silver solder. In some forms of the machine, the individual coils are enclosed between perforated disks of thin German silver. When currents of great strength are required, but not of great electromotive-force, the coils are coupled up in parallel arc, instead of being united in series. In Fig. 217 a small continuous-current machine of vertical pattern, such as was described on p. 193, is shown in action as an exciting machine to furnish the magnetising currents to the stationary field magnets of the alternate-current machine.

Lachaussée's Dynamo.—In a dynamo by Lachaussée, which very strikingly resembles the preceding one, there is iron in the cores of the rotating coils. But the main difference is that the rotating coils are the field magnets, excited by a separate Gramme dynamo, whilst the coils, which are fixed in two crowns on either side, act as armature coils in which currents are induced.

Gordon's Dynamo.—Gordon's dynamo (Fig. 218) is designed on the same lines as the Lachaussée machine; but with many important modifications. In the first place, there are twice as many coils in the fixed armatures as in the rotating magnets, there being 32 on each side of the rotating disk, or, in all, 64 moving coils; while there are 64 on each of the fixed circles, or 128 stationary coils in all. The latter are of an elongated shape, wound upon a bit of iron boiler-plate, bent up to an acute V form, with cheeks of perforated German silver as flanges.[1] The object of thus arranging the coils, so that the moving ones shall have twice the angular breadth of

[1] For further details of the Gordon dynamo, see Mr. Gordon's *Practical Treatise on Electric Lighting* (1884), p. 162.

the fixed ones, is to prevent adjacent coils of the fixed series
from acting detrimentally, by induction, upon one another.
The alternate coils of the fixed series are united together in
parallel arcs, so that there are two distinct circuits, in either or
both of which lamps can be placed ; or they can be coupled
up together. In spite of the great care taken in the construction of this large machine, to guard against the appearance of

FIG. 218.

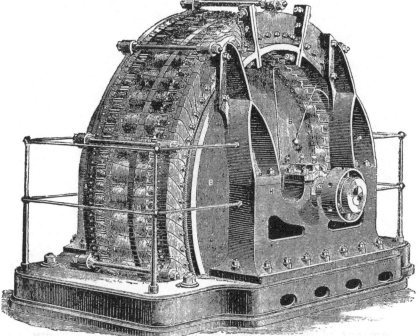

GORDON'S DYNAMO.

eddy currents, by laminating and insulating the cores, frames,
and coils, it appears that in these machines there is a considerable loss from this cause.

Ferranti's Dynamo.—Another alternate-current dynamo,
identical in many respects with the Siemens alternate-current
dynamo, has been brought out, under the name of the Ferranti

machine (Fig. 219).   As in the machines of Wilde and
Siemens, the electromagnets form two crowns with opposing
poles.   The point of difference is the armature, which consists
of zigzags of strip copper folded upon one another.   There
are eight loops in the zigzag (as shown in Fig. 220, which
depicts half only of the arrangement), and on each side are
sixteen magnet poles ; so that the moving parts are twice the
angular breadth of the fixed parts.   The advantage of the

FIG. 219.

FERRANTI-THOMSON DYNAMO.

armature of zigzag copper lies in its strength and simplicity
of construction.   Sir W. Thomson, who is one of the inventors
of this armature, proposed originally that the copper strips
should be wound between projecting teeth on a wooden wheel.
He also proposed to use as field-magnets a form of electromagnet of the kind known as Roberts's, and also used by
Joule, in which the wires that bring the exciting current are
passed up and down, in a zigzag form, between iron blocks
projecting from an iron frame.

The framework of the Ferranti machine is cast in two halves, which are afterwards bolted together. The armature,[1] originally a

FIG. 220.

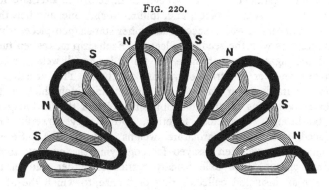

DIAGRAM OF FERRANTI ALTERNATE-CURRENT DYNAMO.

single zigzag piece of copper, has assumed the form shown in Fig. 221, in which it may be seen that the convolutions are

FIG. 221.

ARMATURE OF FERRANTI DYNAMO.

multiplied, and are held in their places by bolts through a star-shaped piece of brass, which also serves to carry to one of the two collectors

[1] See Specification of Patent No. 3702 of 1883, and for later details No. 702 of 1887.

the connexion with one of the zigzag copper strips. There are, in fact, three complete circuits of copper strips in the armature connected in parallel arc. They begin at three of the alternate four bolts of the star-shaped piece, and, folding round one another, they all eventually unite with a second and inner star-shaped piece, which communicates with the second collector. Each strip makes ten turns round the zigzags, so that there are thirty layers, all well insulated from one another by strips of vulcanised fibre. This armature is 30 inches in diameter, and a little more than $\frac{1}{2}$ inch thick in the upper convolutions, so that the opposite poles of the field-magnets can be brought very close together, and a very powerful field produced. The entire armature weighs only 96 lbs. Several arrangements have been essayed for conveying the currents to the external circuit. The axle usually carries on either side of the armature an insulated collector ring of bronze, to which the afore-mentioned star-shaped pieces are respectively connected. Instead of brushes, solid pieces of metal, in the form of hooks, have been employed to collect the current. In another machine mercurial contacts enclosed in cavities in insulated steel bearings are employed instead of the collectors described above. The machine requires a speed of 1400, and weighs $1\frac{1}{2}$ ton.

Dynamos with zigzag conductors have been designed by Rapieff, Matthews, and others.

Mordey's Alternator.—The latest of alternate-current machines is that of Mr. Mordey, brought out by the Anglo-American Brush Corporation. This machine, Fig. 222, differs in two striking features from those previously described. Though, as before, there are two crowns of poles between which the armature lies, all the poles on one side are of one kind, north poles, and all those on the other side are south poles. Hence there is no reversal of the magnetic field through the armature coils ; the number of magnetic lines through any coil simply varying from zero to maximum and back. As a result of this arrangement, there is a great simplification of the means needed to magnetise the field-magnets. It is no longer necessary to wind a separate magnetising coil on each protruding pole. One single coil surrounding a central cylinder of iron suffices to magnetise the whole of the 18 poles. There is indeed only one magnetic circuit, branching into 9 separate branches. The construction of the field-magnet, which is separately shown in Fig. 223, is as follows :—

A short cylinder of wrought iron, through which the shaft passes, forms the core, and is surrounded by the exciting coil. Against the ends of this core are firmly screwed up the two end castings, each of which is furnished with a number of curved horns (nine in this machine), projecting to within

FIG. 222.

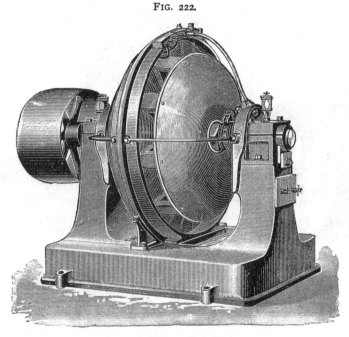

MORDEY ALTERNATOR.

18 millimetres, the narrow polar gap being only just wide enough to admit the armature. The entire field-magnet revolves on the shaft, the exciting coil being supplied with current from a separate machine of small size by means of the two collecting rings on the shaft on the right in Fig. 223. There is no need for the exciting coil to revolve ; but for mechanical reasons it was deemed preferable in the first machine to wind it actually upon the field-magnet core. The

T

armature, which stands still, consists (Fig. 224) of 18 coils of ribbon copper 11 millims. wide wound on porcelain cores, and clamped at the broad end to a light but firm frame, the ends of the conductors of each coil being brought out through porcelain insulators and suitably connected together. All the metal clampings are outside the magnetic field, and they are so arranged that any one coil can be removed in a few moments without dismounting any other part of the dynamo. As the

FIG. 223.

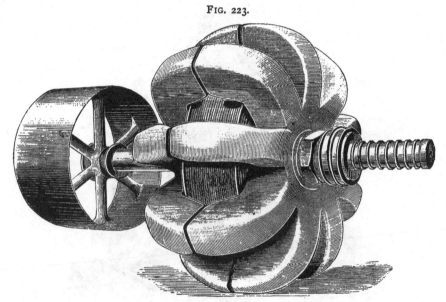

FIELD-MAGNET OF MORDEY ALTERNATOR.

armature is stationary there are no centrifugal forces to be considered, and the coils have to be supported only with a view of resisting the tangential drag of the field. This renders the insulation of the armature coils a very simple matter, and is of great importance in working at high volts. The revolving field-magnet forms an excellent fly-wheel, effectually neutralising any pulsations due to irregularities in the stroke of the engine, and as there are no parts liable to fly out, a high speed of driving presents none of the difficulties that arise with many

other types of machine. In Fig. 222, which gives a view of
the complete machine, the field-magnets are almost hidden by
external covers of sheet metal to prevent too great a disturb-
ance of the air by the rotating pole-pieces. The armature
terminals are at the top of the fixed frame. The shaft is provided at one end with a thrust-bearing resembling that of a
marine propeller, and the bearing blocks are made adjustable
longitudinally, so that the field-magnet may be placed exactly

FIG. 224.

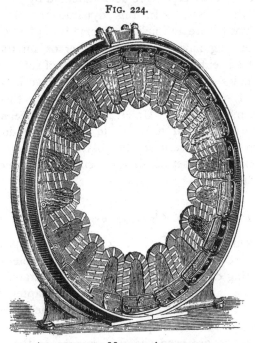

ARMATURE OF MORDEY ALTERNATOR.

symmetrical with respect to the armature. The exciter is a
small 4-pole Victoria 1-HP. machine (p. 173), weighing but $\frac{1}{40}$
of the weight of the alternator. This machine, the first of this
design, gives a current of 17 to 20 ampères at a potential of
2000 volts when running at 650 revolutions per minute. It is
being used for supplying currents to Mordey transformers.
The electromotive-force is 1 volt per $8\frac{1}{2}$ inches of conductor.

The very low resistance of the armature (3 ohms) makes the machine almost self-regulating. One advantage in this type of machine is the little labour required in winding its coils ; and the simplicity of the tooling and fitting of the iron portions is a great gain. Mr. Mordey has designed [1] a considerable number of alternative forms, all characterised by the combination of the two principles of simplicity of magnetic circuit and non-reversal of polarity in the armature. Some designs for machines of kindred type have been patented by W. Main.[2]

Kingdon's Dynamo.—In this dynamo the principle of fixing both armature and field-magnets [3] is applied in a novel fashion. A ring having a large number of internally projecting poles is entirely built up of laminæ of soft iron. The alternate poles are wound with coils to serve as armature part, whilst those between them are wound with other coils to act as the magnet part. Upon an internal wheel are borne masses of laminated iron, which in rotating produce rapid periodic reversals in the magnetic polarity of the cores of the armature parts, and set up alternate currents in the coils that surround them.

Regulation of Alternate-current Dynamos.

A convenient way of regulating the current or potential of alternate-current dynamos is to interpose a variable resistance in the exciting circuit ; the resistance being operated by hand or by some automatic regulator (see Chapter VI. and Appendix VI.). This method is applicable either to separately excited or to self-exciting machines. In the case where separate exciters are used, the performance of the alternate-

---

[1] Specification of Patent 8262 of 1887.

[2] Specifications Nos. 15,858 and 16,032 of 1887. The device of employing field-magnets with a greater number of pole-pieces than of exciting coils had been previously employed by Holmes (Specification 2060 of 1868), and more recently by J. and E. Hopkinson.

[3] This principle, suggested by several early workers (see Historical Notes, p. 7), was revived by the author of this treatise in 1883, in a form which led up to Mr. Kingdon's. Drawings of this machine are given in *Electrical Review*, xxii. 178, 1888.

current machine may be regulated by controlling (by variable resistances, &c.) the exciting circuit of the exciter.

Alternate-current dynamos, when intended for supplying glow-lamps direct at constant potential, are usually constructed with such low resistance in the armature part that they would be almost self-regulating if it were not for the demagnetising influence of the armature currents. This may, as evidenced by the curves given in Chapter XVI., be considerable. More-over, the main leads from the dynamo to the lamps may have

FIG. 225.

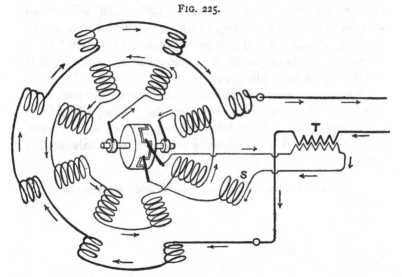

ZIPERNOWSKY'S METHOD OF COMPOUNDING ALTERNATE-CURRENT
DYNAMO.

a sufficient resistance to affect the constancy of the potential at the lamps. Hence it becomes necessary to adopt some system of self-regulation akin to the methods of compound winding adopted for continuous-current machines. Assuming that an initial and constant excitation can be afforded by a separate exciter, it remains to supplement this by a variable excitation which, in the case of constant-potential distribution, must be proportional to the main current, or in the case of constant-current distribution, must be proportional to the resistance in the circuit. The first of these ends has been achieved by Zipernowsky, the second by Kennedy.

Fig. 225 shows Zipernowsky's arrangement, and represents diagrammatically a machine with internal revolving field magnet of 8 poles. The armature consists of eight stationary coils, seven of which are joined in series to feed the main circuit, the eighth coil S being separated from the rest and used to provide excitation for the field magnets, its currents being conveyed in through a suitable commutator on the shaft of the machine. In the main circuit, just beyond the terminals of the machine, is inserted the primary coil of a small transformer T, the secondary of which is connected with the exciting circuit. In this way an additional excitation is procured, always proportional in amount to the currents that are flowing in the main circuit.[1]

Kennedy's constant-current arrangement consists of a somewhat similar device, except that the primary coil of the transformer (the coil in this case being of fine wire in many turns), is placed as a shunt across the terminals of the machine.

[1] Compare Schallenberger's method of regulation described in *Electrical World*, x. 60, 1887.

CHAPTER XIII.

DYNAMOS FOR ELECTROPLATING AND ELECTRO-METALLURGY.

SPECIAL forms of dynamo are needed for the work of electroplating, electrotyping, and the electrolytic treatment of ores and purification of metals. They must, of course, be of the continuous-current type. In general, very low electromotive-forces and very large currents are requisite, for the quantity of metal deposited in the bath depends upon the quantity of ampères of current only, and not on the number of volts of electromotive-force. And though a few volts are requisite to drive the electric current through the resistances of the circuit, the number is in every case small. To decompose water electrolytically requires less than two volts. To deposit metal in a bath in which the anode is of the same metal as the deposit requires usually a very small electromotive-force. In general, if too great an electromotive-force is employed, or if the density of current (i.e. the number of ampères per unit of area of kathode surface) is excessive, the metallic deposits will be uneven or pulverulent. All these circumstances point to the construction of dynamos having at most but four or five volts of electromotive-force, but so designed as to have an exceedingly low internal resistance.

The first application of a dynamo to the purpose of electroplating is due to Mr. J. S. Woolrich, who in 1842 patented this use of a magneto-electric machine.

Wilde's Dynamos.—Wilde, however, was the first to construct machines really fitted for the purpose, when he invented the principle of using a large dynamo, the field-magnets of which were separately excited by the currents of a smaller magneto machine. His first machines, which were used for many years by Messrs. Elkington, had small exciters of the old Siemens type (Fig. 15), mounted upon electromagnets of the form shown in Fig. 72.   Both armatures were of the old shuttle-form introduced by Siemens, and the larger one required to be kept cool by streams of water.   About the year 1867 Wilde introduced another type of machine, to which reference was made on p. 265.   In this dynamo (Fig. 216) the armature consisted of sixteen coils rotating between two crowns of opposed electromagnets, each also consisting of sixteen coils. This machine is shown in the figure as an alternate-current machine, having two collecting rings with a brush pressing against each.   When the machine is used for electro-metallurgical work, the pair of collecting rings are replaced by a second rectifying commutator, of the kind shown in Fig. 205. The armature coils (save the one used for feeding the field magnets) are connected in parallel in the manner shown in Fig. 207.   There are iron cores in the armature coils of this machine, which is, in consequence, prone to heat.

Weston's Dynamo.—The field magnet of this machine (Fig. 226) consists of a cast-iron cylinder, having six internally projecting electromagnets of alternate polarity with steel cores.   The armature is a six-pole arrangement, resembling Fig. 27, but differently connected, the six coils being joined in parallel (as in Fig. 207).   The commutator merely rectifies the alternating currents.   There is added a small centrifugal cut-off to break the circuit when the speed is less than that which will provide the requisite electromotive-force. The object of this device, as well as that of the steel cores, is to prevent the machine from having its polarity reversed by a back-current arising from polarisation in the baths.

Elmore's Dynamo.—Elmore has devised several forms of dynamo for electrolytic purposes.   In the smaller machines the armature is formed of a hollow cast-iron plate carrying six electromagnets on each side.   The use of this cast-iron frame is altogether bad, as it heats and requires special arrangements to keep it cool.   The shaft is perforated, and in the larger machines a supply of cold water is caused to circulate through the armature.   Such a circumstance is sufficient to condemn the design of the armature. The commutator merely rectifies the currents (p. 40), without rendering them continuous. This is a bad feature: for with all electro-chemical work, whether plating, typing, or charging accumulators, there is necessarily much sparking unless the fluctuations of the current are reduced to a minimum by employing a many-part armature with a proper collector. The largest Elmore dynamos, for copper refining, had eighteen electromagnets in each crown, and yielded a current of 3000 ampères

Fig. 226.

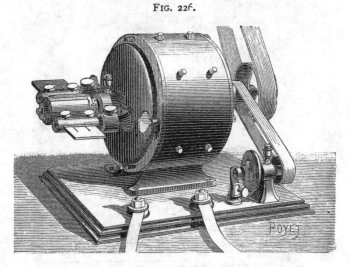

WESTON'S ELECTROPLATING MACHINE.

at a potential of seven to eight volts. Such a machine would deposit over 25 lbs. of copper per hour. The field-magnet coils are included in series with the main circuit. This is a mistake. All electroplating dynamos should be either shunt-wound or else compound-wound. The Elmore machine is now an obsolete form.

Gramme's Dynamo.—When the Gramme dynamo is used for electrolytic purposes, the armature is modified so as to reduce its resistance. The armature shown in Fig. 93, built up of copper strips, is suitable for this purpose. Fig. 227

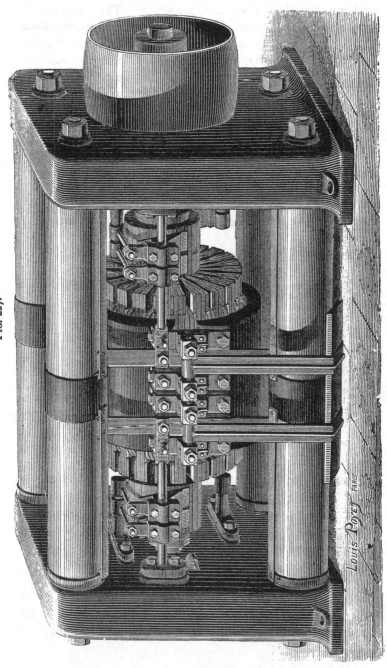

FIG. 227.

GRAMME DYNAMO FOR ELECTRO-METALLURGY (1873).

LOUIS POYET PARIS

illustrates a machine built in 1873 for the Hamburg copper-

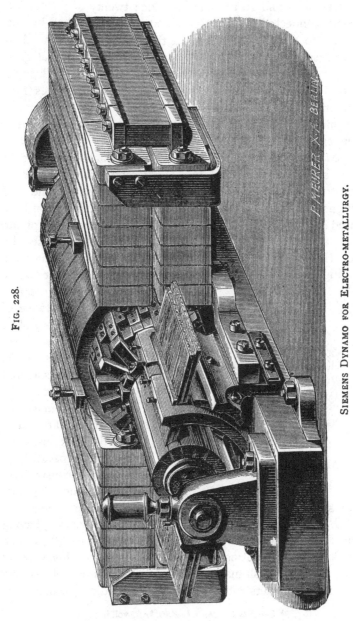

FIG. 228.

SIEMENS DYNAMO FOR ELECTRO-METALLURGY.

refining works. The cylindrical "ring" is constructed in forty

sections, each consisting of seven strips of copper 3 millimetres thick and 10 millimetres wide; twenty of the sections
being connected to a collector at one end, the alternate
twenty to a second collector at the other end of the shaft.
The field magnets are wound with thirty-two turns of sheet
copper, and are in series with the main circuit. When the
two armatures are used in parallel, the machine gives 1500
ampères with an electromotive-force of eight volts. The
machine weighs about one ton, of which about one-third is
copper, two-thirds iron.

Siemens Dynamo.—A special form of machine suitable for
very strong currents and low electromotive-forces has been
constructed by the firm of Siemens. Both armature and field
.nagnets are formed of bar copper of large cross-section insulated by air-spaces which admit of free circulation of air.
The armature is connected up in the manner devised by
von Hefner Alteneck; an end view of the connexions is given
in Fig. 229. These machines are employed at Oker for the

FIG. 229.

ARMATURE OF SIEMENS
DYNAMO.

electrolytic treatment of copper.[1]
The total internal resistance of
this machine is but 0·0007 ohm.
The brushes are solidly mounted,
without spring contacts. The
bars of the armature are soldered
together with silver solder. In
smaller electroplating machines
by the same firm, the armatures
are wound with stout insulated
wires joined four in parallel, as
being more easily constructed.
The unipolar machine (Fig. 203)
was specially designed for electro-metallurgical purposes.

Brush Dynamo.—A variety of the Brush dynamo, having
the armature wound with coils of a low resistance, has been
used for electroplating purposes. It was for this type of
machine that Brush made the important invention of exciting
the field magnets with a compound winding; coarse wire

coils being connected in series, with the addition of a so-called "teazer" coil of finer wire to maintain the magnetism when the main circuit was opened, thus enabling the machine to do either a large or a small amount of work without fear of reversing the current. When objects are to be plated in an electrolytic bath, regularity of deposit can only be ensured when the electrodes are kept at a fairly constant potential, no matter whether a large number of articles are for the

FIG. 230.

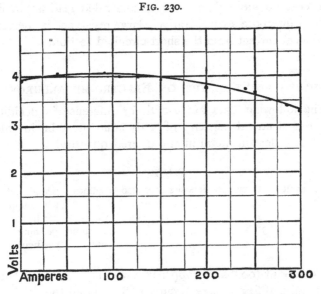

CURVE OF POTENTIAL AT TERMINALS OF BRUSH DYNAMO FOR ELECTROPLATING.

moment undergoing treatment, requiring a large current, or whether the number be small, requiring a small current. The compound winding adopted by Brush answered the desired end with remarkable success. The curve given in Fig. 230 shows how nearly constant the potential at the terminals of this machine is. It varied only from 3·3 to 4·1 volts, whilst the current varied from 300 ampères to zero.

Other dynamos have been designed for electroplating and electro-metallurgical work by the Gülcher Company, and by

Messrs. Stafford and Eaves. The electroplating machine of the latter is solidly built, with a single magnetic circuit; it has a ring armature wound in 18 sections, giving 150 ampères at a pressure of 6 volts, when driven at 640 revolutions per minute. Messrs. Crompton have lately devised a method of dividing the main leads between two pairs of brushes touching adjacent bars of the collector, and are thereby enabled to construct their plating machines with fewer parts in the armature. The divided leads from the dynamo to the plating tanks cost no more than a single undivided lead would do, but they interpose a comparatively large resistance in the path of the local current from the short-circuited section.

NUMERICAL STATISTICS ON ELECTRO-METALLURGY.

The following data are useful for reference in deciding what the electrical capacity of a dynamo must be in order that it may deposit metal in any desired quantity.

Copper.

Current	1	ampère deposits	0·000326 grammes	per second.
,,	1	,, ,,	0·01957 ,,	per minute.
,,	1	,, ,,	1·1739 ,,	per hour.
,,	851·8	,, ,,	1	kilogramme per hour.
,,	386·4	,, ,,	1	pound per hour.

To deposit 100 lbs. of copper in a working day of ten hours will require 3864 ampères of current flowing all the time; or, if conducted in ten baths in series with one another will require 386·4 ampères, but in that case the dynamo will require to be of an electromotive-force ten times as great as for one single large bath. If electrolysis of the crude copper solution is carried on with carbon anodes, there will be required about 1·2 volts for each bath in series, or, at most, 15 volts for the ten baths.

Silver.

Current of	1	ampère deposits	4·025 grammes	per hour.
,,	,, 248·5	,, ,,	1	kilogramme per hour.
,,	,, 112·7	,, ,,	1	pound per hour.

Gold.

Current of 1 ampère deposits 2·441 grammes   per hour.
„   „ 409·7 „   „   1   kilogramme per hour.
„   „ 185·8 „   „   1   pound   per hour.

Nickel.

Current of 1 ampère deposits 1·099 grammes   per hour.
„   „ 910·1 „   „   1   kilogramme per hour.
„   „ 412·8 „   „   1   pound   per hour.

The following statistics as to the various pressures and currents required in various processes of electro-deposition are useful for reference.

PRESSURE AT TERMINALS REQUIRED FOR DIFFERENT KINDS OF BATHS.

	Volts.
Copper (acid bath)	0·5 to 1·5
„ (cyanide bath)	3 „ 5
Silver	0·5 „ 1
Gold	0·5 „ 4
Brass	3 „ 4
Iron (steel-facing)	1 „ 1·3
Nickel on iron, steel, copper, with nickel anode, strike deposit with 5 volts, diminishing to	1·5 „ 2
Nickel on iron, steel, copper, with carbon anode	2 „ 4
Nickel on zinc	4 „ 7
Platinum	5 „ 6

CURRENT DENSITY FOR PROPER DEPOSIT.

	Ampères per 100 sq. in.
Copper Typing—	
Best quality tough deposit	1·5 to 4
Good and tough (for clichés)	4 „ 10
Good solid deposit	10 „ 25
Solid deposit, sandy at edges	25 „ 40
Sandy and granular deposit	50 „ 100
Copper (cyanide bath)	2 „ 3
Zinc (for refining)	2 „ 3
Silver	1 „ 3
Gold	0·5 „ 1
Brass	3 „ 3·5
Iron (steel-facing)	0·5 „ 1·5
Nickel, at first deposit 9 to 10 ampères per 100 square inches, diminishing afterwards to	1 „ 2

CHAPTER XIV.

FIELD-MAGNETS AND MAGNETISM.

1. Magnetic Principles.

INASMUCH as the field-magnets of dynamos are, in the vast·
majority of cases, electromagnets, the iron cores of which are
excited by electric currents circulating in surrounding coils, it
becomes a matter of primary importance to us to know
what is the law that governs the electromagnet. If we once
know the relation that subsists between the exciting current
and the magnetism that is produced by it, we can apply this
knowledge to the design of dynamos : for such knowledge
will enable us to calculate beforehand the size of field-magnet
and the number and gauge of coils that will be required in
a dynamo that is to furnish any given amount of electric
energy. It will be necessary first to define the terms used ;
then we shall give some account of the facts and of the
properties of iron of different kinds ; next will follow an
account of the various algebraic rules that have been suggested to represent approximately the law of the electromagnet ; then some all-important considerations respecting
the magnetic circuit and its theory ; some examples and
useful rules will be given ; and, lastly, the various forms given
to field-magnets will be discussed.

Definitions and General Properties.[1]

Intensity of Magnetic Field.—We have seen in Chapter II.
that every magnet is surrounded by a certain " field," within
which magnetic force is observable. We may completely

---

[1] The paragraphs which follow are taken, with some alterations and additions, from the author's *Elementary Lessons on Electricity and Magnetism* (edition
of 1887).

specify the properties of the field at any point by measuring the *strength* and the *direction* of that force—that is, by measuring the "*intensity of the field*" and the direction of the lines of force. *The "intensity of the field" at any point is measured by the force with which it acts on a unit magnetic pole placed at that point.* Hence, *unit intensity of field is that intensity of field which acts on a unit pole with a force of one dyne.* There is therefore a field of unit intensity at a point one centimetre distant from the pole of a magnet of unit strength. Suppose a magnet pole, whose strength is *m*, placed in a field at a point where the intensity is H, then the force will be *m* times as great as if the pole were of unit strength, and the amount of the force (in dynes) can be calculated by simply multiplying together the strength of the magnetism of the pole and the intensity of the field ; or,

$$f = m \times H.$$

We may also take as a measure of the intensity of the field at any point the number of lines of force that pass through a square centimetre of surface placed across the field at that point. *It follows that a unit magnetic pole will have* 4π *lines of force proceeding from it :* for there is unit field at unit distance away, or one line of force per square centimetre ; and there are 4π square centimetres of surface on a sphere of unit radius drawn round the pole. A magnet, whose pole-strength is *m*, has therefore $4\pi m$ lines of force running through the steel, and diverging at its pole.

Intensity of Magnetisation: Magnetic Susceptibility and Magnetic Permeability.—When a piece of magnetic metal is placed in a magnetic field, some of the lines of magnetic force run through it and magnetise it. The intensity of its magnetisation will depend upon the intensity of the field into which it is put and upon the metal itself. There are two ways of looking at the matter, each of which has its advantages. We may think of the magnetism of the iron or other metal as something resident on the polar surfaces, and expressed therefore in units of magnetism : this is the old way, adopted at a time when magnetism was regarded as a fluid. Or we may think about the internal condition of the piece of metal, and of the number of magnetic lines that are running through it and emerging from it into the surrounding space. This is the more modern way.

The fact that soft iron placed in the magnetic field becomes highly magnetic may then be expressed in the following two ways:—(1) iron when placed in the magnetic field develops strong poles on its end surfaces, being highly *susceptible* to magnetisation ; (2) when iron is placed in the magnetic field, the magnetic lines gather themselves up and run in greater quantities through the space now occupied by iron, for iron is very *permeable* to the lines of magnetic induction, being a good conductor of the magnetic lines. Each of these ideas may be rendered exact by the introduction of coefficients of susceptibility and of permeability. The *coefficient of magnetisation*, or *susceptibility*, is based on unit of pole strength. Suppose a bar magnet to have m units of magnetism on each pole ; then if the length between its poles is l, the product $m \times l$ is called its *magnetic moment*, and the magnetic moment divided by its volume is called its *intensity of magnetisation :* this term being intended, though based on surface-unit of pole strength, to convey an idea as to the internal magnetic state. Seeing that the volume of a bar of uniform section is the product of sectional area into length, it follows that if any such piece of iron or steel of uniform section had its surface magnetism situated on its ends only, its intensity of magnetisation would be equal to the strength of pole divided by the area of end surface. Writing I for the intensity of magnetisation we should have

$$I = \frac{\text{magnetic moment}}{\text{volume}} = \frac{m \times l}{s \times l} = \frac{m}{s}.$$

Now, supposing this intensity of magnetisation were due to the iron having been put into a magnetic field of intensity H, we find that the ratio between the resulting intensity of magnetisation I and the magnetising force H producing it is expressible by a numerical coefficient of magnetisation, or *susceptibility k.* We may write:

$$I = k\,H$$

or

$$k = \frac{I}{H}.$$

This may be looked at as saying that for every magnetic line in the field there will be k units of magnetism on the surface.

In magnetic substances such as iron, steel, nickel, &c., the susceptibility k has positive values ; but there are many substances such as bismuth, copper, mercury, &c., which possess feeble negative coefficients. These latter are termed " diamagnetic " bodies, and are repelled by the poles of magnets. The values of k vary very much in iron, not only in the different qualities of iron, but in every specimen with the stage of magnetisation. When a piece of iron has become well magnetised it is no longer as susceptible to magnetisation as it was at first ; it is becoming " saturated." Barlow found the value of k for iron to be $32\cdot8$; Thalen found it from 32 to 44 ; Archibald Smith, 80 to 90 ; Stoletow, 21 to 174 ; Rowland found it in Norwegian iron to go as high as 366 ; Ewing found *thin* soft iron wires go up to 1300 or 1400. Stoletow showed that iron in a weak magnetic field showed a small susceptibility, which greatly increased as the magnetising force in the field was strengthened, but again fell off with still greater forces as the iron got saturated. When very intense magnetising forces are used, so that the intensity of magnetisation is very great, the susceptibility is practically reduced to zero. It appears that the maximum intensity of magnetisation that can be given to iron and steel is ·about 1500 (units, per square centimetre of cross section). According to Rowland the maximum for cobalt is 800, for nickel 494. Steel does not retain all the magnetisation that can be temporarily induced in it, its maximum permanent intensity being, according to Weber 400, according to von Waltenhofen 470, according to Rowland 785, according to Hopkinson 878.

Everett has calculated (from Gauss's observations) that the intensity of magnetisation of the earth is only 0·0790, or only $\frac{1}{17600}$ of what it would be if the globe were wholly iron. In weak magnetic fields the susceptibility of nickel exceeds by about five times that of iron; but in strong fields iron is more susceptible.

The *coefficient of magnetic induction*, or *permeability*, is based on the lines of magnetic induction. The number of magnetic lines that run through unit area of cross section, at any point, is called "the magnetic induction" at that point; it is denoted by the letter B. The ratio between the magnetic induction and the magnetising force producing it is expressed by a numerical coefficient of induction, or *permeability*, μ. We therefore write

$$B = \mu\,H$$

or

$$\mu = \frac{B}{H}.$$

This coefficient it always positive : for empty space it is 1, for air it is practically 1 ; for magnetic materials it is greater than 1, for diamagnetic materials it is slightly less than 1. The student may think of it in the following way: Suppose a certain magnetising force to act in a certain direction, there would naturally result from its action induction along a certain number of lines of induction (or so-called lines of force), and in a vacuum the number of lines would numerically represent the magnetising force. But if the space considered were occupied by iron the same magnetising force would induce many more lines. The iron has a sort of multiplying power or specific inductive capacity, or conductivity for the magnetic lines. This permeability is easily calculated from the susceptibility. It was shown above that there are 4π magnetic lines proceeding from each unit of pole magnetism. Hence if, as shown above, each line of force of the magnetising field produces k units of magnetism there will be $4\pi k$ lines added by the iron to each 1 line in the field, or the multiplying power of the iron μ is equal to $1 + 4\pi k$. It follows at once that the iron examined by Barlow, mentioned above, had for the value of μ, its permeability, $4\pi \times 32\cdot8 + 1$, or 413; that examined by Thalen, from 402 to 552; Stoletow, 264 to 2186. The fact is that not only do the magnetic permeabilities of different kinds of iron vary enormously, but the magnetic permeability in any one given piece of iron varies under varing amounts of magnetising force. The values of the permeability, like those of susceptibility, decrease as the magnetisation of the iron gets increased towards saturation. In the following table two sets of values are given from the researches of Stoletow, and the more recent ones of Bidwell :—

OBSERVATIONS OF STOLETOW.

H	k	I	μ	B
0·43	21·5	9·24	275·6	118·5
0·44	30·5	13·45	390·5	171·8
3·20	174·0	556·6	2222·	7113·
30·6	39·4	1206·	504·2	15427·

OBSERVATIONS OF BIDWELL.

3·9	151·0	587	1899·1	7390
10·3	89·1	918	1121·4	11550
40·	30·7	1226	386·4	15460
115·	11·9	1370	150·7	17330
208·	7·0	1452	88·8	18470
427·	3·5	1504	45·3	19330
585·	2·6	1530	33·9	19820

The figures in the first column (H) may be regarded as the number of magnetic lines that there would be to the square centimetre if all iron were removed, whilst the figures in the last column (B) give the number which the same magnetising force would create if the space were filled up with iron. If the figure in the last column be divided by the figure opposite in the first column the quotient gives the multiplying power of the iron — in other words, its magnetic permeability μ, which is set down in the fourth column. It will be noticed that the highest value obtained by Bidwell for the induction B is 19,820. Other experimenters have found other values. Rowland gives 16,600 for wrought iron. Kapp gives 16,740 for wrought iron, 20,460 for charcoal iron (sheet), and 23,250 for charcoal iron (wire). Hopkinson (see p. 300) gives 18,250 for wrought iron, and 19,840 for mild Whitworth steel.

It appears then that about 20,000 magnetic lines per square centimetre of cross section is the utmost to which the induction can be pushed in wrought iron, on the average. This result is of great importance in the designing of dynamos, as will appear hereafter.

According to Hopkinson the induction B for cast iron is about 11,000, in a field H of 220 : the residual induction being about 5000. In a field of 240 the induction in grey cast iron reached 10,783 ; in malleable cast iron, 12,408 ; in mottled cast iron, 10,546. Bosanquet finds maximum induction B for charcoal iron and wrought iron from 16,800 to about 19,000, but has succeeded in magnetising a wrought-iron bar so that the induction in the middle bit of the bar reached 29,388. Professor Ewing, by special devices for producing an enormous magnetising force, has driven the induction in Lowmoor wrought iron up to 31,560 magnetic lines to the square centimetre, the permeability being reduced to about 3 ; and still more recently to over 40,000 under abnormal circumstances, the permeability being reduced below 2. Steel containing 12 per cent. of manganese is curiously non-magnetic. Hopkinson found its maximum induction only 310. The author has found that the peculiar "mitis" metal, which is a cast iron containing a small percentage of aluminium and is wonderfully malleable, possesses high magnetic powers, and can be magnetised up to 13,000, or even 13,500. It should be well adapted for use in dynamo-machines.

Curves of Magnetisation.

In every electromagnet we have to deal with an iron core, which is magnetised by a current flowing in a coil around it. Experiment shows that the magnetising power of a current is proportional not only to the strength of the current (that is to say to the number of *ampères*), but also to the number of turns in the coil round which it circulates ; for it is found that the magnetising power of 1 ampère flowing 10 times round is exactly equal to that of 10 ampères flowing once round, or to that of 2 ampères flowing 5 times round. The number of ampères multiplied by the number of turns in the coil is called for convenience the number of *ampère-turns.* *The magnetising power of the circulating current is proportional to the number of ampère-turns.*[1]

But though the magnetising force is thus strictly proportional to the ampère-turns, the magnetism produced in the iron core is by no means proportional to the ampère-turns; the effect is not proportional to the cause simply because the properties of the iron alter. As the iron gets more and more magnetised it gets less susceptible to further magnetisation, less permeable toward any additional magnetic line: additional ampère-turns produce only a small effect when the iron core is getting well saturated with magnetism. The simplest way to study this set of facts is to make experiments and plot out the results as a curve.

The apparatus which is required in addition to the electromagnet comprises the following pieces: a sufficiently powerful battery; a reliable galvanometer or ampère-meter to measure the strength of the current; a set of adjustable resistances to vary the current; and a tangent magnetometer, or failing this a compass with a short needle having an index moving over a scale of degrees to serve as a magnetometer.

Fig. 231 shows in what manner these pieces may be arranged, the electromagnet E being placed either due east or due west (magnetically) of the magnetometer, and " end-on " towards it, at a convenient distance. A number of observations are made with different strengths of current,

---

[1] It can be shown mathematically that if a current of strength i (absolute C. G. S. units) flows in a coil of S turns around the iron that forms part of a magnetic circuit, then the total magnetising power of this circulating current— that is to say, its total power to drive magnetic lines round the magnetic circuit, or, in precise language, the line-integral of the magneto-motive forces—is equal to 4π S i (see the author's *Elementary Lessons on Electricity and Magnetism*, edition of 1887, pp. 291 to 296). If the strength of the current i is given in ampères (instead of C. G. S. units), the value must be divided by 10, because one ampère is only the tenth part of the absolute C. G. S. unit of current. In this case the magneto-motive force will be equal to $1 \cdot 2566 \times$ S i.

and the corresponding magnetic forces are read off upon the magnetometer. For each particular value of the exciting current there will correspond a certain value of the magnetic force. These may be plotted out in a graphic diagram,

FIG. 231.

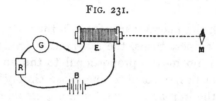

EXPERIMENT ON SATURATION OF ELECTROMAGNET.

the strength of the current being plotted out horizontally and the magnetic forces vertically. The curve will take, in general, the form shown in Fig. 232, and is seen to consist of two parts—a part which rises at a more or less steep angle,

FIG. 232.

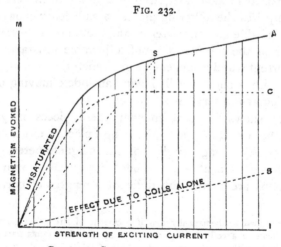

CURVE OF SATURATION OF ELECTROMAGNET.

and which for some distance continues nearly straight from the origin, and another part also nearly straight but inclined at a much smaller angle, the two parts being joined by a curved portion. The former part corresponds to the state of things when the iron core is unsaturated, the latter part to the state when the iron core is more than half saturated and the curved intermediate portion corresponding to the intermediate state when the iron core is approaching saturation. In this curve two effects are in reality simultaneously blended : the effect of the magnetism of the iron core and the effect of the magnetic action of the coils through which the current is flowing. It is possible to separate these effects ; for if the iron core be removed, and the magnetic effect of the coils alone be observed, a new set of data are obtained which when plotted out will yield the gently sloping line O B. From this line two conclusions may be drawn. It slopes at a small angle, for (1) the magnetic effect of the coils is small compared with that of the iron core. It is quite straight, for (2) the magnetic effect of the current in the coil is exactly proportional to the strength of the current in the coil. Having thus found what part of the whole effect is due to the current in the coil alone, we can find what part is due to the core alone. For if we subtract from the heights of the points on the curve O A the heights of the corresponding points on the line O B, we shall get the curve O C which will represent the effect due to the iron core alone. It will be noticed that in this curve the second part rises very slowly : in other words, the core when once saturated is much less susceptible to any additional magnetisation. The subsequent rise of the curve O A after attaining the diacritical point of semi-saturation S,[1] is very slow. It practically requires an infinite current to double the magnetism from that point.

The curve O C tells us, then, in a graphic way, the state of saturation of the core when magnetised by currents of different strengths ; whilst the curve O A gives us a graphic representation of the total magnetic effect due to currents of various strengths.

It is possible to obtain the curve for the effect due to the core alone by a direct process, by introducing between the electromagnet and the magnetometer a compensating coil to

[1] The meaning of the diacritical point on the curve of magnetisation is explained on p. 307.

balance exactly that part of the total effect which is due to the direct magnetic action of the coils. Arrangements of this kind have indeed been made by Weber and by Hughes.

But in an actual dynamo machine we do not want to know what part of the effect is due separately to the core, because the field produced by the field-magnets is due to core and coils acting together. Moreover, in the actual dynamo the important thing to be considered in the magnetisation of the field-magnets is not the strength of the exciting current, but the product of that strength into the number of turns of wire in the coil. The important thing to know, in considering the saturation or otherwise of the field-magnets, is therefore the number of *ampère-turns* rather than the number of ampères.

The following set of experiments, made with a coil of exactly 500 turns on an iron core 10 centimetres long and 1 centimetre in diameter, illustrate the matter. The figures in the column marked N are the resulting numbers of the magnetic lines as calculated from the deflexions produced in a magnetometer :—

Ampères.	Ampère-turns.	Magnetic lines.
i	$S\,i$	N
0·00	0	39
0·22	110	368
0·39	195	576
0·98	490	1383
1·33	665	1778
3·65	1825	5242
4·6	2300	6326
9·2	4600	8363
9·4	4700	8474

The above values when plotted out give a saturation curve like Fig. 232.

It will be noticed that the iron core possessed a slight residual magnetism, for even with no current in the coil there was a small magnetic moment. Whenever there is residual magnetism, the curve will start from a point above the origin O, not from the origin itself.

The method illustrated in Fig. 231, where the electromagnet —a bar in form—acts on a magnetometer at a distance, is not applicable in cases where (as in all well-designed dynamos) the electromagnet forms part of a closed, or nearly closed magnetic circuit. In such cases a second method of measuring the magnetism must be resorted to, namely, the induction method. A small coil of wire in circuit with a suitable galvanometer is placed over the electromagnet at some suitable point; if this coil is suddenly removed from the magnetic circuit there will be a momentary induction current, propor-

FIG. 233.

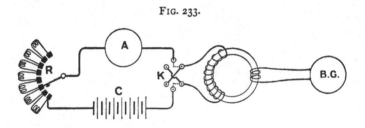

tional to the number of magnetic lines that traversed the coil; hence the first swing of the galvanometer needle is a measure of the magnetism. A third and somewhat similar method consists in leaving the measuring coil *in situ*, while the magnetism of the iron is reversed by reversing the current in the exciting coil. This is the method followed by Stoletow, Rowland, and Bosanquet. Fig. 233 illustrates this method of investigation. The iron to be examined is made in the form of a ring. Around this are wound two coils—one to magnetise the iron; the other, usually consisting of a few turns, is for measuring the magnetic induction. A variable resistance R is inserted in the battery-circuit along with an ampère-meter A, so that various strengths of magnetising current may be used. B G is a ballistic galvanometer, and K is the switch for suddenly reversing the exciting current.

In Fig. 234 are given curves of magnetisation for *annealed wrought iron*, obtained by the second method of experimenting by Dr. J. Hopkinson.[1] The black line shows the relation between the intensity of the magnetising force H and the induction B during the process of increasing the magnetising

FIG. 234.

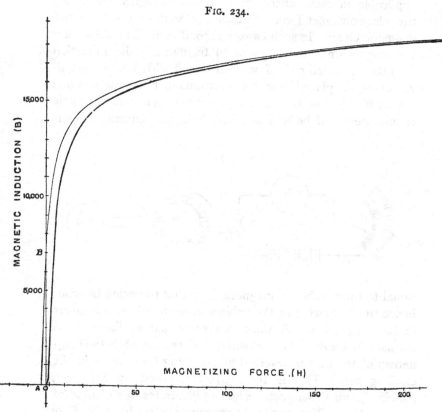

force from zero to about 220; and the light line shows the same relation during the process of decreasing the magnetising force to zero and then reversing it so as to remove the residual magnetic induction. Fig. 235 gives similar curves for a sample of *grey cast iron* such as is used by Messrs. Mather and Platt in building parts of dynamo machines.

[1] Hopkinson, in *Phil. Trans.*, 1885, pt. ii. p. 455.

Every sample of iron will show, on being tested, a similar set of facts which can be plotted down as a curve that is characteristic of the relation in question ; but the curves for cast iron and steel always lie lower than those for wrought iron. Moreover, it will usually be noticed that when a fresh piece of iron or steel is subjected to a gradually increasing magnetising force, the lowest part of the curve presents near its origin a small concavity (see Fig. 234), showing that there is a certain stage where under small magnetising forces the permeability

FIG. 235.

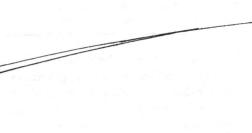

is greater than at the initial stage. This concavity is more pronounced in the case of hard iron and of steel than in the case of soft iron. But the curves differ in detail even in different specimens of the same sort of iron. In designing dynamos it is convenient to keep for reference a series of curves such as Figs. 234 and 235, made by careful observation on samples of the same iron as it is intended to use in construction.

As we shall require to refer again to these curves when dealing with calculations of the magnetic circuit, we may give in brief tabular form the values of B and H corresponding to the above curves :—

OBSERVATIONS OF HOPKINSON.

Annealed Wrought Iron.			Grey Cast Iron.		
B	H	μ	B	H	μ
5,000	2	3,000	4,000	5	800
9,000	4	2,250	5,000	10	500
10,000	5	2,000	6,000	21·5	279
11,000	6·5	1,692	7,000	42	133
12,000	8·5	1,412	8,000	80	100
13,000	12	1,083	9,000	127	71
14,000	17	823	10,000	188	53
15,000	28·5	526	11,000	292	37
16,000	52	308			
17,000	105	161			
18,000	200	90			
19,000	350	54			

The values of B here taken are the mean values between those found [1] with ascending and descending magnetisations. The values of μ are got by dividing B by H. In order the better to observe how the permeability μ diminishes as the induction is carried toward saturation, we may examine Fig. 236, in which the values of μ are plotted out vertically, and those of B horizontally. To the curves relating to Hopkinson's results is added a curve which refers to Bidwell's results on p. 293. It will be noted that in the case of the annealed wrought iron, between the values of 7000 and 16,000 the values of μ lie almost on a straight line, and might be calculated from the equation $3 \cdot 5 \times \mu = 17,000 - B$.

2. Formulæ used for Electromagnets.

Many suggestions have been made for equations connecting the strength of an electromagnet, or its magnetic moment, with the strength of the current which excites it. Most of these are purely empirical. Only those of Weber and of

---

[1] Hopkinson's paper unfortunately gives no tabular values, only the curves, from which these numbers have been measured off with as much precision as such a method allows.

Lamont are based upon abstract theories of magnetism, and Weber's does not represent the actual facts so satisfactorily as some of the other arbitrary formulæ. Space only admits of very brief enumeration of the various suggestions; but as it is most important to be able to calculate from the dimensions of an electromagnet the amount of magnetism that will be

Fig. 236.

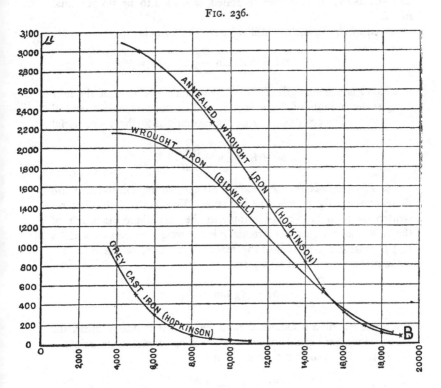

excited in it by a given current, it is worth while to note some of the attempts that have been made to find a formula to fit the facts. The student who does not interest himself in the historical treatment of the subject may pass on at once to the next section on the Theory of *the Magnetic Circuit.*

1. *Lenz and Jacobi's Formula.*—According to the experiments of these early investigators (1839), the magnetism of the electromagnet

is simply proportional to the strength of the current and to the number of turns of wire in the coil.   This may be written as:

$$m = k\,S\,i,$$

where i is the strength of the current, S the number of turns in the coil, k a constant depending on shape and quality of iron, and m the strength of the pole.   Joule (1839) showed this rule to be incorrect, and that as the iron became saturated, m ceased to be proportional to Si.

2. *Müller's Formula.*—Müller[1] gave the formula

$$S\,i = A\,d^{\frac{3}{2}} \tan \frac{m}{B\,d^2},$$

in which i is the strength of the current, d the diameter of the iron core, m the strength of the pole produced in the core, and A and B constants.   The equation may be transformed into the more useful form—

$$m = B\,d^2 \tan^{-1} \frac{S\,i}{A\,d^{\frac{3}{2}}}.$$

3. *Von Waltenhofen's Formula.*—Müller's formula introduced into the expression the diameter of the core.   Von Waltenhofen has re-written the formula to make it apply to the case where the weight of the core is given.   In this formula g is the weight in grammes and α and β constants.   M is now the magnetic moment of the core, and x the magnetic moment of the coil.

$$M = \beta\,g \tan^{-1} \frac{x}{\alpha\,g^{\frac{4}{3}}}.$$

From this equation it follows that, when x is indefinitely great, the magnetic moment of the saturated bar is proportional to its weight.

Similar formulæ of the general form

$$M = b \,.\, \tan^{-1} \frac{S\,i}{a}$$

have been used by Dub, Cazin, and Breguet, but they are only varieties of Müller's equation.   The defect of all these "arc-tangent" formulæ is, that they lend themselves so little conveniently to use in the equations of dynamos.   Moreover, they do not accord with the

---

[1] See Wiedemann's *Die Lehre von der Elektricität*, vol. iii. p. 414; also see Müller-Pouillet's *Lehrbuch der Physik*, vol. iii. p. 482 (ed. 1881); and S. P Thompson's *Electricity and Magnetism*, art. 330.

observations of Lenz, Jacobi, Scoresby, Sturgeon, and hundreds of others, that for small values of the magnetising current the magnetism evoked is very exactly proportional to the strength of the current. In the saturation curve, such as is drawn in Fig. 232, p. 296, the first part of the curve is for a long way nearly straight. But if this "arc-tangent" rule were true, this portion of the curve would be very decidedly convex.

4. *Kapp's Formula.*—Kapp has also proposed to employ a tangent formula in calculating the magnetic resistance of iron cores; it is that taking Z as the maximum number of lines which could be forced through the iron core with an unlimited magnetising power, and z as the actual number forced through, and writing σ for $z \div Z$, then the magnetic resistance at any stage can be found by taking the initial resistance and multiplying it by the function,

$$\frac{\tan \frac{\pi}{2} \sigma}{\frac{\pi}{2} \sigma}.$$

This gives us the formula of the electromagnet, using c as a constant:

$$M = c\,S\,i\,\frac{\frac{\pi}{2} \sigma}{\tan \frac{\pi}{2} \sigma},$$

which, since M is proportional to z, is capable of being reduced to Müller's formula.

5. *Weber's Formula.*—Weber's formula [1] is framed on the supposition that in every unit of volume of the iron there exist n molecules, each of magnetic moment m; so that the magnetic moment of unit volume (which numerically equals the intensity of magnetism), if all were set with axis parallel, would be $I = m\,n$. It also supposes that every molecule is set in a certain arbitrary direction, and tends to remain there, under the coercion of a molecular force, the value of which is called D. If X, the magnetising force, act on a molecule at an angle u, the effective magnetic moment of that molecule will become

$$m' = m\left(\frac{X + D \cos u}{\sqrt{D^2 + X^2 + 2\,D\,X \cos u}} - \cos u\right).$$

[1] *Elektrodynamische Maasbestimmungen*, p. 572, where a saturation curve is also given; see also Maxwell, *Electricity and Magnetism* (2nd edition), vol. ii. p. 78, for a discussion of Weber's theory.

X

This formula was duly integrated and developed by Weber. Maxwell has reduced it to simpler forms for particular cases, according to the relative magnitudes of X and D. In the equations, I stands for magnetic moment per unit volume.

When X is *less* than D,

$$I = m\,n\left(1 - \frac{1}{3}\right)\frac{X}{D};$$

When X *equals* D,

$$I = m\,n\left(1 - \frac{1}{3}\right);$$

When X is *greater* than D,

$$I = m\,n\left(1 - \frac{1}{3}\frac{D^2}{X^2}\right);$$

When X is infinitely great,

$$I = m\,n.$$

Maxwell gives the results of these formulæ in curves, which also fail to agree with the observed facts.

6. *Frölich's Formula.*—Frölich [1] used an interpolation formula to express the effective magnetism M of a dynamo-machine in terms of the current *i*, and of a set of arbitrary constants, *a*, *b*, and *c*. His most complete form was—

$$M = \frac{i}{a + b\,i + c\,i^2}.$$

Finding the term in i^2 unnecessary, he adopted the simpler form

$$M = \frac{i}{a + b\,i}.$$

Here *b* equals the reciprocal of the maximum value of M; as will be seen by assigning to *i* any very large value.

A very similar formula had been used twenty-five years previously by Robinson [2] to express the lifting power of electromagnets. Similar formulæ have been used by Lamont, Oberbeck, [3] Fromme, [4] Deprez, Clausius, Ayrton and Perry, and by Rücker. The advantage of the formula is its adaptability to use in other equations. A modification of it was used by the author in the former editions of this

[1] Frölich, *Elektrotechnische Zeitschrift*, pp. 90, 139, 170, 1881; and p. 73, 1882.

[2] Robinson, *Trans. Roy. Irish Academy*, vol. xxii. p. 1, 1835.

[3] Oberbeck, *Pogg. Ann.*, cxxxv. pp. 74–98, 1868.

[4] Fromme, *Pogg. Ann.*, clv. p. 305, 1875.

book for expressing the intensity of the field due to the electromagnet, viz. :—

$$H = G \frac{\kappa S i}{1 + \sigma S i};$$

where G is a coefficient depending only on the geometrical form of
the magnet and the position of the place where H is measured, κ the
coefficient of initial magnetic permeability of the core, σ the saturation coefficient, and S the number of turns in the coil. In fact, the
author found it necessary to introduce ampère-turns instead of
ampères into the equation, because in compound-wound dynamos
the magnetising power is partly due to the main current flowing a
few times round the core, and partly to a weak shunt current flowing
many times round the core. The author also investigated the meaning of the "saturation-coefficient" σ, and found that it was the
reciprocal of that number of ampère-turns which would reduce the
magnetic permeability to half its initial value, or in other words, it
was the reciprocal of that number of ampère-turns which would bring
up the magnetism to half its highest possible value, *i. e.* to half-
saturation. This number of ampère-turns he named the *diacritical*
number; and the current producing half-saturation he called the
diacritical current. Dr. Frölich has independently made use of this
conception, and has applied it to simplify the formula of the electromagnet. The argument is his; the notation here used is, however,
the author's. Writing $(S i)'$ for the diacritical number of ampère-
turns, we have $(S i)' = \frac{1}{\sigma}$. Then suppose we use the formula (giving
the appropriate value to G), to express not the average strength of the
field, but the total number N of magnetic lines evoked in the magnet,
we may write

$$N = \frac{G \kappa S i}{1 + \sigma S i} = \frac{G \kappa}{\sigma} \cdot \frac{S i}{\frac{1}{\sigma} + S i}.$$

and putting for the maximum value of N at saturation,

$$\overline{N} = \frac{G \kappa}{\sigma},$$

we have

$$N = \overline{N} \frac{S i}{S i + (S i)'}.$$

But, further, if S is known, then we may cancel it out and write
simply

$$N = N \frac{i}{i + i'};$$

where i' is the diacritical or half-saturating current. This very simple equation is approximately true for every electromagnet excited by a single current. Two observations made on any electromagnet will suffice to determine the two constants $\overline{N}$ and i'. Further, if r be the resistance of the magnetising coil, since $i\,r = e$ (the potential required to send the current i through the coil), we may obviously write the equation

$$N = \overline{N}\,\frac{e}{e + e'};$$

where e' is the diacritical difference of potential, namely that difference of potential which, applied to the coil having resistance r and convolutions S, will half-saturate the core. This last form is most convenient when calculations are to be made about shunt-dynamos, whilst the one preceding it is the more useful for series dynamos.

7. *Sohncke's Formula.*—Professor Sohncke[1] has lately proposed an exponential formula which expresses the facts very accurately: though it is not so convenient as the preceding. It is

$$M = \frac{1}{a}i \cdot \epsilon^{-bi}.$$

8. *Lamont's Formula.*—Lamont deduces[2] a formula of great value from certain theoretical considerations which are of special significance. He assumes that for every bar there is a certain maximum of magnetisation to which it would attain only under the influence of an infinitely great magnetising force; and that the permeability of the bar is at every stage of the magnetisation proportional to the difference between the actual magnetisation and the possible maximum magnetisation. This is, in other words, as if in every bar there were room for only a certain limited number of magnetic lines of force, and that when any lesser number have been induced in it, the susceptibility of the bar to the introduction of additional lines is proportional to the room yet left for them in the bar. A very similar hypothesis has lately been put forward by Mr. R. H. M. Bosanquet.[3] Lamont's theory may be expressed as follows. Let the magnetism present at any stage be called m, and let the maximum magnetism be called M. Then the amount of magnetism which the bar can still take up is $M - m$; and the permeability $\frac{d\,m}{d\,x}$ is pro-

[1] *Elektrotechnische Zeitschrift*, April 1883, p. 160.
[2] Lamont, *Handbuch des Magnetismus*, 1867, p. 41.
[3] *Philosophical Magazine*, S. v. vol. xix. p. 85, Feb. 1885. See also Bosanquet, in *Electrician*, Feb. 14, 1885.

portional to M − *m*. Here *x* may be understood as the number of ampère-turns (or S *i*), in the exciting current; and we may write

$$\frac{d\,m}{d\,x} = k\,(\mathrm{M} - m),$$

where *k* is a constant depending on the units employed. This equation we may arrange as

$$\frac{d\,m}{\mathrm{M} - m} = k\,d\,x,$$

and, integrating

$$\log_\epsilon (\mathrm{M} - m) = -\,k\,x + \mathrm{C};$$

or,

$$\mathrm{M} - m = \mathrm{A}\,\epsilon^{-k x};$$

where A is a constant of integration. But when *x* = 0, *m* = 0 also; hence A = M, giving us the formula

$$m = \mathrm{M}\,(1 - \epsilon^{-k x}).$$

This formula is probably much more nearly true for soft-iron electromagnets than any of the preceding.

Lamont further points out that this expression may be expanded in terms containing ascending powers of *k x* as

$$m = \mathrm{M}\,k\,x\left(1 - \frac{k\,x}{1 \cdot 2} + \frac{k^2\,x^2}{1 \cdot 2 \cdot 3} - \&c. \dots\right).$$

If the magnetising force *x* is small, we may neglect all terms after the first; which virtually reduces the formula to that of Lenz and Jacobi, the magnetism being simply proportional to the ampère-turns. Lamont also points out that a first approximation to the formula above is given by an expression of the form

$$m = \frac{\mathrm{M}\,k\,x}{1 + k\,x},$$

which is identical with the formulæ of Fromme, Oberbeck, Frölich, and Clausius. This formula, as Bosanquet observes, can be derived by taking *m/x*, instead of *d m/d x*, as proportional to M − *m*.

None of these formulæ account for the phenomenon alluded to on p. 301 as being observed in the magnetisation of many pieces of iron and steel (and especially in closed rings of iron and steel), namely, that there is an apparent increase in the permeability after a certain early stage in the magnetisation has been reached. Lenz first noticed this in 1854. Wiedemann, Dub, Stoletow, Rowland, Chwolson, Bosanquet, and Siemens have all investigated the matter; Rowland, Bosanquet, and Ewing in particular having given careful numerical determinations of the variations of permeability under varying degrees of ascending magnetisation. The researches of Chwolson and of Siemens seem, however, to show that the apparent increase of the permeability is due to the want of homogeneity in the iron, and to the presence of a certain proportion of particles having the properties of hard steel and requiring a certain minimum of magnetising force to be applied before they become sensibly magnetised, there being an apparent more rapid growth of the magnetism when this stage is reached. According to this view the permeability due to temporary magnetism begins by being a maximum, and diminishes as the magnetising force is increased; whilst the permeability due to permanent magnetism is zero at first and until a certain stage, when it rises rapidly to a maximum of its own, and thereafter dies gradually away. That which Stoletow, Rowland, and Bosanquet have measured with so much care, is the sum of these two effects. Siemens makes the valuable remark that the harder the piece of iron or steel, the later is the stage at which this apparent maximum of magnetic permeability is observed.

The result of this superposition of effects in the dynamo machine is that when the " characteristic " is taken *with ascending strengths of current*, there may be observed—and this is not marked, save in dynamos in which the iron constitutes very nearly a closed circuit on itself—a concavity in the first part of the characteristic, which, as explained on p. 296, is usually taken as an oblique straight line. But if the characteristic is taken with descending strengths of current, no such concavity is observed, the magnetism of the field-magnets, and also the electromotive-force, having values considerably higher, for the same value of exciting current, than in the ascending curve. The presence of permanent magnetism in the core is therefore detrimental to the steadiness of the field. Even with the softest Swedish iron, differences may be observed in the electromotive-force, with the same speed and same exciting current, before and after the exciting current has been increased to a high degree. For this reason the approximate formula known as Frölich's is probably quite as near to the truth as the more perfect formula of Lamont. Neither of them take into account the presence of the apparent increase in permeability or the retardation of the apparent maximum in cores having greater coercitivity. For further information on the differtnces between the ascending and descending curves of magnetism, he reader is referred to the researches of Warburg, Ewing, and Hopkinson.

Most of the formulæ set forth above are not entirely satisfactory, because they involve the determination of constants upon the actual electromagnet after it has been built. For example, the valuable approximate formula of Frölich on p. 307 (bottom) cannot be directly applied until one has ascertained the values of $\overline{N}$ and i'. Some have thought this a fatal objection to the use of these simple formulæ; but this objection is no longer valid, since it is shown on p. 409 how these two constants can be predetermined. A formula is required by the aid of which an engineer can calculate from the known properties of iron and from the dimensions of the various parts the actual number of magnetic lines that it will furnish when excited by a known current; or by which, *vice versâ*, when the number of magnetic lines that are required is known, he may be able to calculate the dimensions of the various parts that will give the desired result. The consideration of the principle of *the magnetic circuit* has been found to furnish the formulæ that were desired.

3. The Magnetic Circuit.

The most important recent step in the treatment of dynamo problems has been the use of the conception of the magnetic circuit. The principle is briefly as follows : that *a minimum expenditure of magnetising power will give a maximum of effective magnetism when the iron parts of the dynamo, the cores of the field-magnets, the yoke, the pole-pieces and the core of the armature are so disposed that they constitute together a magnetic circuit having a minimum magnetic resistance.*

It is no novelty to regard the magnetism of a magnet as something that traverses or circulates around a definite path, flowing more freely through such substances as iron, than through other relatively non-magnetic materials. Analogies between the flow of electricity in an electrically-conducting circuit, and the passage of magnetic lines of force through circuits possessing magnetic conductivity are to be found abundantly in the literature of the science. So far back as 1821, Cumming[1] experimented on magnetic conductivity. The idea of a magnetic circuit was more or less familar to

[1] *Camb. Phil. Trans.*, April 2, 1821.

Ritchie,[1] Sturgeon,[2] Dove,[3] Dub,[4] and De la Rive,[5] the last-named of whom explicitly uses the phrase "a closed magnetic circuit." Joule[6] found the maximum power of an electromagnet to be proportional to "the least sectional area of the entire magnetic circuit," and he considered the resistance to induction as proportional to the length of the magnetic circuit. Faraday[7] considered that he had *proved* that each magnetic line of force constitutes a closed curve; that the path of these closed curves depended on the magnetic conductivity of the masses disposed in proximity; that the lines of magnetic force were strictly analogous to the lines of electric flow in an electric circuit. He spoke of a magnet surrounded by air being like unto a voltaic battery immersed in water or other electrolyte. He even saw the existence of a power, analogous to that of electromotive-force in electric circuits, though the name *magneto-motive force* is of more recent origin. The same idea is more or less implicitly recognised in the latter half of the magnetic papers in Sir William Thomson's collected volume on Electrostatics and Magnetism. The notion of magnetic conductivity is to be found in Maxwell's great treatise (Vol. II. p. 51), but is only briefly mentioned. Rowland[8] in 1873 expressly adopted the reasoning and language of Faraday's method in the working out of some new results on magnetic permeability, and pointed out that the flow of magnetic lines of force through a bar could be subjected to exact calculation; the elementary law, he says, "is similar to the law of Ohm." Writing R for the "resistance to lines of force," M for "mag-

[1] *Phil. Mag.*, series iii., vol. iii., p. 122.

[2] *Ann. of Electr.*, xii., p. 217.

[3] *Pogg. Ann.*, xxix., p. 462, 1833.   See also *Pogg. Ann.*, xliii., p. 517, 1838.

[4] *Dub. Elektromagnetismus* (ed. 1861), p. 401 ; and *Pogg. Ann.*, xc., p. 440, 1853.

[5] De la Rive.   *Treatise on Electricity* (Walker's translation), vol. i., p. 292.

[6] *Ann. of Electr.*, iv., 59, 1839 ; v., 195, 1841 ; and *Scientific Papers*, pp. 8, 34, 35, 36.

[7] *Experimental Researches*, vol. iii., art. 3117, 3228, 3230, 3260, 3271, 3276. 3294, and 3361.

[8] *Phil. Mag.*, series iv., vol. xlvi., August 1873.   'On Magnetic Permeability and the Maximum of Magnetism of Iron, Steel, and Nickel.'

netising force of helix," and Q for number of "lines of force in bar at any point," he wrote, for a particular case (a ring-magnet, having therefore a closed magnetic circuit), the equation,

$$Q = \frac{M}{R};$$

an equation for magnetic circuits which every electrician will recognise as precisely like Ohm's law. He applied the calculations to determine the permeability of certain specimens of iron, steel, and nickel. In 1882 [1] and again in 1883, Mr. R. H. M. Bosanquet [2] brought out at greater length a similar argument, employing the extremely apt term " Magneto-motive Force," to connote the force tending to drive the magnetic lines of induction through the " magnetic resistance" of the circuit. In these papers the calculations are reduced to a system, and deal not only with the specific properties of iron, but with problems arising out of the shape of the iron. Bosanquet shows how to calculate the several resistances of the separate parts of the circuit, and then add them together to obtain the total resistance of the magnetic circuit.

Prior to this, however, the principle of the magnetic circuit had been seized upon by Lord Elphinstone and Mr. Vincent, who proposed to apply it in the construction of dynamo-electric machines. On two occasions [3] they communicated to the Royal Society the results of experiments to show that the same exciting current would evoke a larger amount of magnetism in a given iron structure, if that iron structure formed a closed magnetic circuit than if it were otherwise disposed. They embodied their ideas in a form of dynamo, [4] which however, on account of mechanical difficulties, did not establish itself as a permanent type of machine. The work

[1] *Proc. Roy. Soc.*, xxxiv., p. 445, December 1882.

[2] *Phil. Mag.*, series v., vol. xv., p. 205, March 1883. ' On Magneto-motive Force.' Also *ib.*, vol. xix., February 1885, and *Proc. Roy. Soc.*, No. 223, 1883. See also *Electrician*, xiv., p. 291, February 14th 1885.

[3] *Proc. Roy. Soc.*, xxix., p. 292, 1879, and xxx., p. 287, 1880. See *Electrical Review*, viii., p. 134, 1880.

[4] Specification of Patents 332 of 1879 and 2893 of 1880. See also p. 211 of this work.

of Lord Elphinstone and Mr. Vincent was not however lost, for the principle thus introduced by them into the construction of dynamos took root and bore fruit. In June 1884, the author of this book, in the preface to its first edition, writing on the imperfection of our knowledge of the laws of magnetic induction, wrote : *We want some new philosopher to do for the magnetic circuit what Dr. Ohm did for the voltaic circuit fifty years ago.* In the same year at the conference of electricians at Philadelphia, Professor Rowland read a paper[1] in which he proposed a formula for the number of lines of force in a field-magnet. His notation is different from that used in this book. He writes B for the total number of lines, N the number of magnetising coils, C the current. B will be proportional to N and C, and inversely proportional to the resistance to the lines of force in the circuit. The resistance to the lines of force is proportional to L, the length of the iron of the system, divided by S, the cross-section of the magnet (supposing it uniform), and by μ, the magnetic permeability of the iron (or conductivity of the iron for the lines of force). To this we must add something for the resistance of the air-gap. Let l be twice the width of the air-gap between armature and pole-piece, and A the area across which the lines of force flow : then we have to add another quantity p which depends upon the resistance to those lines of force which escape in all directions, and represents the loss due to that cause. Then we have as the equation,

$$B = \frac{N\,C}{\dfrac{L}{S\mu} + \dfrac{l}{A + p}}$$

An important development of the foregoing ideas was made in 1885 by Mr. Gisbert Kapp, in a series of papers contributed to the *Electrician*[2] between February 1885 and April 1886, in which the question of the design of dynamos

---

[1] Report of the Electrical Conference at Philadelphia in 1884, p. 77. See also *Electrical Review*, xv., p. 368, 1884, and *Electrician*, xiii., p. 536, 1884.

[2] *Electrician*, xiv., pp. 259, 307, 347, 390, 431, 511 ; xv., p. 23, 190, 250 ; xvi., pp. 7, 406.

was discussed from this point of view, with many illustrations touching the forms of field-magnets, and some calculations and formulæ of great interest. While these papers were in progress, the author was preparing for the press the second edition of this work, and drew up the sketches of the magnetic circuits of a large number of types of dynamo machines, chiefly of dynamos of Class I., that are given in a collected form in Fig. 240. Three months later a sheet of figures, very closely resembling these, was given by Mr. Kapp, who had prepared them in entire independence, in an important paper[1] which he communicated on Nov. 24th, 1885, to the Institution of Civil Engineers. In this paper Kapp adopted a new unit of lines of force, 6000 times greater than the C.G.S. unit. The reason of this being that he preferred to have to divide by 10^6 instead of 10^8, and to speak of the number of revolutions per minute instead of the number of revolutions per second. Using Z for the number of lines of force (in Kapp units, where $Z = N$ (of our notation) C.G.S. units $\div$ 6000), and n for revolutions per minute, Kapp puts,

$$E \text{ (volts in armature)} = Z \cdot Nt \cdot n \cdot 10^6 ;$$

the symbol Nt standing for number of turns of wire around the armature as counted all round the periphery (the same for which symbol C is used in this book hereafter). Kapp then further wrote for the magnetic circuit the formula

$$Z = \frac{P}{R_a + R_A + R_F},$$

where P is the exciting power in ampère-turns, R_a the magnetic resistance of the air-space, R_A that of the armature core, R_F that of the field-magnets. Now let δ represent distance across the span between armature core and polar surface, b breadth of armature as measured parallel to axis, λ the length of arc embraced by polar surface, so that λb is the polar area out of which magnetic lines issue, a radial depth of armature core, so that ab is area of section of

[1] *Proc. Inst. Civil Engineers*, lxxxiii. (1885-6), pt. i, "Modern Continuous-Current Dynamo-Electric Machines and their Engines."

armature core (space actually occupied by iron only being reckoned), AB area of field-magnet core, *l* length of magnetic circuit within armature, L ditto in field-magnet; all the dimensions being measured in inches or square inches respectively. Then, in case of an ordinary single magnetic circuit, Kapp writes

$$R_a = 1440 \frac{2\delta}{\lambda b'}$$

$$R_\lambda = \frac{l}{ab'}$$

$$R_\tau = 2 \frac{L}{AB}.$$

The coefficient 1440 arises from the peculiar units, inch for length, minute for time, and 6000 C.G.S. for unit of lines of force; and this number may be taken as representing in this mixed system of units the specific magnetic resistance of air. The coefficient 2 relates to the specific resistance of wrought iron or charcoal iron; the figure 3 must be substituted in case of cast-iron magnets.[1] Moreover, on account of magnetic leakage in the formula for Z, 0·8 P has to be used instead of P. In the debate on this paper, Prof. G. Forbes gave a further modification of the formula. Still more recently Mr. Kapp[2] has returned to the subject and has given an empirical rule (No. 4, on p. 305) for calculating the variations of the magnetic resistance of the iron at the various stages of magnetisation.

In May, 1886, Drs. John and Edward Hopkinson communicated to the Royal Society[3] a very complete and

[1] These formulæ are explained in *Electrician*, xv., p. 250, August 14th, 1885.
[2] *Journal Soc. Telegraph Engineers and Electricians*, xv., pp. 524–529, November 11th, 1886. "On the Predetermination of the Characteristics of Dynamos"; a very valuable paper marred by mixed units.
[3] *Phil. Trans.*, 1886, part i., p. 331. "Dynamo-Electric Machinery." This important paper is reprinted *in extenso*, but without the plates, in *Electrical Review*, xviii., p. 471, November 12th, 1886, and subsequent numbers. See also *Electrician*, xviii., pp. 39, 63, 86, and 175 in issues of November 19th, November 26th, December 3rd, and December 31st, 1886, where the figures of the plates are printed in the text.

elegant investigation of the problem of magnetic circuits for the purpose of finding a suitable and accurate expression for the electromotive force of the dynamo in terms of the magnetising current ; the most important part of their investigation being directed to constructing the curve characteristic of magnetisation of the dynamo from the ordinary laws of magnetism and from the known properties of iron. The process in its first approximation resembles the process adopted by Kapp; but it stands alone in an important respect, namely, that its authors plot a separate characteristic curve for the relation between the magnetising force and the induction for each separate part of the magnetic circuit, and then sum up the separate curves so as to obtain a final resultant characteristic curve. This is done first on the assumption that there is no magnetic leakage, and with other assumptions for the sake of simplifications. After a first approximation has thus been obtained, the theoretical result is compared with the actual result of experiment, thereby affording a means of estimating the corrections that must be introduced, the magnitude of the correcting factors being thus known, and the theory perfected to a second degree of approximation.

In describing the Hopkinsons' formulæ, we shall employ not their notation, but the same that is adopted in the rest of this book (see p. 335).

Let N be the whole magnetic flux through the armature, that is to say, the total number of lines of magnetic force that pass into the armature core on one side and out of it at the other. Assume (to simplify matters) that there is no waste or leakage of magnetic lines. Then remember that the magneto-motive force (or line-integral of the magnetising forces acting round the circuit) is [1] equal to $4 \pi S i \div 10$, where S is the number of turns of wire and i the ampères. Now the magnetic resistance of any magnetic conductor is proportional to its length and inversely proportional to its

[1] See Maxwell's *Electricity and Magnetism*, vol. ii., Art. 499, or S. P. Thompson's *Elementary Lessons on Electricity and Magnetism* (edition of 1887), pp. 291-296. See also p. 295 of this book.

sectional area, and also inversely proportional to its permeability. Suppose then, as shown in Fig. 237, that the
magnetic circuit of the dynamo is made up simply of three
parts : an iron armature-core, the two air-gaps, and the iron
field-magnets, then the iron armature-core, in which the
average length of path for the magnetic lines is l_1, the average sectional area A_1, and of permeability μ_1, will have a

magnetic resistance equal to $\dfrac{l_1}{\mu_1 \, A_1}$. Similarly for the two

FIG. 237.

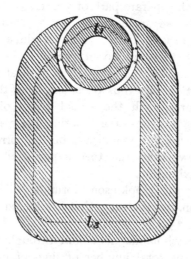

gaps : if the length of path across each, from iron to iron, be
called l_2, and the area of either polar surface be A_2, and the
permeability of the matter that occupies the gaps be called
μ_2, then the magnetic resistance these two offer to the

magnetic flux will be $2 \dfrac{l_2}{\mu_2 \, A_2}$; or, remembering that for air,

copper, and all ordinary non-magnetic substances $\mu = 1$, this

may be written simply $2 \dfrac{l_2}{A_2}$. Similarly for the iron field-

magnet ; writing l_3 for length of path through the iron from
pole to pole, A_3 for sectional area (supposed to be equal

throughout), and μ_3 for permeability, the magnetic resistance will be $\dfrac{l_3}{\mu_3 A_3}$. Adding these three resistances together, we get as the total resistance of the magnetic circuit the value

$$\frac{l_1}{\mu_1 A_1} + 2 \cdot \frac{l_2}{A_2} + \frac{l_3}{\mu_3 A_3}.$$

Whence we may immediately write down as the appropriate formula,

total number of magnetic lines $= \dfrac{\text{magneto-motive force}}{\text{magnetic resistance}}$; or

$$N = \frac{4 \pi S i}{10 \left\{ \dfrac{l_1}{\mu_1 A_1} + 2 \dfrac{l_2}{A_2} + \dfrac{l_3}{\mu_3 A_3} \right\}}.$$

As a matter of fact, the Hopkinsons stated their formula a little more generally. In the first place, instead of adding up the separate magnetic resistances, they calculated the magnetic forces needed in the separate parts and then added these together. In the second place, instead of assuming the existence of μ for the different parts, they contented themselves with saying that the magnetic induction in each part must be some function of the magnetic force acting in that part. Now, if there be N magnetic lines passing through sectional area A square centimetres, the number of lines per square centimetre, which we call "the induction," will be $\dfrac{N}{A}$, or, as it is often written (see p. 292), B. Accordingly, we may write for the magnetic force acting in the armature part of the magnetic circuit $f\left(\dfrac{N}{A_1}\right) \times l_1$, which "function" may be examined and plotted out as a curve. In fact, the curves of magnetisation, such as are given on p. 300, are nothing else than curves which show the relation between the magnetising forces and the amount of magnetism they induce. There will be a similar expression $f\left(\dfrac{N}{A_3}\right) \times l_3$ for the magnetic force that acts in the field-magnet part; whilst for the gaps the magnetic force is simply $\dfrac{N}{A_2} \times 2l_2$, for the function for air.$= 1$. Now, if we know the separate amounts of magnetising force required to produce these magnetic inductions in the separate parts, it is clear that the whole or integral magnetising force in question will be got by adding them together, giving

$$l_1 f\left(\frac{N}{A_1}\right) + 2\, l_2\, \frac{N}{A_2} + l_3 f\left(\frac{N}{A_3}\right) = 4\,\pi\, \mathrm{S}\, i \div 10.$$

There are three advantages in this mode of stating the matter :—(1) The use of the function, of which the value is to be found by reference to a curve or tabulated set of observations (such as those given in Figs. 234, 235, and 236, or statistically on p. 302), instead of merely using the symbol μ, makes the expression more general; (2) the separate terms being differently affected by leakage of the magnetic lines, it is easy to apply a correction to any one of them separately ; (3) this form of the formula is convenient, in the case of a given iron carcase, to calculate the number of ampère-turns required to excite the working quantity of magnetism. For we have now three terms, the first telling us how many ampère-turns of excitation are required to drive N magnetic lines through the resistance of the armature core, the second telling how many are required to drive N lines through the gap resistance, and the third telling how many are required to drive N lines through the iron cores of the field-magnets. In a well-designed dynamo the second term is numerically the most important of the three, and it is not complicated by any question of saturation ; for in air the magnetic induction is always proportional to the inducing magnetic force. To represent this graphically upon a diagram (Fig. 238) plot out as ordinates the numbers of magnetic lines that are forced round the circuit, and plot horizontally as abscissæ the corresponding values of the requisite magneto-motive force $4\pi\,\mathrm{S}\,i \div 10$. For the second term which relates to the gaps, the relation will be represented simply by a sloping line such as O B. For example, the number of magnetic lines N that are to be forced across the gaps between the armature and field-magnet being represented to scale by the length of the line O N, the corresponding value of $2\,l_2\dfrac{N}{A_2}$, which is the corresponding part of the magneto-motive force, when plotted out on the horizontal scale as $O\,x_2$, gives b as a point through which the line passes. Similarly we may construct a curve to represent the first term by the curve A ; for if we know from experiments made on iron of the same kind as is used in the armature-core, the values of the function f (or, what comes to the same thing,

FIG. 238.

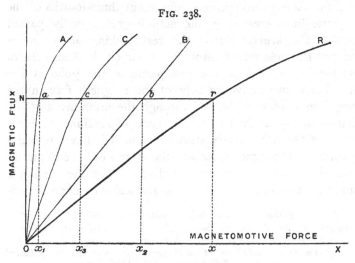

MAGNETIC FLUX

N a c b r

MAGNETOMOTIVE   FORCE

O x_1 x_3 x_2 x X

the various corresponding values of μ_1) we can calculate the value of the quantity $l_1 f\left(\dfrac{N}{A_1}\right)$ and set it off as $O\,x_1$, giving a as a point on this curve. Similarly the curve C is calculated point by point for the third term, by using knowledge derived from experiments made on iron of the same kind as is used in the field-magnets. Now, from these three curves A, B and C, which represent the various values of the three terms on the left-hand side of the equation, we can at once get the resultant curve that is characteristic of the whole magnetic circuit of the dynamo. For if we draw a level line across at the point N, we know that the three overlapping' lengths N a, N b, N c

Y

(which are respectively equal to the three lengths $O x_1$, $O x_2$, and $O x_3$,) represent the three corresponding separate parts of the magneto-motive force. Adding these three lengths together, we get $O x$ or $N r$ as the total magneto-motive force: and this gives us r as a point on the resultant curve, which we may complete by finding other points in a similar way and sweeping the curved line $O R$ through them. Having in this way built up a curve characteristic of the magnetisation the Hopkinsons then proceeded to correct it by considering the leakage. They found that in the dynamo experimented upon (an Edison-Hopkinson) only about three-fourths of the magnetic lines created in the field-magnet actually passed through the armature-core, the rest leaking across either between the pole-pieces through the air or the bed-plate, or else turning back from the pole-pieces to the yoke at the top. Experiment gave the ratio of the magnetic flux in the upright iron cores to the flux through the armature as 1·32. That is to say, in this particular type of machine, to force 100 lines through the armature-core one will have to excite 132 in the field-magnet cores, and therefore will have to put on more magneto-motive force accordingly. Let the symbol v stand [1] for this ratio. Then in the particular dynamo experi-

[1] Professor Forbes has shown how to calculate this ratio from theoretical considerations. See *Journ. Soc. Teleg. Engineers*, xv., No. 64, p. 555, November 25th, 1886. He states three propositions or lemmas:

(i.) The magnetic conductivity of the air space between two parallel planes of nearly equal area is the mean of the two areas divided by the distance between the planes (the measurements being made in centimetres). This is useful for calculating the waste field between the parallel limbs and between pole-pieces and bed-plate.

(ii.) If the induction in an air space between two equal rectangular areas similarly situated in the same plane be assumed to lie along semi-circles, with the medial straight line as the line of centres, the magnetic conductivity of that space is equal to $\frac{a}{\pi}$ nat. log $\frac{r_2}{r_1}$; where r_1 and r_2 are the distances from the medial line to the nearest or furthest edges respectively, and a is the width of each rectangle. This is used for calculating side leakage between limbs and across pole-pieces.

(iii.) If the induction in an air space between two equal rectangular areas, which are similarly situated in the same plane, be assumed to be along quadrants with neighbouring edges of the planes as centres, the quadrants being connected by straight lines, the magnetic conductivity of that space is equal to $\frac{a}{\pi}$ nat. log $\frac{\pi r + b}{b}$; where r is width and a the depth of each rectangle, and b their

mented on there was a yoke at the top through which the length of (curved) path was l_4, and which had cross-section A_4. There were also solid pole-pieces, for which the corresponding quantities were called l_5 and A_5. Inserting these additional matters into the equation, it now becomes

$$l_1 f\left(\frac{N}{A_1}\right) + 2\, l_2 \frac{N}{A_2} + l_3 f\left(\frac{\nu N}{A_3}\right) + l_4 f\left(\frac{\nu N}{A_4}\right) + 2\, l_5 f\left(\frac{N}{A_5}\right)$$
$$= 4\,\pi\,\mathrm{S}\,i \div 10.$$

There are now five terms to be calculated, giving five curves. Moreover, as is well known, with descending magnetising forces the curve of magnetisation is different from the curve with ascending magnetising forces. Fig. 239, which is taken from the Hopkinsons' paper, shows how they plotted out both for ascending and descending magnetisations the five curves. Of these A relates to the armature, B to the two interstitial gaps, C to the field-magnet cores, G to the yoke, and H to the two pole-pieces. The resultant ascending and descending curves are also shown. They agree remarkably well with the crosses and points which were plotted out from actual experiment. The dotted curves and the crosses surrounded with circles relate to descending magnetisation. The student should not fail to consult the original paper, which though brief is full of important matter, and in which applications of this method to two dynamos of different types are given.

The magnetisation curves, Figs. 234 and 235, pages 300 and 301, being the same used by the Hopkinsons, may be used by the student. The ordinates being values of $N \div A$, the abscissæ give the corresponding values of $f\,(N \div A)$ or of

$$\frac{1}{\mu}\,(N \div A).$$

Example.—In one particular machine examined by the Hopkinsons, the same Edison-Hopkinson dynamo described on p. 203,

---

distance apart. This is used for calculating leakage from pole-pieces back to the yolk.

According to Ravenshaw, who has used these lemmas in designing the recent machines for Messrs. W. T. Goolden and Co. (Fig. 157), they enable the designer to predetermine the performance of the machine within two per cent. They enable the perturbing effects of iron masses in bearings, fly-wheels, &c. to be calculated. Forbes by them reckons the leakage coefficient of the Edison-Hopkinson machine to be 1·40 instead of 1·32.

Figs. 144, 146, the following were the dimensions :—l_1 13 centimetres, $_2$ 1·5 centimetre, l_3 91·4 centimetres, l_4 49 centimetres, l_5 11 centimetres ; A_1 810 square centimetres ; A_2 on core 1410, on polar face 1513 square centimetres, or, allowing for spreading, say 1600 square

FIG. 239.

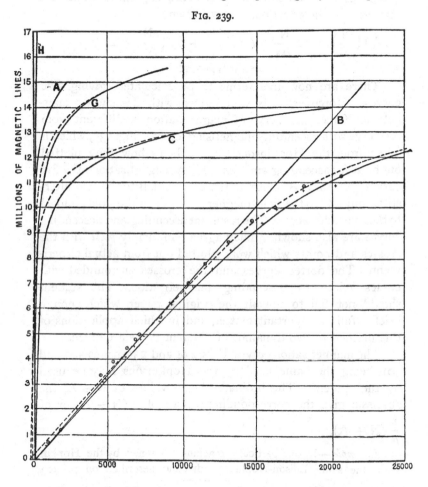

centimetres ; A_3 930 square centimetres ; A_4 1120 square centimetres ; A_5 1230 square centimetres ; $v = 1\cdot32$; N, when the machine was at full work, a little over 10,850,000 ; S = 3260 ; $r_s = 16\cdot93$.

Some calculations made by a process of the same nature as Hopkinson's, but here applied for the purpose of predetermining the two constants contained in the useful approximate equation of the series dynamo $E = E\dfrac{i}{i + i'}$ are given on p. 408.

4. Forms of Field-Magnets.

In Chapter III., on the organs of dynamos, much was said about the general principles which should be followed in the design of field-magnets. The reader will doubtless have noted in the descriptive matter which followed, how these principles were observed, and how also many of them were violated, in the actual forms of machines which are to be found in commerce. A brief review of the leading forms will present the matter in a clearer light. With the principle of the magnetic circuit to guide him, the reader will have little difficulty in judging of the relative value of the various designs; for he will remember that the magnetic circuit of highest conductivity will have the most compact form, greatest cross section, softest iron, and fewest joints. It has, moreover, been pointed out by Rowland that theoretically it is better that there should be one such magnetic circuit than that there should be two: though for practical structural reasons the author thinks the double circuit preferable in many cases. These points should be borne in mind in considering the forms depicted in the accompanying figure, and which relate almost exclusively to dynamos of Class I., that is to say, those in which the armature rotates in a simple field. No. 1 of these illustrations shows the form adopted by Wilde for use with the shuttle-wound armature of Siemens. Two slabs of iron are connected at the top by a yoke, and are bolted below to two massive pole-pieces. There are four joints in the magnetic-circuit, in addition to the armature-gaps, and the yoke is insufficient. No. 2 shows the form adopted in the latest Edison dynamos (American pattern). The upright cores are stout cylinders. The yoke is of immense thickness: the pole-pieces are massive, but their useless corners are cut away. There are as many joints as in Wilde's form; but such a circuit would possess a far higher magnetic conductivity than Wilde's owing to the greater cross-section. One difficulty with such single-circuit forms is how to mount

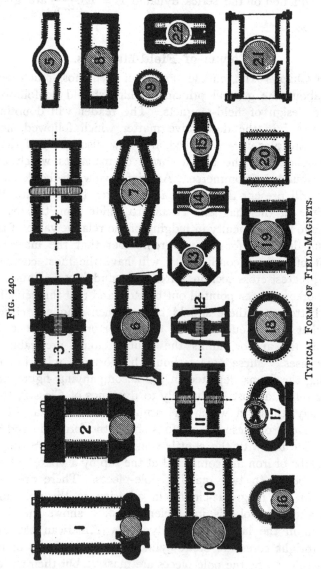

FIG. 240.

TYPICAL FORMS OF FIELD-MAGNETS.

them upon a suitable bed-plate. If mounted on a bed-plate of iron, a considerable fraction of the magnetism will be short-circuited away from the armature, even though an intermediate bed-plate of zinc some inches deep be interposed. In the larger form No. 10, used by Edison in his steam-dynamos, this difficulty is only partially obviated by turning the magnets on one side.

The favourite type of field - magnet, having a double magnetic circuit with consequent poles, is represented in No. 3; it was introduced by Gramme. It may be looked upon as the combination of two such forms as No. 1, with common pole-pieces. Nos. 3 to 9 may be looked upon as modifications of a single fundamental idea. No. 4 gives the form used in the Brush dynamo (plan), the two magnetic circuits being separated by the ring armature. The diagram will serve equally for many forms of flat-ring machine; but in most of these the poles at the two flanks of the ring are joined by a common hollow pole-piece, embracing a portion of the periphery of the ring. No. 5 shows the well-known form of Siemens, with arched ribs of wrought iron, having consequent poles at the arch. The circuit is here of insufficient cross-section. No. 6 depicts the form adopted by Weston : and very similar forms have been used by Crompton, and by Paterson and Cooper. There is a better cross-section here. No. 7 is a form used by Bürgin and Crompton, and differs but slightly from the last. It has one advantage—that the number of joints in the circuit is reduced. No. 8 is a form used by Crompton, Kapp, and by Paterson and Cooper. No. 9 is the form adopted in the little Griscom motor. No. 18 is a further modification due to Kapp. No. 19, which also has consequent poles, is used by Mc Tighe, by Joel, and by Hopkinson ('Manchester' dynamo), by Clark, Muirhead & Co. ('Westminster' dynamo), by O. E. Brown (Oerlikon), by Blakey, Emmott & Co., and in some of Sprague's motors, but with slight differences in proportions of the details. The main difference between No. 19 and No. 6, lies in the position selected for placing the coils, No. 19 requiring two, No. 6 four. No. 20, which is the design of Elwell and Parker, is a further modification of No. 3, and would be improved by having a greater cross-section. In No. 3 (Gramme) it is usual to cast the pole-pieces and end-plates, but to use wrought-iron for the longitudinal cores. The requisite polar surface must be got by some means, and when the core was made thin, the two courses open were either to fasten upon the core a massive pole-piece (Nos. 1, 3, 4, 6, 7, 19, 20), or else to arch the core No. 5 so that its lateral surface was available as a pole. Now, however, that it is known that massive cores are of advantage, the requisite polar surface can be obtained without adding any polar expansion or " piece," but by merely shaping the core to the requisite form (No. 8). This must not be regarded as a mere thinning of the magnet ; for though mere reduction of cross-section at any part of the circuit would reduce the magnetic conductivity, reduction of the thickness for the purpose of bringing the armature more closely into the circuit will have quite the opposite effect. Nos. 11 to 15 illustrate forms of field-magnet having *salient*, as distinguished from *consequent* poles. No. 11 is the double Gramme machine designed by Deprez. Nos. 12 and 13 are two of the innumerable patterns due to Gramme himself. These are both of cast iron ; and it will be noticed that in No. 13 there are no joints, it being cast in one piece. No. 14 is the form used by Hochhausen, and is practically identical with 21 save in the position of the axis of rotation. The iron flanks of No. 14 tend to produce a certain short-circuiting of the magnetism by their proximity to the poles ; and their sectional area is insufficient. No. 15, used by Van de Poele, is similar. No. 16 is the form used by the author in small motors, and is cast in one piece. The semi-circular form adopted for the core was intended to reduce the magnetic circuit to a minimum length. No. 17 illustrates the form used by Jürgensen, having salient poles reinforced by other electro-magnets within the armature. No. 21 shows in section the double tubular magnets of the Thomson-Houston dynamo, the spherical armature being placed, as in Nos. 12, 14 and 15, between two salient poles. There is a curious analogy between Nos. 21 and 19 ; but they entirely differ in the position of the coils. No. 22 is a design by Kapp, in which there are two salient poles of similar polarity, and two consequent poles between

them, one pair of coils sufficing to magnetise the whole
quadruple circuit. Almost identical forms have been em-

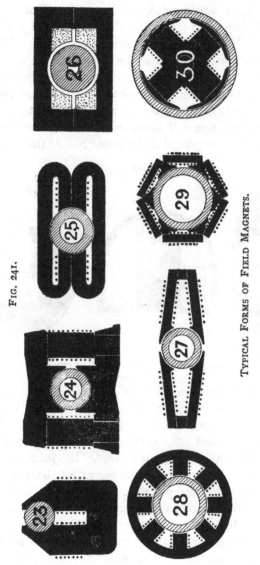

FIG. 241.

TYPICAL FORMS OF FIELD MAGNETS.

ployed by Kennedy ("iron-clad" dynamo), and by Lahmeyer,
and by Wenström. No. 23, Fig. 241, is a type which, used

long ago by Sawyer and by Lontin, has recently become a
favourite one, having been revived almost simultaneously by
Gramme ("type supérieur"), by Kapp, by Siemens ("F"
type), by Cabella ("Technomasio"), and lately by Paterson
and Cooper.   No. 24 is Brown's very massive form.   No. 25 is
a design by Kennedy, known as the "iron-clad" dynamo : the
iron cores are forged to shape.   No. 26 is designed by Professor Geo. Forbes.   The ironwork is in two halves ; the coils,
which are entirely enclosed, are so placed as to magnetise the
armature directly, one coil occupying all the available space
between the field-magnet and the upper half of the armature ;

FIG. 242.   FIG. 243.   FIG. 244.

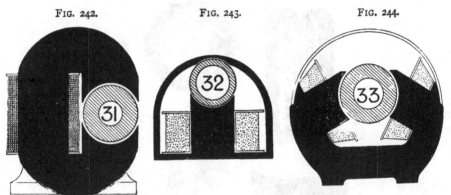

TYPICAL FORMS OF FIELD MAGNETS.

the other the similar space around the lower half.   No. 27 is
the four-pole form adopted by Elwell and Parker in some of
their larger machines.   No. 28 is a multipolar form used by
Wilde, Gramme, and others, the poles which surround the
ring being alternately of opposite sign.   In No. 29 a modification of this design by Thury, for use with a drum armature,
the six inwardly-directed poles are magnetised by coils wound
upon the external hexagonal frame.   No. 30 is a sketch of the
latest form adopted by Siemens and Halske, wherein an
external ring rotates outside a very compact and substantial
four-pole electromagnet.   A similar six-pole machine has
been designed by Ganz, of Buda-Pesth, and a four-pole also
by Fein.

Another recent form of field-magnet is shown in No. 31. This, which is a single horse-shoe with but one coil upon it, was a design by the author of this work early in 1886 ; and a similar form was independently designed by Messrs. Goolden and Trotter about the same time. One-coil machines have also been recently designed by Messrs. Schorch, of Darmstadt, and by Mr. R. Kennedy, of Glasgow, by Mr. Immisch, and by Messrs. J. G. Statter and Co. No. 32 represents also a machine requiring but one coil, and is of the iron-clad type. It was devised by McTighe in 1882, and has been recently revived by Messrs. Stafford and Eaves. No. 33 represents the latest machine of Messrs. Fein, of Stuttgardt, with inward-pointing poles.

The amount of magnetic leakage that takes place in the various forms of field-magnet differs greatly in different forms. No doubt there is least waste field in those machines which have the most compact magnetic circuits, fewest joints, and fewest protruding edges and corners. The magnetic lines of the waste field sometimes take curious forms, which have been experimentally explored,[1] in various types of machines, by Professor Carl Hering.

It was stated by the author, on p. 53, that theoretically the best cross-section for field-magnet cores was circular, as this gave the greatest area for least periphery, and therefore presumably would for a given length of wire in the coil give the largest amount of iron to be magnetised. This, of course, means that if the length of wire and the number of turns be given, a core of this section will, of all possible shapes of core, take the greatest number of ampères to bring it to the diacritical point of semi-saturation. Now, it was the author's discovery, in 1884, that either the electromotive-force or the current of every dynamo is proportional to that number of ampère-turns which will bring its core to this diacritical point. This discovery renders it more than ever needful in designing dynamos to adhere as closely as possible to the author's previous advice to make the core of circular section whenever

---

[1] See *Electrical Review*, xxi. 186 and 205, 1887.

the construction will admit of it.   Again, as was pointed out
by Hopkinson, it is a mistake to construct a field-magnet with
two or more parallel cores uniting at a common pole-piece ; for
not only is the wire between the two cores useless, it is worse,
because it offers wasteful resistance.   To divide the iron
that might be in one solid cylindrical core into two parallel
cylindrical cores, implies, of course, that for every turn of
wire two turns must be used, each of which is more than
half as long as the original one, the total length being
increased as $\sqrt{2} : 1$, while the magnetising power is actually
reduced.   The following calculations are therefore added,
which show the area (in square centimetres) enclosed in a
number of different forms of section, the total periphery of
each being one metre.

Circle	796
Square	625
Rectangle, 2 : 1	555
„ 3 : 1	469
„ 4 : 1	400
„ 10 : 1	236
Oblong, made of square between 2 semicircles	675
„ „ 2 squares „ „ „	548
Two circles (section of 2 parallel cores as in Edison " L " and Siemens " F. 34 " machines)	398
„ „ but assuming wire to be wound right around both cores at once	594
Three circles (section of 3 parallel cores, as in Edison " K " and early Weston dynamo)	265
Four circles (section of 4 parallel cores, as in Gramme vertical dynamo, Fig. 92)	199
Eight circles (section of 8 parallel cores, as in Edison's steam dynamo, Fig. 142)	99

CHAPTER XV.

ELEMENTARY THEORY OF THE DYNAMO.

THE theory of the dynamo which is developed in this chapter is treated in a somewhat different way from that adopted in the two former editions. In the earlier editions the field-magnet was regarded apart from the armature, or its core, as having the function of creating a magnetic field; and the armature was regarded simply as a complex coil having a certain "effective area," rotating in that field. The average intensity of the field was denoted by the symbol H, and the equivalent effective area by A : the electromotive-force of the dynamo being proportional to the product of the two, and the speed n; or as it was written

$$(\text{average}) \; E = 4 \, n \, AH.$$

There was nothing incorrect in this way of regarding the matter ; but neither A. nor H were quantities that could be easily determined during the different stages of action of the machine. But experience has shown that the main problems that require to be considered in the design of dynamos, are best solved by reference to the *magnetic circuit of the machine as a whole,* the iron core inside the armature being regarded as a constituent part of that circuit, and not as something which merely increases the effective area of the armature coils. For this reason the manner of treating the subject has been changed. In all that follows the armature is regarded merely as consisting of a certain number C of conductors, grouped in a particular way around an axis of rotation, their function being to cut across a certain actual number of magnetic lines of force, that are furnished by the magnetic circuit. And instead of the average intensity of the magnetic field,

the actual whole number N of magnetic lines of force that
are forced by the field-magnet circuit to traverse the space
swept out by the armature coils, is now taken as the other
factor.   As will presently be seen, the fundamental equation
is now written

$$\text{(average) } E = n\text{CN} \quad \ldots \ldots \ldots \text{[I.]}$$

The iron core of the armature, in the present mode of treating the subject, is now regarded as having nothing to do with
C, the armature factor, but as playing its part in the magnetic
circuit in determining how many magnetic lines, namely N,
shall be cut by the armature in its rotation.   In brief, instead
of trying to find an average intensity, we endeavour to think
of the magnetic field as a whole—we endeavour to think of
the *quantity* of it.   In all that follows the symbol N stands
for *the whole number of lines of magnetic force that traverse
the armature*, entering it on one side and again leaving it on
the other ; it is called by some writers the " total induction "
through the armature, by others the " total magnetic flux."

   In the present chapter an expression is first found for the
average electromotive-force, which expression serves as the
fundamental equation of all dynamos.   In subsequent
chapters, by introducing approximate formulæ for the law of
magnetisation, equations are then deduced for the various
kinds of series-wound, shunt-wound, and compound-wound
dynamos.

   In the former editions of this work the geometrical treatment of the subject was kept separate from the algebraic.
Experience has shown that students derive in general more
benefit from a mixed treatment of the subject, than from one
that is either purely algebraic or exclusively geometrical.
In this edition therefore the geometrical illustrations are set
beside the algebraic proofs.

   It may be well to point out that in this and the succeeding chapters
the following symbols are used in the following significations :—

   A   area, expressed in *square centimetres.*
   B   the magnetic induction, or number of magnetic lines per square
           centimetre.

b   number of external wires in a section of the armature.

β   angular breadth of a section of armature coil or of segment of collector.

C   number of conductors on the armature, counted all round the external periphery.

c   number of segments of collector or commutator.

E   entire electromotive-force generated in an arma-
      ture,

ε   difference of potential from brush to brush,

e   difference of potential from terminal to terminal,

&   electromotive-force of some external supply of
      electricity,

} expressed in *volts.*

η   economic coefficient (see p. 361).

F   force (i. e. push or pull), expressed in either *dynes, poundals, grammes'* weight, or *pounds'* weight.

G   a geometric coefficient, pertaining to field-magnets, and depending only on their size and shape, or on the size and shape of their coils and pole-pieces (see p. 307).

H   intensity of magnetic field.

i   current in external circuit,

i_a current in armature,

i_s current in shunt coil,

i_m current in series coil of field-magnet,

} expressed in *ampères.*

L   coefficient of self-induction.

λ   average length of one turn of wire in a coil.

μ   coefficient of magnetic permeability of iron.

N   whole number of lines of magnetic force that traverse armature core.

n   number of revolutions *per second.*

ω   angular velocity (expressed in *radians*-per-second).

R   resistance of external circuit,

r_a resistance of armature coils,

r_s resistance of shunt coils,

r_m resistance of series coil on field-magnets,

r   internal resistance of dynamo ; equal to $r_a + r_m$
      or to $r_a + r_s$ according to circumstances,

ρ   resistance per unit of length,

} expressed in *ohms.*

S   number of turns of wire in field-magnet coil in series with armature.

σ   saturation-coefficient of iron (see p. 307).

T   torque, or turning-moment, or angular force, or couple, or "effort statique," or "statisches Moment," expressed in *dyne-centimetres, gramme-centimetres, kilogramme-metres,* or *pound-feet,* according to circumstances.

T   is also used in the section on alternate-current dynamos for the periodic time of the alternating current, measured in *seconds.*

t time, measured in *seconds*.

W } activity, or work-per-second, expressed in *watts* or in *horse-*
w }   *power*.

Z   number of turns of wire in shunt field-magnet coil.

Fundamental Equation of Dynamo.

To find the average electromotive-force of a moving conductor we must remember that, by definition, see page 20, this is (in absolute C.G.S. units) numerically equal to the number of magnetic lines that are cut in one second by the conductor. Also the practical unit, the *volt* being (see page 21), by definition equal to 10^8 absolute C.G.S. units of electromotive-force; it will be necessary to divide the number of C.G.S. units by 10^8 in order to reduce the number to volts. Further, when there are, as in the armatures of dynamos, a number of conductors in series with one another, the total electromotive-force of the dynamo will be equal to the sum of the electromotive-forces of those conductors that are in series with one another.

We will deal with an ordinary two-pole dynamo, having an armature in which the number of "sections" is denominated by the symbol c; the number of "segments" or "bars" in the commutator or collector will also be c. Let there be in each section b external wires or conductors, as counted on the outside of the armature core. (In ring-armatures there will be the same number of external wires as there are loops or windings in the section; in drum armatures there are twice as many external wires as there are loops or windings in the section). Then the number of external conductors or wires, reckoned all round the armature will be bc; it will be more convenient to use the single symbol C for this number. The number of external conductors or wires that are in series with one another electrically from brush to brush will be $\frac{bc}{2}$ or $\frac{1}{2}$C. Now let the armature rotate with a speed of n revolutions per second. (Engineers usually count the revolutions made in one minute, necessitating division by 60 to get n.) Then one revolution will take $\frac{1}{n}$ part of 1 second. We are now ready to calculate the electromotive force.

No. of lines cut by 1 external wire in 1 revolution .. .. '.. .. .. = 2 N ;

(because each wire cuts all the lines where they go in at one side of the armature, and where they come out on the other).

No. of lines cut by 1 external wire in 1 second = 2 *n* N ;

No. of lines cut by ½C external wires in series in 1 second .. .. = 2*n* N × ½C ;

No of lines cut by ½C external wires in series in 1 second .. .. = *n* CN.

Average electromotive-force (in C.G.S. units) = *n* C N ;

Average electromotive-force (in *volts*) .. .. = $\dfrac{n \, C \, N}{10^8}$ .. [I*a*.]

As an example, a certain "Manchester" dynamo (*see p.* 165), having field-magnets so constructed as to force four millions of magnetic lines through the armature core, when at work, had 160 external conductors on its armature. At a speed of 17·5 revolutions per second—*i.e.*, 1050 per minute) this machine would have the following electromotive-force :

$$E = \frac{17\cdot5 \times 160 \times 4{,}000{,}000}{100{,}000{,}000}.$$

$$E = 112 \text{ volts.}$$

It will be unnecessary in every case to write the divisor 10^8 in the formula, because it is easily remembered that, if omitted for the sake of brevity, the numbers obtained can be transformed at once to volts [1] by so dividing down.

[1] Another way of lightening the formula would be to adopt, as the unit quantity of magnetic flux, one hundred million lines, and in that case the symbol N would be used for the number of bundles of lines containing a hundred million each. But very few dynamos have been constructed in which the magnetic flux attained so great a quantity as one such bundle. The awkwardness of having the values of N always fractional would scarcely be compensated for in the gain of having a unit which fits so conveniently to the other practical international units. No difficulty is likely to arise from the omission of the divisor in the formulæ.

Z

For many purposes it is more convenient to have the fundamental equation in terms of the angular velocity. Let the symbol ω represent the angular velocity. Then

$$\omega = 2\pi n;$$

for, in each revolution, the angle described is 2π *radians* or 360 *degrees*. Consequently $n = \dfrac{\omega}{2\pi}$, which gives;

$$\text{(Average)} \ E = \frac{\omega}{2\pi} C N \ \ldots \ldots \ldots \ldots [Ib.]$$

It will be observed that this electromotive-force is simply an average; and it depends on the construction of the arma-

FIG. 245.

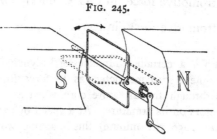

IDEAL SIMPLE DYNAMO.

ture how much fluctuation there is in the value during a rotation.

If, as in Fig. 245, the armature had but two external conductors forming a simple loop, then the electromotive-force would fluctuate between zero and a maximum. Calling the lowest point of the rotating loop in its vertical position 0°, then the position on the left of the dotted line will be 90°, if we reckon the angle of rotation in the clockwise direction. The top point will be 180°, and the point on the extreme right 270°. Then the induced electromotive-force will be zero as the coil passes through 0° and 180°, for at the positions 0° and 180°, the conductors will be sliding along, rather than cutting, the magnetic lines, and a maximum as the coil passes through 90° and 270°. The *rate* of enclosing or "cutting" will be a *maximum* when the *actual number* of lines enclosed is a *minimum*, and *vice versâ.* (See p. 36).

At any intermediate angle, if the field is uniform, the actual number of lines of force enclosed is proportional to the cosine of the angle through which the coil has turned from its zero position, and the electromotive-force will be proportional to the sine of that angle. Strictly speaking, we ought to take the sine with a *negative* value to represent the electromotive-force, because as usually defined the induced electromotive-force is proportional to the rate of *decrease* in the number of lines of force enclosed. We need not, however, trouble about signs, because, if the commutator is properly set, all the induced electromotive-forces are thereby made to act in the same direction through the external circuit. The exact expression at any particular angle θ, for the electromotive-force of the loop may be calculated as follows :—The number of lines of force inclosed when the loop has turned through angle θ is $= N \cos \theta$; hence the rate of cutting will be $\omega N \sin \theta$, or $2\pi n N \sin \theta$. Now, since the average value of $\sin \theta$, between the limits $\theta = 0°$ and $\theta = 90°$, is $\frac{2}{\pi}$, the *average* electromotive-force *per loop* may be obtained by substituting this value, giving us

$$\text{Average E per loop} = 4\,n\,N.$$

And since the number of loops that are in series between brush and brush is $\frac{1}{4}C$, we have finally

$$(\text{Average}) \; E = n\,C\,N.$$

If the coil consisted of many turns all wound in one group, like the Siemens shuttle-wound armature, p. 38, the same expressions would obviously hold good on substituting the proper number for C.

We next proceed to study the fluctuations produced in the electromotive-force as the result of various groupings of the coils.

Fluctuations of Electromotive-force in a One-coil Armature.

As explained on the previous page, the actual induced electromotive-force is proportional to the sine of the angle through which the coil has turned, or

$$E = 2 \pi n N \sin \theta \times \tfrac{1}{4}C,$$

whence

$$E = \frac{\pi}{2} n C N \sin \theta \dots\dots\dots [II.]$$

As θ increases from 0° to 360°, the value of the sine goes from 0 to 1, then from 1 to 0, from 0 to − 1 and from − 1 back to 0.

FIG. 246.

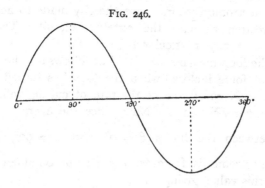

The values of the sine are depicted in Fig. 246. The same curve may serve then to show how the electromotive-force would fluctuate if there were no commutator. But the action of the commutator is to commute the negative inductions

FIG. 247.

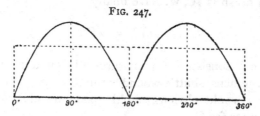

into positive ones: the brushes being so arranged as to slide from one part of the commutator to the other at the moment when the inverse induction begins. This gives the curve the form of Fig. 247, which therefore represents how the current pulsates in the circuit of a simple old-fashioned shuttle-wound Siemens armature. Now if we could level these hills, and change our undulating induction into a steady one, we should get a single straight line, shown in Fig. 247 as a dotted line, enclosing below it a rectangular area equal to the sum of the areas enclosed by the sinuous curves, and therefore at a height which is the average of the heights of all the points along the curves: in fact, since each sinuous curve is part of a curve of sines, the *average* height will be $\frac{2}{\pi}$, or about $\frac{7}{11}$ of the maximum height.

Fluctuations in a Closed-coil Armature divided into Sections.

As shown in the argument on pp. 40 and 42, it is, for reasons of construction, usual to wind armature coils in two sets connected in parallel arc. The two halves of the Pacinotti ring, the two halves of the windings on the Siemens drum, meet at the brushes in parallel arc. If each of the two coils consisted of 100 turns, their joint effect in inducing electro-motive-force will be no greater than that of either of them separately, but the internal resistance of the armature will be halved. From this point onwards in the argument it will be assumed that the armature windings consist of *pairs* of coils. Thus, instead of one coil of 200 turns as shown in Fig. 248, we

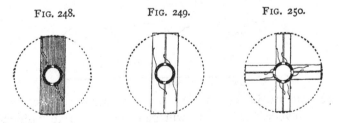

| FIG. 248. | FIG. 249. | FIG. 250. |

shall take it that there is a pair of coils each of 100 turns as in Fig. 249.

Now suppose that, in order to get a less fluctuating effect, we divide each of our original single pair of coils into two parts, and set these at right angles to one another. To take a numerical case, suppose there were originally 100 turns in each coil and we split each into two coils of fifty turns, but set them across one another so that one comes into the best position in the field as the other is going out of it. (This arrangement is indicated in Fig. 250, which may be contrasted

FIG. 251.

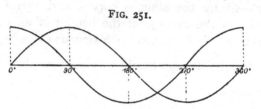

with Fig. 248 representing the undivided coil.) In this case we shall have two sets of overlapping curves—each of them will have to be but half as high as before, because the equivalent

FIG. 252.

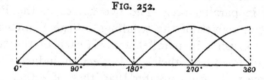

area of each coil is only half what it was for the whole coil. Then, if there were no commutator, the induced electromotive-force in the two sets of coils would fluctuate as shown by the two curves of Fig. 251. But if the ends of the two "sections"

FIG. 253.

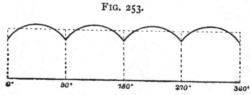

of the coil are joined to a proper commutator or collector, all the "inverse" inductions will be commuted into "direct" ones by the sliding of the brushes at the right moment, and the two curves would then become as in Fig. 252. The next process is to ascertain what the joint result of these over-lapping electromotive-forces will be : it is evident that from 0° and 90° the two inductive actions are assisting one another, and that at 45° they are equal. The nett result here is therefore double either of them ; and, in fact, the curve representing the *sum* of the two curves is given in Fig. 253. This curve shows at once a step towards *conti-nuity*, as the fluctuations are far less than those of the single coil, Fig. 247. If, as before, we level the undulating tops by a dotted line, we get precisely the same height as before. The *total* amount of induction (the total number of lines of force cut) is the same, and the *average* electromotive-force is the same. There is no gain, then, in the total electric work resulting from rearranging the armature coils in two sets at right angles to each other : but there is a real gain in the greater continuity and smoothness of the current.

Fig. 254.

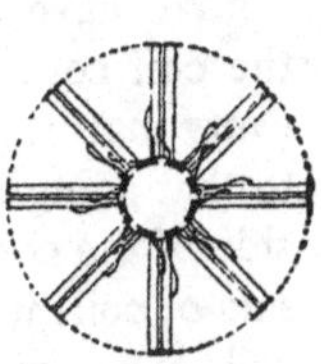

If we again split our coils and arrange them as shown in Fig. 254 at angles of 45°, in four sets of pairs of coils of twenty-five turns each, and connect them up to a proper commutator, we shall get an effect which is very easily represented by constructing two curves, each similar to the

Fig. 255.

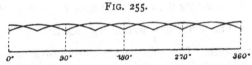

last but each of half the height, and compounding them together (Fig. 255). One of them will of course have the maximum heights of crests occurring 45° further along than those of the other curve ; and when these are compounded to-

Fig. 256.

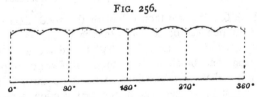

gether we get for a resultant a curve shown in Fig. 256, which has exactly the same *average* height as before, but which has still less of fluctuation. It is easily conceived that this process of dividing the coil into sections, and spacing these sections out at equal angles symmetrically, would give us a result approaching as near as we choose to an absolutely continuous one. If our original pair of coils of 100 turns each were split into twenty sets of pairs of five turns each, or even into ten sets of pairs of ten turns each, the approach to continuity would be very nearly truly attained. It only remains to calculate the continuity algebraically; which, though not difficult, is rather tedious.

Calculation of Fluctuations of the Electromotive-force in Closed-coil Armatures.

We have seen in Chap. III. that in every armature a section of the coil connected with any two commutator-bars is undergoing at every instant an inductive effect exactly similar, but opposite in sign, to that going on in the section connected with the two bars on the side of the commutator diametrically opposite. We likened the two sets of coils in the two halves of the armature to two sets of galvanic cells arranged in parallel. Suppose the armature had in all thirty-six sections, then in reality there are two sets of eighteen, and the electromotive-force induced in each set is alike. We will take the case of a ring armature as being less complicated than the drum. Let the symbol c stand for the total number of sections in the arma-ture. There will be therefore $\frac{c}{2}$ sections in each half of the armature from brush to brush. Let each section consist of b turns of wire. The whole armature will consist of $b\,c$ turns. If these c sections are set symmetrically round, the angle between the plane of each section and the next one to it will be $\frac{360}{c}$ degrees or $\frac{2\pi}{c}$ radians. This may for some purposes be written $\frac{\pi}{\frac{1}{2}c}$; and for shortness we will call this angle $= \beta$. We will then calculate the total electromotive-force induced in one set of sections, that is to say, in one of the rows of $\frac{1}{2}c$ sections of coils extending half round the armature and commutator from one brush to the other. Referring to page 339 and remembering that only $\frac{1}{2}$ N lines, at most, go through any one section at any instant, we see that in the first section, when it has turned through angle θ, the induced electromotive-force e_1 will be

$$e_1 = \omega\,\frac{N}{2}\,b\sin\theta,$$

where ω is the angular velocity. In the second section the electro-motive-force will be

$$e_2 = \omega \frac{N}{2} \sin (\theta + \beta),$$

because this section has a position differing by an angle β from the first section. In the third section we have similarly

$$e_3 = \frac{N}{2} b \sin (\theta + 2\beta);$$

and so on, until we come to the last section of the set, for which the electromotive-force will be

$$e_{\frac{1}{2}c} = \omega \frac{N}{2} b \sin (\theta + \overline{\tfrac{1}{2} c - 1} \, \beta).$$

But the whole electromotive-force of the set is the sum of all these separate electromotive-forces; so we have

$$E = \omega \frac{N}{2} b \times$$

$$\left\{ \sin \theta + \sin (\theta + \beta) + \sin (\theta + 2\beta) \ldots \ldots \sin (\theta + \overline{\tfrac{1}{2}c - 1} \, \beta) \right\}.$$

We can get, however, no information with respect to the maximum and minimum values of this fluctuating electromotive-force as long as the expression for E is in the form of a long series of values. We must proceed to sum this series within the brackets. Call its sum S. Then, in order to get its separate terms added together, we will multiply each by $2 \sin \frac{\beta}{2}$; so that;

$$\sin \theta \qquad \times 2 \sin \frac{\beta}{2} = \cos \left(\theta - \frac{\beta}{2} \right) \quad - \cos \left(\theta + \frac{\beta}{2} \right),$$

$$\sin (\theta + \beta) \times 2 \sin \frac{\beta}{2} = \cos \left(\theta + \frac{\beta}{2} \right) \quad - \cos \left(\theta + 3 \frac{\beta}{2} \right),$$

$$\sin (\theta + 2\beta) \times 2 \sin \frac{\beta}{2} = \cos \left(\theta + 3 \frac{\beta}{2} \right) - \cos \left(\theta + 5 \frac{\beta}{2} \right),$$

&c. &c. &c.

$$\sin (\theta + \overline{\tfrac{1}{2}c - 1}\beta) \times 2 \sin \frac{\beta}{2} = \ldots \ldots \ldots - \cos \left(\theta + \overline{\tfrac{1}{2}c - 1} \, \beta + 1 \frac{\beta}{2} \right).$$

It will be noticed that on adding up the terms on the right-hand side they cancel out in pairs, leaving only the first and last : thus—

$$S \times 2 \sin \frac{\beta}{2} = \cos \left(\theta - \frac{\beta}{2} \right) - \cos \left(\theta + \overline{\tfrac{1}{2}c - 1}\,\beta + \frac{\beta}{2} \right).$$

But $\tfrac{1}{2}c\,\beta = \pi$, and $\overline{\tfrac{1}{2}c - 1}\,\beta = \pi - \beta$; and therefore the last term may be written $-\cos \left(\theta + \pi - \frac{\beta}{2} \right)$, which is same as $+\cos \left(\theta - \frac{\beta}{2} \right)$.

Then the expression becomes

$$S \times 2 \sin \frac{\beta}{2} = 2 \cos \left(\theta - \frac{\beta}{2} \right)$$

and

$$S = \frac{\cos \left(\theta - \frac{\beta}{2} \right)}{\sin \frac{\beta}{2}}.$$

Inserting this value, we get at once

$$E = \omega \frac{N}{2} b \frac{\cos \left(\frac{\beta}{2} - \theta \right)}{\sin \frac{\beta}{2}}.$$

The amount of fluctuation implied in this formula depends on how the brushes are set. They slide, of course, from one bar of the commutator to another while the commutator moves through the angle β. So, if $\theta = 0$ at the beginning, when the commutator-bar is just beginning to touch the brush, then $\theta = \beta$ just as the bar leaves contact with the brush. And when the brush touches the middle of the bar $\theta = \frac{\beta}{2}$. Now the cosine is a maximum when the angle is a minimum. Therefore E will be a maximum when $\frac{\beta}{2} = \theta$, *i. e.* when $\frac{\beta}{2} - \theta = 0$: and E will be a minimum when either $\theta = 0$, or $\theta = \beta$. We have, consequently, the following results as the bar of the commutator passes under the brush :—

(1.) At beginning ($\theta = 0$),

$$E \text{ (a minimum)} \quad \ldots \quad \ldots \quad = \omega \, \frac{N}{2} \, b \, \frac{\cos \dfrac{\beta}{2}}{\sin \dfrac{\beta}{2}}$$

$$= \omega \, \frac{N}{2} \, b \, \cot \frac{90°}{\frac{1}{2} c} \, .$$

(2.) At middle of bar $\left(\theta = \dfrac{\beta}{2} \right)$,

$$E \text{ (a maximum)} \quad \ldots \quad \ldots \quad = \omega \, \frac{N}{2} \, b \, \frac{1}{\sin \dfrac{\beta}{2}}$$

$$= \omega \, \frac{N}{2} \, b \, \operatorname{cosec} \frac{90°}{\frac{1}{2} c}$$

(3.) At end ($\theta = \beta$),

$$E \text{ (again a minimum)} \quad \ldots \quad = \omega \, \frac{N}{2} \, b \, \frac{\cos \dfrac{\beta}{2}}{\sin \dfrac{\beta}{2}}$$

$$= \omega \, \frac{N}{2} \, b \, \cot \frac{90°}{\frac{1}{2} c}$$

The greatest fluctuation therefore that can occur, will be the difference between $\operatorname{cosec} \dfrac{90°}{\frac{1}{2} c}$ and $\cot \dfrac{90°}{\frac{1}{2} c}$; and, since each bar as it passes under the brushes comes into the position just occupied by the bar preceding it, there will be as many fluctuations in every revolution as there are bars in the commutator or sections in the armature, namely c. Further than this, if we could increase the number of sections indefinitely, so that $\dfrac{90°}{\frac{1}{2} c}$ or $\dfrac{\beta}{2}$ was practically $= 0°$, then both $\operatorname{cosec} \dfrac{90°}{\frac{1}{2} c}$, and $\cot \dfrac{90°}{\frac{1}{2} c}$ would be equal, and would be equal to $\dfrac{c}{\pi}$; for $\dfrac{90°}{\frac{1}{2} c} = \dfrac{\pi}{c}$, and for small angles the arc is sensibly equal to either the sine or the tangent. We will, however, calculate the actual amount of fluctuation in certain cases. Many dynamos are built with armatures having a 36-part collector, and thirty-six sections in the armature coil. We want to know the fluctuations in this case, and in other cases with fewer or more segments. The following table gives the results of the calculations; the number of sections of the armature and commutator being given in the first column, and their angular breadth in the second. The fluctuation is the difference between columns 3 and 4 :—

c	β	$\dfrac{\operatorname{cosec}\frac{\beta}{2}}{c}$	$\dfrac{\cot\frac{\beta}{2}}{c}$	The Fluctuation.	Percentage Fluctuation.
2	180°	·5	0·	·5	±50·00
4	90	·3479	·2500	·0979	14·04
10	36	·3236	·3077	·0159	2·38
12	30	·3220	·3110	·0110	1·70
15	24	·3206	·3136	·0070	1·10
20	18	·3196	·3157	·0039	·61
24	15	·3192	·3165	·0027	·42
30	12	·3189	·3171	·0018	·28
36	10	·3187	·3175	·0012	·19
40	9	·3186	·3177	·0009	·14
45	8	·31857	·31780	·00077	·12
60	6	·31846	·31802	·00044	·07
90	4	·31838	·31819	·00019	·03
360	1	·31832	·31830	·00002	·003
5400	0° 4'	·3183099	·3183098	·0000001	·00001

From these figures it is apparent that the fluctuations become practically insignificant when the number of sections is increased ; as indeed the curves of Figs. 247 to 256 showed. With a 20-part commutator the fluctuations of the electromotive-force in the armature are less than 1 per cent. of the whole electromotive-force. With a 36-part commutator, they are less than one-fifth of 1 per cent. So far as mere fluctuations are concerned then, it is practically a useless refinement to employ commutators of more than thirty-six parts. But there are other reasons, as we shall see in considering the self-induction in the separate sections, for making the number of sections as great as possible.

Now, assuming that we wind our coil in a large number of sections, so that the fluctuations may be negligible, what will the total electromotive-force be? We may write, as we have seen, $\dfrac{c}{\pi}$, for either cosec $\dfrac{\beta}{2}$ or for cotan $\dfrac{\beta}{2}$, giving us $E = \omega \dfrac{N}{2} b \dfrac{c}{\pi}$. Now $\omega = 2\pi n$, and $bc = C$, so that our formula once more becomes $E = n\,C\,N$, as before

Measurement of Fluctuation.

The relative amount of fluctuation in the current furnished by a dynamo may be observed by noticing the inductive effect on a neighbouring circuit. Let a coil be introduced into the circuit, and let a second coil, wholly disconnected from the first, be laid coaxially with it, so that the coefficient of mutual induction between the coils shall be as great as possible. Introduce into the circuit of the second coil a Bell telephone receiver. If the main-circuit current is steady there will be no sound heard. If it fluctuate, each fluctuation will induce a corresponding secondary current in the telephone circuit, and the amount and frequency of the fluctuations may be estimated by the loudness and pitch of the sound in the telephone. The fluctuations of the current of a Brush dynamo are in this manner readily detected. Professor Ayrton has suggested the introduction into the secondary induction circuit of an electro-dynamometer to serve as a " discontinuity-meter."

Effect of Non-simultaneous Commutation.

If the brushes are not so set that the sliding of contact under one brush is not accomplished at the same instant as that under the other brush, then it is clear that there will be slightly unequal electromotive-forces in the two halves of the armature circuit. This momentary inequality will die out, to be succeeded by another inequality (of opposite sign) when commutation occurs at the other brush. The effect will be the same as though a small alternating current having $2\,n\,c$ periods of alternation per second were made to act around the circuit of the armature. Such effects may be occasioned in armatures by various causes; if the number of sections in the armature be an odd number; if the numbers of conductors in all the sections are not alike or their connections are unsymmetrical; or lastly, if the contact-edges of the brushes do not lie exactly at opposite ends of a diameter. These effects are independent of a still smaller alternation arising in every armature in consequence of mutual inductions between the currents in the commutated coils and those adjacent to them.

Measurement of N.

An important problem is how to measure the actual number of magnetic lines that pass through the armature. This number is really best ascertained by calculation from the performance of the machine itself. The speed being observed by aid of a suitable speed-counter, the number of conductors round the armature being known, and the whole electromotive-force generated in the machine being measured by proper electrical methods, then it only remains to apply the fundamental formula, transformed so as to calculate back to N :—

$$N = \frac{E \times 10^8}{n\,C}$$

To measure E the means adopted must depend upon the construction of the machine. If it is a continuous current machine, either magneto or separately-excited, then a simple voltmeter applied at the open terminals or brushes will suffice. If the machine is series-wound or shunt-wound, the same method may apply, provided the magnetising coils can be disconnected and separately excited up (for example, by using accumulators) to the right degree while the machine is run. In all these cases the result will not correctly represent the working values, because in work there are the reactions due to considerable currents in the armature coils. To measure E while the machine is running, it must either be run upon known resistances (so as to enable E to be calculated from Ohm's law) ; or E may be calculated by measuring (see p. 68) the difference of potentials at the brushes with a voltmeter, and then calculating from the resistance of and current in the armature the volts lost internally, which, added to the measured volts, make up the whole E.

Another way, which does not involve the running of the machine—and is therefore less satisfactory—is to wind closely around the armature, exactly in the diameter of commutation, a single turn of fine insulated wire, the ends of which are connected through two insulated wires lying close together, or, better, twisted together, to a ballistic galvanometer of slow period and of appropriate sensitiveness. This being done, arrangements are then made to separately excite the field-magnets, by some auxiliary current, to an amount equal to the working value. On turning on the exciting current, a current of short duration, proportional in its integral value to the number of magnetic lines which have thus been introduced through the wire loop, is excited in the circuit and produces a throw on the galvanometer. On breaking the exciting circuit or short-circuiting the exciting coils, another throw is observed in the opposite direction, and of equal amount. To ascertain the absolute number of lines indicated by this throw, a comparative experiment must be made with some sort of apparatus for introducing a known number of magnetic lines into the same circuit. A good way to do this is to use a hand-coil of large size (resembling on a large scale the coil of a tangent galvanometer), wound with a known number of turns of fine wire, the average area of these turns being known from careful measurement. Such a coil should be included in the circuit of the aforesaid galvanometer, but with leading wires enabling it to be placed at a sufficient distance away from the dynamo and from all other magnets and iron, to allow of no error arising from such causes of perturbation. It should be laid level in some spot for which the value of the vertical compound of the earth's magnetic field is known. (In London this is 0·43 C.G.S. units.) On suddenly inverting this coil, another throw is produced in the ballistic galvanometer. The result may be calculated out as follows :—Let δ_1 be the throw due to introducing the N lines through the loop, and δ_2 the throw given on inverting the coil. Calling the number of turns in the coil S, the (average) area of each of them A, and V the intensity of the vertical component of the earth's field, then the number of lines cut by inverting the coil is 2 S A V. And N is found by solving the simple rule-of-four sum :—

$$N : 2\,S\,A\,V :: \delta_1 : \delta_2.$$

CHAPTER XVI.

The Magneto-machine and the Separately-excited Machine.

In the equations hitherto considered it was assumed that the armature rotated in a magnetic field the quantity of which was specified by the symbol N. Nothing was specified as to the kind of field-magnets : and the general formula deduced is of course applicable for all kinds, provided their magnetic power is known. In magneto-dynamos, in which the field is due to permanent magnets of steel, N depends both on the magnetism of the steel and on the iron core of the armature. The number of lines that find their way through the armature is, however, lessened by the reaction of the current in the armature when this becomes strong. If the magnetism of the field-magnets were so overpoweringly great, as compared with that due to the armature coils, that this reaction were insignificantly small, then, since our fundamental formula is

$$E = n \, C \, N,$$

E would, for any given magneto machine, be directly proportional to n, the speed of rotation. But we know in practice that this is not the case. Suppose we turn a magneto machine at 600 revolutions per minute ($n = 10$, for then there will be 10 revolutions per second) and get, say, 17 volts of electromotive-force from it, then, if there were no reactions from the armature, turning it at 1200 revolutions per minute ought to give exactly 34 volts. This is never quite attained ; though in many machines, as, for example, in the laboratory-pattern magneto-Gramme machines made by Breguet (Fig. 86), the direct proportion is very nearly attained even with much higher speeds.

Relation between Speed and Electromotive-force.

Theory indicates that in a magneto-electric machine the electromotive-force induced in the armature should be exactly proportional to the speed of rotation, provided the magnetic field in which the armature rotates is of constant intensity. This is only true in practice if the current in the armature is kept constant by increasing the resistances of the circuit in proportion to the speed, because currents of different strength react unequally on the intensity of the field. In some experiments made by M. Joubert at different speeds the electromo-

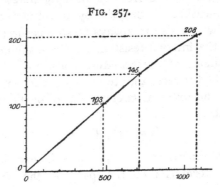

FIG. 257.

CURVE SHOWING RELATION BETWEEN SPEED AND ELECTROMOTIVE-FORCE.

tive-force was measured by an absolute electrometer, which allowed no current whatever to pass. The only possible reactions were those due to possible eddy currents in the core : and the theoretical law was almost exactly fulfilled. The observations are given below, and plotted in Fig. 257, in which the straightness of the " curve " shows how nearly truly the theoretical condition was attained.

Speed	500	720	1070	revolutions per minute.
Electromotive-force	103	145	208	volts.

The student may here with advantage study the lines given in Fig. 185, p. 236, obtained from a certain separately-excited Brush machine.

2 A

Potential at Terminals of Magneto-dynamo.

The potential at terminals of the magneto machine—and indeed of every dynamo—is, when the machine is doing any work, less than E, the total induced electromotive-force, because part of E is employed in driving the current through the resistance of the armature. The symbol e may be conveniently used for the difference of potential between terminals. Only when the external circuit is open, so that no current whatever is generated, $e = E$. It is convenient to have an expression for e in terms of the other quantities, seeing that when any current is being generated it is impossible to measure E directly by a voltmeter or by an electrometer, whereas e can always be so measured.

Let r_a be the internal resistance of the machine, that is to say the resistance of the armature coils, and of everything else in circuit between the terminals; and let R be the resistance of the external circuit. Then, by Ohm's law, if i be the current,

$$E = i \, (r_a + R).$$

But by Ohm's law also, if e be the difference of potential between the terminals of the part of the circuit whose resistance is R,

$$e = i \, R \, ;$$

whence

$$\frac{e}{E} = \frac{R}{r_a + R} \, ; \qquad\qquad \text{[III.]}$$

from which we have

$$e = \frac{R}{r_a + R} \, E,$$

or

$$e = \frac{R}{r_a + R} \, n \, C \, N. \qquad\qquad \text{[IV.]}$$

It is also convenient to remember that

$$E = \frac{r_a + R}{R} \, e \, ; \qquad\qquad \text{[V.]}$$

for this formula enables us to calculate the value of E from observations of *e* made with a voltmeter.

Relation between whole Electromotive-force and Difference of Potentials at the Terminals.

The essential distinction pointed out above between the whole electromotive-force E, and that part of it which is available as a difference of potentials at the terminals *e*, may be further illustrated by the following geometrical demonstration which is due to Herr Ernst Richter.[1]

In a machine (such as are chiefly dealt with later) in which *e* is constant, E will not be constant, except in the unattainable

FIG. 258.

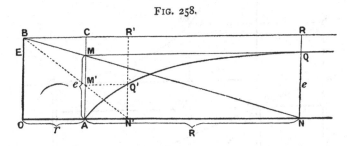

case of a machine which has no internal resistance. Let *r* represent the internal resistance of the machine, including that of the armature and of any magnet-coils that are in the main circuit ($r = r_a + r_m$); then,

$$E = e + ir.$$

If E is constant, then *e* cannot be constant when *i* varies ; and if *e* is constant, E cannot be. We have then two cases to consider :—

(1) E *constant.* Take resistances as abscissæ and electromotive-forces as ordinates, and plot out (Fig. 258) O A = *r*, A N = R, O B = E. The line B N represents the fall of potential through the entire circuit. Of the whole electromotive-force O B, a part equal to C M is expended in

[1] *Elektrotechnische Zeitschrift*, iv. p. 161, April 1883.

driving the current through the resistance r, leaving the part A M available as the difference of potentials at the terminals, when the total resistance of the circuit is represented by the length from O to N. Accordingly at N erect a vertical line N Q equal to A M. Take a less external resistance R' = A N', and by a similar process we find that the corresponding value of e is A M' or N' Q'. Similarly, any number of points may be determined; they will all lie on the curve A Q Q', which therefore shows how, as the external resistance is increased,

FIG. 259.

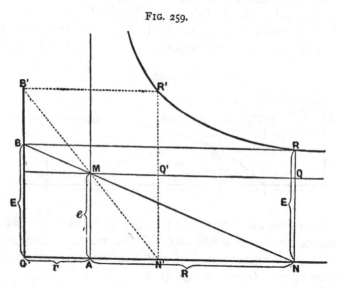

the terminal potential rises, whilst the whole electromotive-force remains constant and is represented by the horizontal line B R. The equation of this curve is given by the condition

$$\frac{E - e}{E} = \frac{r}{R + r}$$

whence $(E - e)(R + r) = E r$ = constant; which equation is the equation of an equilateral hyperbola having O B and B R as asymptotes.

(2) *e constant.* As in the preceding case, O A = r; A N = R; and A M = e. From N (Fig. 259) draw the line

N M and produce it backwards to B. Then O B represents that value of E which will give *e* volts at terminals when R = N M. Accordingly set off at N the line N R = O B In a precisely similar way draw N' B', to correspond with any other value of R, and make N' R' equal to O B'. N' R' represents the value of E when the value of the external resistance R is equal to A N'. By determining other values we obtain the successive points of the curve R R', which shows how the whole electromotive-force must vary in order to maintain a constant difference of potentials at the terminals, as represented by the horizontal line M Q. The equation to this new curve is given by the condition

$$\frac{E - e}{r} = \frac{e}{R}$$

or $(E - e) R = e r = $ constant.

This curve is also an equilateral hyperbola.

The Separately-excited Dynamo.

For separately-excited dynamos the same formulæ hold good as for magneto-dynamos ; but in this case N depends upon the strength of the independent exciting current.

In estimating the nett (or commercial) efficiency of a separately-excited dynamo, the energy spent per second in exciting the field-magnets ought to be taken into account.

Characteristic of Magneto Machine, and of Separately-excited Dynamo.

In the magneto-dynamo the magnetism of the steel magnets is approximately constant. This has given rise to a common idea that in such machines the electromotive-force depends on the speed alone. This is not true. For owing to the cross-magnetising and demagnetising tendency of the currents in the armature coils, the number of magnetic lines that actually traverses the armature core diminishes when the currents in the armature are strong. The stronger the current in the armature the stronger the reaction. As will be explained later (p. 368), it is convenient to plot out certain curves, known as *characteristics*, to exhibit the relation that subsists between the electromotive-force and the current under different condi-

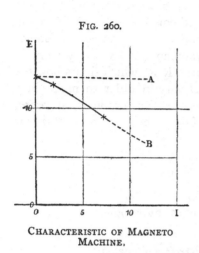

FIG. 260.

CHARACTERISTIC OF MAGNETO
MACHINE.

tions of speed, resistance, etc. Usually one of the conditions assumed is that the speed is constant. Such curves are particularly useful for studying the various reactions that exist between the field-magnet and armature. Fig. 260 gives the characteristic of a small magneto machine, of the laboratory type, having a Gramme ring. It was capable of lighting two small Swan lamps of about 5 candle - power. On open circuit the electromotive-force was 13·1 volts at a speed of 1400 revolutions per minute. The value of E, the total electromotive-force generated in the armature, fell from 13·1 to 12·4 volts when a current of 1·8 ampères was taken from the machine ; and when it was short-circuited to give 6·1 ampères, the value of E fell to 9·2 volts. This was, however, an exceptionally bad case.

The reaction of the armature current was here very strongly marked. Were there no reaction the characteristic would follow the dotted line to A instead of dropping down to B. As explained on p. 92, the demagnetising tendency of the armature current increases with the forward lead given to the brushes.

It ought not to be forgotten that beside the prejudicial reactions in the magnetic field there are prejudicial reactions in the armature coils themselves, due to mutual-induction between those parts of the coils in which the current is increasing or decreasing in strength and those in which a steady current is flowing.

The characteristics of separately-excited dynamos exhibit a similar decline in the electromotive-force ; and the reasons are exactly similar. A care-
ful study of these machines was made in 1884 by Mr. W. B. Esson,[1] who gives the following curve for a sepa-rately-excited dynamo having a modified Pacinotti ring ar-mature. The line E (Fig. 261) represents the total electro-motive-force if there were no reactions. The line ε repre-sents the values of the poten-tial between the brushes of the machine (called ε in this book in contradistinction to E the whole electromotive-force) as it would be if there were no reaction. The curved

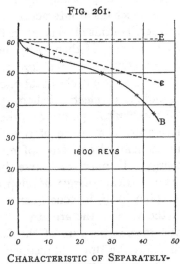

FIG. 261.

CHARACTERISTIC OF SEPARATELY-EXCITED DYNAMO.

line B gives the actually-observed values of ε when different currents were taken from the machine. The great drop at the lower end of the curve is probably due to the greater demagnetising effect of the armature current when there is (as with strong currents) a considerable lead at the brushes. The characteristic always shows such downward curvature more when the field-magnets are weakly excited. (See Fig. 315.)

Efficiency and Economic Coefficient of Dynamos.

Suppose that we know the actual mechanical horse-power applied in driving a dynamo. This can be measured directly either by using a "transmission dynamometer," or by taking an indicator diagram from the steam-engine that is driving it, or, in certain special cases where the field-magnets can be

---

[1] *Electrical Review*, vol. xiv. p. 303, April 1884. See also papers by M. Marcel Deprez, *Comptes Renaus*, xciv. pp. 15 and 86, 1882.

pivoted or counterpoised, by applying the method originally pursued by the Rev. F. J. Smith, and later described by M. Marcel Deprez and by Professor Brackett, in which the actual mechanical interaction between the armature and field-magnets is utilised to measure the horse-power used in driving the machine. If, then, we know the horse-power applied, and if we measure the "activity" of the dynamo, that is to say its rate of giving out electrical energy, or, as some people call it, its output of electrical horse-power, we have by comparing the mechanical power absorbed with the electrical activity developed, a measure of the "efficiency" of the dynamo as an economical converter of mechanical energy into electrical energy. It must, of course, be borne in mind that part of the electrical energy developed is inevitably wasted in the machine itself, in consequence of the unavoidable resistance in the wire of the armature, and, in the case of self-excited dynamos, in the wire of the field-magnet coils. There must, therefore, be drawn a distinction between the gross efficiency of the machine, or as it is sometimes called, its " efficiency of electric conversion," and its nett efficiency, or "useful commercial efficiency."

To express efficiency, whether gross or nett, we must, however, have the means of measuring the electric "activity" of the dynamo, or of any part of its circuit.

As is well known, the energy per second of a current can be expressed, provided two things are known, namely, the number of ampères of current, and the number of volts of potential between the two ends of that part of circuit in which the energy to be measured is being expended. The number of ampères of current is measured by a suitable ampère-meter; the number of volts of potential by a suitable volt-meter. The product of the volts into the ampères expresses the electric energy expended per second in terms of the unit of activity denominated "watts." As 1 horse-power is equal to 746 watts, the number of volt-ampères (i. e. of watts) must be divided by 746 to give the result in horse-power. If i represents the current in ampères, and e the difference of potential in volts, then the "activity" or " electric energy per second," for which we may use the symbol w, may be written

$$w = \frac{ei}{746}.$$

Now we know, in the case of every dynamo, that the electric energy developed usefully in the external circuit is *not* the whole of the electric energy of the machine, part being absorbed (and wasted in heating) in the resistances of the armature and magnet coils. The ratio of the useful electrical energy realised in the external circuit to the total electric energy that is developed is sometimes called, though not very happily, the "electrical efficiency" of the machine. I prefer instead to call this ratio the "economic coefficient" of the machine. It may be expressed algebraically as follows :—If a series-wound or a magneto machine is giving a current of i ampères, and its total electromotive-force be E volts, then its total electric activity will be

$$= E\,i \text{ watts.}$$

If the electromotive-force between the terminals of the dynamo be e volts, then the useful activity is

$$= e\,i \text{ watts.}$$

Using the symbol η for the "economic coefficient," or so-called "electrical efficiency," we have

$$\eta = \frac{\text{useful activity}}{\text{total activity}} = \frac{e\,i}{E\,i},$$

or,

$$\eta = \frac{e}{E}.$$

But we know that the ratio $\frac{e}{E}$ depends on the relation of the internal and external resistances, for

$$\frac{e}{E} = \frac{R}{r + R} \text{ (see equation [III.]),}$$

where R is the resistance of the external circuit, and r the internal resistance (armature, magnets, &c.) of the machine. Hence, for a series dynamo or a magneto machine,

$$\eta = \frac{R}{r + R}. \qquad [\text{VI.}]$$

Obviously, this coefficient will approach more and more nearly to *unity* the more that the value of r can be diminished. For if a machine could be constructed of *no* internal resistance there would be none of the energy of the current expended in driving the current through the armature and wasted in heating its coils.

We shall see later on how the expression for the economic coefficient η must be modified in the case of shunt dynamos and compound dynamos.

Returning now to the real efficiency of the machine, let us use the symbol W for the mechanical work-per-second, or horse-power, actually used in driving the machine. And, remembering that the gross electric activity of the machine is $\frac{E\,i}{746}$, we have for the *gross efficiency*, or efficiency of electric conversion,

$$\frac{E\,i}{W \times 746},$$

and for the *nett efficiency*, or useful commercial efficiency,

$$\frac{e\,i}{W \times 746}.$$

It will be seen that, as the first of these expressions contains E, and the second e, the nett efficiency can be obtained from the gross efficiency by multiplying by η, the economic coefficient.

It must be noticed before passing from this topic that since i, the strength of the current, enters into each of the expressions for efficiency as a factor, and as i depends not only on the resistance of the machine itself, but on that of the lamps, or other parts of the system which it is used to feed, it is somewhat misleading to talk of the efficiency of the *dynamo*, as if the efficiency was a property of the dynamo.

On the contrary, not only the gross efficiency, but also economic coefficient, and therefore *a fortiori* the nett efficiency, depend on the external resistance, that is to say on the number of lamps that may happen to be alight ! But still there is a sense in which these expressions are justifiable. Every dynamo is designed to furnish a certain quantity of lamps, and therefore to carry a certain average current. Its efficiency and coefficient of economy ought therefore to be expressed in terms of that current (and of that external resistance) which may be considered the fair working load of the machine.

Variation of Economic Coefficient with Current.

It will be noticed that in the case of the series machine considered above, the value of η will be different when R the external resistance is varied. When R is very great compared with r, then the value of η is very nearly $= 1$; but for small values of R, the value of η diminishes indefinitely. But when R is large the current is small, and when R is small the current is large. It appears, therefore, that a series dynamo has its maximum value for the economic coefficient when it is doing its minimum of work. A curve showing the relation is given in Chapter XXII., Fig. 312.

Relation of Size to Capacity and Efficiency.

For some years past the author has been the advocate of large dynamos, not because he has any admiration for mere bigness, but because, as in steam-engines, so in dynamos, the larger machines may be made more efficient than the small, in proportion to their cost. There has been a considerable amount of controversy upon the relation that subsists between the linear dimensions of similar machines and their permissible output and their efficiency ; the divergence of views arising mainly from difference of opinion as to the assumptions that are suitable at the outset. In the first edition of this work the author laid down the proposition that if the speed of rotation remains the same and the intensity of the magnetic field (per square centimetre) is also maintained unchanged, then the output of a machine n times as great, in all linear dimensions, as any given machine, would be increased in the proportion of n^5, and the coefficient of waste would be in the proportion of n^{-3}. Now though this may be *a priori* true in the abstract, the very assumptions made for the sake of simplification are opposed to actual conditions of working; for it is inexpedient to drive large machines at the same speed as small ones; and to procure equal magnetisation with large magnets wastes more energy in proportion on account of the relatively greater difficulty of getting rid of the waste heat from the magnetising-coils, the energy requisite to be spent on magnetising being nearly proportional to the volume of iron to be magnetised and the power of getting rid of the heat being only proportional to the surface. Amongst those who have discussed the problem are Hopkinson, Frölich, Ayrton, Mascart and Joubert, Kapp, Storch, Rechniewsky, and Pescetto. According to Hopkinson[1] the capacity of similar machines is proportional to the *cube* of their linear dimensions; the work wasted in magnetising the field-magnets proportional to the linear dimensions, whilst that wasted in heat in the armature conductors is proportional to the square of the linear dimensions. Mascart and Joubert[2] place the capacity as low as the *square* of the linear dimensions, and draw the conclusion that small machines are preferable to large ones. Pescetto[3] arrives at similar conclusions. Rechniewsky[4] follows Hopkinson in assigning n^3 as the proportion. Frölich[5] assigns the value n^4; and criticises the rule of the fifth power given by the author of this work and by Deprez as involving an increase of n^3 in the current, whilst there is an increase of only n^2 in the section of the conducting wires; which further involves that the density of the current in the wires must increase with the size of the machine, which is clearly

[1] *Proc. Instit. Civil Engineers*, April 1883.
[2] *Leçons sur l'Électricité* (1886), ii. 815.
[3] *L'Électricien*, xi. 357, 1887.
[4] *La Lumière Électrique*, xxii. 311.
[5] *Die dynamoelektrische Maschine* (1886), p. 168.

impracticable. Storch [1] considers constant-current machines to be in a different category from constant-potential machines. Assuming equal intensity of magnetic-field, equal peripheral velocity, and equal permissible current-density, he finds that in all machines the ampère-turns requisite for excitation vary as the linear dimensions. For constant-current machines the capacity is proportional to n^3, that is to say to the weight of the machine, or to the volume of copper on the armature. For constant-potential machines he finds the total length of wire on the armature to be independent of the dimensions of the machines; the number of external armature conductors to vary inversely as the linear dimensions; whilst the capacity of the machines is found to vary as n^4, though with undue heating, unless the volume of copper on the armature is also increased as n^4. Storch and Rechniewsky agree with Hopkinson that the work lost in field-magnets decreases, relatively to that lost in armatures, with an increase in the linear dimensions.

The most recent contributions to the discussion come from Mr. Kapp [2] and Professor Ayrton. [3] Kapp proposes that the speeds of rotation shall be assumed to vary inversely as the linear dimensions so as to put all machines into equal conditions as regards strains from centrifugal force, and that all the similar machines shall be considered as being worked up to the same safe limit of heating. This involves that the work wasted internally in heat shall be proportional to surface or as $1 : n^2$. The resistances, both magnetic and electric of the field-magnets will be proportional to n^{-1}, and the exciting powers to $n^{\frac{3}{2}}$. The intensities of field will be proportional to $n^{\frac{1}{2}}$, and the electromotive-forces to $n^{\frac{3}{2}}$. The diameters of wires allowed are as n^1 on the magnets and $n^{\frac{3}{4}}$ on the armatures: the resistances of armatures will be proportional to n^{-2}, and the permissible current to n^2. It follows at once that the capacities of the machine (in watts) will vary as $n^{3\frac{1}{2}}$, whilst the work wasted will vary as n^2: hence the economic coefficient

[1] *Centralblatt für Elektrotechnik*, viii. pp. 544, 594, and 743, 1886.
[2] *Proc. Instit. Civil Engineers*, lxxxiii. p. 36, 1886.
[3] *Ibid.*, p. 116.

will increase with the size of the machine. Kapp gives the cost of machines as proportional to $n^{2\frac{1}{2}}$, whence it follows that the cost of a dynamo per unit of output (say per lamp) varies inversely as its linear dimensions. He gives the following illustrative table :—

Diameter of armature (inches)	10	15
Revolutions per minute	1000	670
Number of glow-lamps	150	620
Weight (in hundredweights)	10	34
Price	100*l.*	276*l.*
Price per lamp	13*s.* 4*d.*	8*s.* 11*d.*
Electrical (efficiency per cent.)	80	89

Professor Ayrton assumes that the speeds of similar machines may be safely put as inversely proportional to the square-roots of the linear dimensions or as to $n^{-\frac{1}{2}}$ instead of n^{-1}, for the number and the strength of the binding wires could easily be increased in the larger machines. In the larger machines the smaller relative space required for clearance makes admissible the increase of the current in proportion to n^2. But this increased current would magnetise the iron more highly in proportion, and the electromotive-force would be greater than $n^{\frac{1}{2}}$, probably nearer $n^{1\cdot7}$; bringing up the capacity to be proportional to $n^{3\cdot7}$.

The common opinion of dynamo constructors appears to be that the capacity of dynamos is, for similar machines, a little greater than in proportion to the weight.

CHAPTER XVII.

THE SERIES DYNAMO, AND ITS CHARACTERISTIC.

IN the series dynamo (see Fig. 262, also Fig. 73), there is but one circuit, and therefore but one current, whose strength i depends on the electromotive-force E and on the sum of the resistances in the circuit. These are :—

FIG. 262.

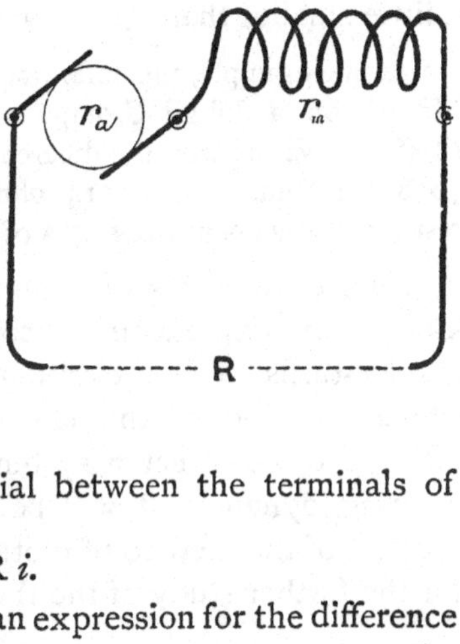

R = the external (variable) resistance.

r_a = the resistance of the armature.

r_m = the resistance of the field-magnet coils.

By Ohm's law—

$$E = (R + r_a + r_m)\, i.$$

Also e, the difference of potential between the terminals of the machine, is

$$e = R\, i.$$

It is also convenient to find an expression for the difference of potential between the brushes of the machine ; the volts measured here being greater than e, because of the resistance of the field-magnets ; and less than E, because of the resistance of the armature coils. For this difference of potential between brushes we will use the symbol ϵ. Then, by Ohm's law, remembering that the current running through r_m and R is of strength i, we have

$$\epsilon = (R + r_m)\, i\,;$$

whence, also,

$$e = E - (r_a + r_m)\, i.$$

Economic Coefficient of Series Dynamo.

From Joule's law of energy of current it follows that the economic coefficient η, which is the ratio of the useful electric energy available in the external circuit to the total electric energy developed, will be

$$\eta = \frac{\text{useful work}}{\text{total work}} = \frac{i^2 \, R \, t}{i^2 \, (R + r_a + r_m) \, t} = \frac{e}{E},$$

or

$$\eta = \frac{R}{R + r_a + r_m}. \qquad [\text{XII.}]$$

This is obviously a maximum when r_a and r_m are *both* very small. Sir W. Thomson recommends that r_m be made a little smaller than r_a. A good proportion is two-thirds.

As an example, the following is taken from the tests made at Munich on a Bürgin-Crompton dynamo of which the economic coefficient varied with the different external resistances from 62·5 to 70·8 per cent. :—$r_a = 2 \cdot 14$ ohms, $r_m = 1 \cdot 78$. Here the magnet resistance was about five-sixths of that of the armature.

Further than this we cannot go without introducing some kind of an expression to connect E with the number of ampère-turns in the exciting coils. But this involves the whole question of the law of magnetisation. Before we attempt to construct a systematic set of equations for self-exciting dynamos, it will be more convenient to give some account of the method of plotting out the characteristic curves for the further study of the theory of the machine.

Characteristic Curves.

So many practical problems in the construction of dynamo-electric machines are in the present state of science solved by the use of graphic diagrams, and particularly by the use of certain curves technically called *characteristics*, that the method of constructing and using them forms an important part of the theory of the dynamo. For many practical purposes no other method is half so useful.

The characteristic curve stands indeed to the dynamo in a

relation very similar to that in which the indicator diagram stands to the steam engine. As the mechanical engineer, by looking at the indicator diagram of a steam engine, can at once form an idea of the qualities of the engine, so the electrical engineer, by looking at the characteristic of the dynamo, can judge of the qualities and performance of the dynamo. The comparison may even be said to reach further than this.

The steam-engine indicator diagram serves two purposes which though not unconnected with one another are yet distinct. When the scale on which the diagram is drawn is known, it gives direct information as to the horse-power at which the engine is working, depending on the total area enclosed by the curve, and quite irrespective of its form. But even though the actual scale be not known, the details of the form of the curve at its various points give very definite information to the engineer as to the working of the engine, the perfection of the exhaust, the setting of the valves, the efficiency of the cut-off, and the adequacy of the supply pipes and port-holes of the valves.

So also the characteristic curve of the dynamo may serve two functions. When the scale on which it is drawn is known it tells the horse-power at which the dynamo works; nay, can indicate at what horse-power the dynamo may be worked to the greatest profit. But even though the actual scale be not known, the details of the form of the curve afford definite information as to the conditions of the working of the machine; the degree of saturation of its magnets, the sufficiency of the field-magnets in proportion to the armature, and the goodness of the design in several respects.

The suggestion to represent the properties of a dynamo machine by means of a characteristic curve is due to Dr. Hopkinson, who in 1879 described such curves to the Institution of Mechanical Engineers, and gave the curve of the Siemens dynamo reproduced in Fig. 263. The name of "characteristic" was assigned in 1881 by M. Marcel Deprez [1]

---

[1] Vide *La Lumière Électrique*, Dec. 3, 1881 ; where, however, Deprez gives a method of observation that is open to the objection that it neglects the armature reactions.

to Hopkinson's curves; and the excellence of the name has been attested by its general adoption.

Dr. Hopkinson's object was to represent the relation subsisting between the electromotive-force and the current; he therefore constructed from observations a curve in which the abscissæ measured horizontally represent the number of ampères of current flowing, and the vertical ordinates the corresponding values of the electromotive-force. The following table (taken with some trifling modifications from Dr. Hopkinson's paper in the *Proceedings of the Institution of Mechanical Engineers*, 1879, p. 249) gives the observed values of i the strength of the current, and E the electromotive-force, of a certain series-wound dynamo :—

EXPERIMENTS ON SIEMENS DYNAMO AT SPEED OF 720 REVOLUTIONS PER MINUTE.

Current (in ampères). i	Resistance (in ohms). R	Electromotive-force (in volts). E
0·0027	1025	2·72
0·48	8·3	3·95
1·45	5·33	7·73
16·8	4·07	68·4
18·2	3·88	70·6
24·8	3·205	79·5
26·8	3·025	81·1
32·2	2·62	84·4
34·5	2·43	83·8
37·1	2·28	84·6
42·0	2·08	87·4

It may be remarked that the electromotive-force E is the total electromotive-force generated in the machine, and must not be confounded with e the difference of potential between the terminals as measured by a voltmeter, or other similar instrument. In many cases we now prefer to plot e instead of E; but that was not Hopkinson's original method. He determined E by measuring i and multiplying it by the total resistance of the circuit; for by Ohm's law $i\,\mathrm{R} = \mathrm{E}$. It should also be remarked that the dynamo was a "series dynamo," shunt-wound machines not having at that date come into vogue.

Before entering into other points, it may be worth while to consider the meaning of the curve. It begins at a point a little above the origin. This shows that there was a small amount of residual magnetism remaining permanently in the

FIG. 263.

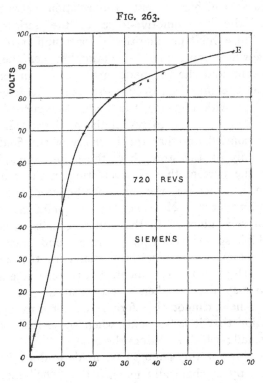

CHARACTERISTIC CURVE OF A SERIES DYNAMO.

field-magnets. The curve ascends at first at a steep angle, then curves round and eventually assumes a nearly straight course, but at a gentler slope than before. Whence arise these typical forms? It is known that the electromotive-force of a dynamo depends not only on the speed of running and on the number of coils of wire in the armature, but also on the intensity of the magnetic field. Now if the speed is constant—it was maintained at 720 revolutions per minute in Hopkinson's experiments— the only variable of importance is the intensity of the magnetic field. As the magnetism of the field-magnets rises and grows toward its maximum, the intensity of the field also rises and grows toward a maximum, and so does the induced electromotive-force. We might therefore expect, as Hopkinson points out, that the curve which represents the relation between the current and the electromotive-force of the series-wound dynamo should exhibit peculiarities of form similar to those of the curve which represents the relation between the magnetising current and the magnetic moment of an electro-magnet; and a comparison of Fig. 263, the "characteristic" of the dynamo, with Fig. 232, the "saturation curve" of an electromagnet, will suffice to reveal the analogy. It must, however, be pointed out that the intensity of the field does not depend only on the strength of the field-magnet, but is affected by the current that is circulating in the armature coils, in consequence of the reaction between the field-magnets and armature. Moreover, certain prejudicial effects arising from self-induction in the armature coils come into play with high speeds and strong currents, and prevent the electromotive-force from being proportional to the strength of the field. Hopkinson's statement, that the characteristic of the dynamo may also be taken to represent the intensity of the magnetic field, cannot therefore be accepted except with the reservation that it is true only when these reactions are negligibly small ; which is seldom the case.

It is possible for a dynamo to be made to draw its own characteristic by mechanically moving the pencil relatively to the paper (as in steam indicators) by means of two electro-magnets, one of them being excited by the main current, the other being connected as a shunt to the terminals of the machine.

Dr. Hopkinson in the paper alluded to, and in a second one published in the *Proc. Inst. Mech. Engin.*, in April 1880, p. 266, pointed out a great many of the useful deductions to be drawn from a consideration of these curves. Some other deductions have been made by M. Marcel Deprez for which the reader is referred to *La Lumière Électrique* of Jan. 5th,

1884. Dr. Frölich has also published several important papers on the subject in the *Elektrotechnische Zeitschrift* for 1881 and 1885. Dr. Hopkinson has returned to the subject in a lecture before the Institution of Civil Engineers, " On Some Points in Electric Lighting," April 1882.

Horse-power Characteristics.

As mentioned at the beginning of this chapter, if the characteristic curves are drawn to scale the activity of the dynamo may be read off from them in horse-power. The product of the current into the potential is proportional to the rate at which the electric energy is being evolved. The product of one volt of potential into one ampère of current is sometimes called one *volt-ampère*, and has also been called by the special name of one *watt*. One watt or volt-ampère is equal to $\frac{1}{746}$ of a horse-power. To calculate the horse-power (electrical) evolved in the circuit when the dynamo is running at any particular speed, with a particular number of lamps in circuit, two measurements have ordinarily to be made—the volts of electromotive-force and the ampères of current. These must then be multiplied together and divided by 746 to obtain the horse-power. But if the characteristic of the dynamo at the particular speed be known, a reference to the curve will show at once what the electromotive-force is that corresponds to any particular current. For example, in the Siemens dynamo examined by Hopkinson, the characteristic of which is given in Fig. 263, p. 371, suppose the dynamo was working through such a resistance as to give 30 ampères when running at 720 revolutions, we see at once that the corresponding electromotive-force is 83. Hence

$$\frac{83 \times 30}{746} = 3\cdot3 \text{ horse-power.}$$

Now to obviate such calculations we may plot out on the diagram some additional curves crossing the characteristics and mapping them out into equal values of horse-power. These "horse-power lines" are nothing else than a set of rectangular hyperbolas. For example, the 1-horse-power line will pass through all the points for which the product of volts and ampères is equal to 746. It will therefore pass through the point corresponding to 74·6 volts and 10 ampères ; through 37·3 volts and 20 ampères ; through 14·92 volts and 50 ampères, &c., because the products in each of these cases is equal to 746 watts or 1 horse-power. The 2-horse-power line will pass through points whose product values are equal

FIG. 264.

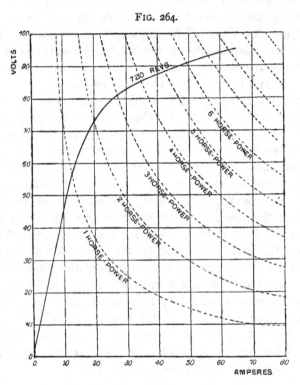

CHARACTERISTIC WITH HORSE-POWER LINES.

to 746 × 2, and the other lines in the same way.   Fig. 264 shows the characteristic of the Siemens machine, reproduced from Fig. 263 above, but with the horse-power lines added.

In this case the volts plotted are the total electro-motive-force, "E," of the dynamo, and therefore the horse-power represents the total electric energy converted per second in the circuit of the dynamo.   If instead of E we had plotted the values of "*e*," the difference of potentials between the terminals, we should have had a slightly different curve, representing the amount of electric energy appearing per second in the external circuit and available for useful purposes.

A horse-power characteristic of a shunt-wound dynamo is given further on, in Fig. 273.

If the vertical and horizontal scales are not chosen equal, the horse-power lines, though hyperbolæ, are of course distorted.

"External" Characteristics or Terminal Potential Curves.

For many purposes it is more useful to know the relation between the current and the "external" difference of potential at the terminals than to know the relation between the current and the whole electromotive-force induced in the armature : and it is mostly easier to measure *e* than to measure E ; seeing that while the former can be directly measured with a voltmeter, the latter can only be got at indirectly. The name *external characteristic* may be given for the sake of distinction to those curves which exhibit the relation between the potentials and the currents of the external circuit. In the series dynamo it is a simple matter to derive one of these curves from the other, provided the internal resistance of the machine (armature and field-magnets) is known. In the Siemens dynamo examined by Hopkinson in 1879, and of which Figs. 263 and 264 give the total characteristic, the total internal resistance was 0·6 ohm. The curve is reproduced for a third time in Fig. 265, where it is marked "E." Now to force a current of 10 ampères through a resistance of 0·6 of an ohm would require a difference of potential of 6 volts between its terminals. Looking at the curve, we see that the whole electromotive-force corresponding to 10 ampères was about 46·5 volts. Of this number, six were employed as mentioned in overcoming the internal resistance, leaving 40·5 volts as the available potential between terminals. Further, when the current was running at 50 ampères, there must have been no less than 30 volts employed in overcoming the internal resist-

ance of 0·6 ohm; and as the value of E for this current is
90·5 volts, there remain 60·5 volts for *e*.   There are now two
ways open to us of representing these matters on our diagram.
They are both shown in Fig. 265.   The line J is drawn through
the origin, and through the values of 6 volts for 10 ampères
and 30 volts for 50 ampères.   (The tangent of the slope of the

FIG. 265.

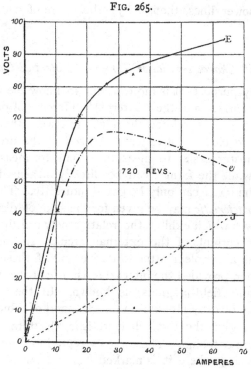

TOTAL AND EXTERNAL CHARACTERISTICS.

line J is equal to 6 ÷ 10 = 0·6.   We shall see later that this
slope represents the internal resistance.)   Then if the heights
of the ordinates from the base line up to the line E represent
total volts induced, and if the heights of the ordinates from
the base line up to the line J represent the corresponding volts
used in overcoming internal resistance, it follows that *the difference of potentials at the terminals will be represented by the*

differences of the ordinates between the lines J *and* E. This is the first way of representing those differences of potential. The second way is to cut off from the tops of the ordinates portions equal to those of the line J. This amounts to subtracting the internal volts, which as shown in the algebraic theory are equal to $i\,(r_a + r_m)$, from E, and so obtaining the values of e. These are plotted out in the curve marked "e" in the figure ; and as this curve represents the available electromotive-force in the external circuit it obtains the name of *external characteristic* or terminal potential curve.

Characteristics of Series Dynamo.

The Siemens dynamo of which the characteristic is given in Fig. 263 was a series dynamo. For the sake of comparison the characteristic is given in Fig. 266 of an "A" Gramme

EXPERIMENTS ON GRAMME SERIES-WOUND DYNAMO.

Current (in ampères).	Electromotive-force (in volts).	
	Speed 1440.	Speed 950.
5	72	45
10	107	70
15	122	77
20	127	79
25	129	79
30	128	79
35	128	79
40	127	78
45	125	76
50	123	74
55	120	72
60	116	..
65	110	..
70	101	..

machine also series-wound. This machine had, when it was measured by M. Marcel Deprez, 0·41 ohm resistance in the armature coils and 0·61 ohm in the coils of the field-magnets. Two characteristics are given : one corresponding to a speed of 1440, the other to a speed of 950 revolutions per minute. The horse-power lines are shown in dot also. The figures are given in the preceding table.

In the series dynamo the magnetisation of the magnets increases with the current, and therefore, at first, the electromotive-force increases also, giving the first straight portion of the curve. As the magnets approach saturation the curve turns, and, as the reactions due to the current in the armature

FIG. 266.

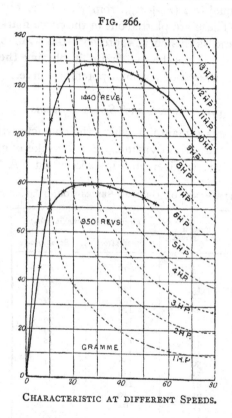

CHARACTERISTIC AT DIFFERENT SPEEDS.

now become of relatively great importance, flattens itself and ultimately turns down again.

One circumstance that contributes to the diminution of the electromotive-force when large currents are passing through the armature, is that if the field-magnets are relatively not powerful, the cross-magnetising effect of the armature current causes a great displacement of the neutral point, and obliges a considerable lead being given to the brushes, with the result that the demagnetising effect of the armature on the field-magnets (p. 92) becomes greatly increased. The dip in the characteristic is always greater in the case of weak field-magnets. It also occurs most in those machines in which the core of the armature is more nearly saturated than the cores of the field-magnets; for when with large currents the armature cores get saturated, the magnetic leakage from the pole-pieces becomes relatively greater.

One more curve of a series-wound dynamo is given in Fig. 267. This is a small Brush machine (intended to supply a single arc light), of the old pattern with solid iron ring, in which, owing to the peculiar arrangement of the coils, (see p. 231), the reactions of the armature make themselves known by a very extraordinary down-bending of the characteristic. This is partly due to the arrangements for cutting out a pair of coils as they approach the neutral point. It will be noticed that the maximum horse-power of this small machine is $1\frac{3}{4}$ horse; and that this value is only obtained when the reactions have already set

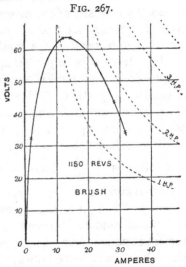

FIG. 267.

DESCENDING CHARACTERISTIC.

in. In the first edition of this work this circumstance was alluded to in terms which were needlessly condemnatory. The remarkable diminution in the electromotive-force which takes place when the machine is so treated as to demand from it an output which it was never intended to give, is in practice a real advantage. Should the machine be accidentally short-circuited while running, the reactions of the armature prevent the production of an injuriously large current, which might overheat the coils. It is considered an advantage in machines for arc-lighting, where a nearly constant current is required, to employ machines with drooping characteristics, and to work them at this part of the curve.

Relation of Characteristic to Speed.

We know that the electromotive-force generated in a rotating coil or armature, would be strictly proportional to the intensity of the magnetic field, were it not for the reactions of the current in the armature. Now in a series dynamo, the intensity of the field depends on the strength of the current ; and, if the current is kept constant (by adjusting the resistances), the intensity of the magnetic field will also be constant even though the speed of the armature be varied. If therefore the characteristic of a machine be known at any speed, its characteristic for any other speed can be found by the very simple process of increasing the ordinates of the curve in a similar proportion. Take, for example, the case of the Gramme dynamo, of which a characteristic at the speed of 950 revolutions is given in Fig. 266. The characteristic at 1440 could be calculated from it by increasing the ordinates in the proportion of $\frac{1440}{950}$. Thus we see from the lower curve that when the current was 20 ampères the electromotive-force was 79 volts. Then $79 \times 1440 \div 950 = 119\cdot7$ volts. The actual electromotive-force observed at the speed of 1440 and with current at 20 ampères was 127 volts. There is a slight discrepancy here ; and indeed always ; for dynamo machines behave invariably as if a certain number of the revolutions did not count electrically. If the number of " dead turns " were here reckoned as 140, the number of volts calculated by theory would agree very exactly with that observed.

Resistance in the Characteristic.

In the characteristic we have volts plotted vertically and ampères horizontally. Now by Ohm's law, volts divided by ampères give ohms. How can this be represented in the characteristic? Suppose, for example, it is required to represent the resistance of the circuit corresponding to some particular current. Let Fig. 268 be the characteristic of the dynamo in question, and it is desired to know what is the resistance corresponding to the state of things at the point marked P. Draw the vertical ordinate P M, and join P to the origin O. The line P O has a certain slope, and the angle of its slope is P O M. Now P M is equal to the electromotive-force under consideration, and O M is the current. Therefore, by Ohm's law,

$$\text{Resistance} = \frac{\text{electromotive-force}}{\text{current}} = \frac{P M}{O M};$$

but

$$\frac{P M}{O M} = \tan P O M;$$

therefore the resistance = tan P O M. Put into words, this is :—*The resistance corresponding to any point on the characteristic is represented in the characteristic by the trigonometrical tangent of the angle made by joining the point to the origin.*

An easy way of reckoning these tangents is shown in Fig. 268. At the point on the horizontal line corresponding to 10 ampères erect a vertical line. A line drawn from the origin at an angle whose tangent is = 1 (namely 45°) would cross this vertical line at a point opposite the 10-volt mark. This point may then be called 1 ohm, and equal

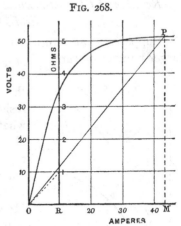

Fig. 268.

METHOD OF REPRESENTING RESISTANCE GRAPHICALLY.

distances measured off on this line will constitute it a scale of resistances. In Fig. 268 the resistance corresponding to point P of the characteristic is seen to be about 1·2 ohms on the scale of resistances. Now P is placed at 51·3 volts, and the current is 43·2 ampères. Dividing one by the other, we get

1·18 ohms. Such calculations are then obviated by the graphic construction.

If in the actual dynamo the resistance of the circuit were gradually increased, we should have the point P displaced along the curve backwards towards the origin, the volts and ampères both falling off, and the steepness of the line O P increasing. When O P arrived at a certain steepness it would practically form a tangent to that part of the characteristic which is nearly straight, and then any very small increase in the resistance would cause the dynamo to lose its magnetism, from lack of current to magnetise the magnets. The resistance may be similarly represented on the characteristics of shunt dynamos (see Fig. 273); but in this case if the characteristic is drawn for the external current and the external difference of potential, then the resistance so represented will be the external resistance.

Relation of Characteristic to Winding of Armature and Field Magnets.

Suppose the armature of a machine to be re-wound with a larger number of turns of proportionally thinner wire. What will be the result when rotated at the same speed as before? The resistance will be increased somewhat, and the electromotive-force also will be higher. Let Fig. 269 represent the characteristic of the machine as it was when there were X turns of wire on the armature. How must it be drawn when the number is increased to X'? Let P represent a point corresponding to a certain strength of current. Taking the new armature, let the resistance be varied until the current once more comes to the same value. The magnets are now magnetised exactly as strongly as before; but there are X' turns

FIG. 269.

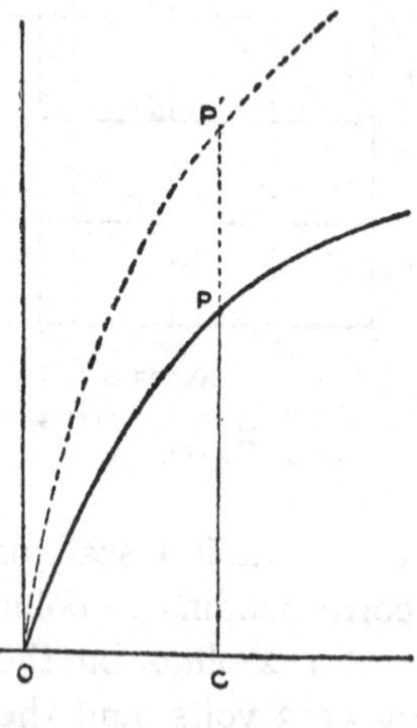

of wire cutting the lines of magnetic force instead of X. The electromotive-force will therefore also be greater in the proportion of $\frac{X'}{X}$. Draw therefore P′ C so as to have the proportion P′C : PC :: X′ : X. All other points on the new characteristic can be obtained by similarly enlarging the ordinates in the same ratio.

It will be evident from this that increasing the number of turns of wire in the armature has the same effect as increasing the speed of driving. This shows that *slow-speed* dynamos (as required for use on ships, &c.) may be made to give the requisite electromotive-force provided the number of turns of wire be relatively increased. This involves, however, a sacrifice of economy because of the increase of resistance in the armature.

The effect of altering the number of turns of wire on the field-magnets can also be traced out on the characteristic diagram. Suppose the number of turns in the magnetising coil be S, and that we re-wind the machine, increasing the number to S′ turns. What will the result be? In this case we shall get the same electromotive-force when driving at the same speed as before, provided the magnets be equally magnetised. But if the current goes S′ times round instead of S we shall want a current only $\frac{S}{S'}$ as strong as before, to pro-

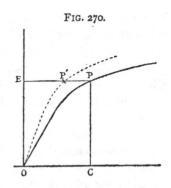

Fig. 270.

duce the same magnetisation. To get the new characteristic then (see Fig. 270) draw P E horizontally. P E = C O = the current corresponding to electromotive-force E. Find P′ such that P′E : PE :: S : S′; then the new characteristic will pass through P′. Similarly all other points of the new characteristic may be determined by reducing their abscissæ in a similar proportion.

It must be noted that these two processes are not admissible for the characteristics of shunt-wound machines.

Critical Current of Series Dynamo.

From the fact that the characteristics for different speeds differ only in the relative scale of the ordinates, an important consequence may be deduced. The first part of every characteristic for any speed is nearly straight up to a point where for that speed the electro-motive-force is nearly two-thirds of its maximum value. When the current is such that the electro-motive-force has attained to this value, any very small change either in the speed of the engine or in the resistance of the circuit produces a great change in the electro-motive-force, and therefore in the current ; therefore, since this critical case occurs always with the same current (see Fig. 271), this current—corresponding to the point on all the curves where the straight line begins to turn—may be called the "critical current" of the dynamo.

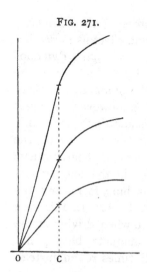

Fig. 271.

Each dynamo has its own critical current, and it will not work well with a less one ; for a less one will not adequately excite the field-magnets. It will further be seen that since with each speed the characteristic rises with a corresponding slope, there will be one particular resistance at each value of the speed, at which the critical current will be obtained, and the higher the speed the higher may be the resistance. There is no such thing as a critical resistance in a series dynamo ; for whether a resistance is critical or not depends upon the speed. *Neither is there any such thing* per se *as a critical speed for a series dynamo ;* for whether the speed is critical or not depends on the resistance of the circuit.

CHAPTER XVIII.

The Shunt Dynamo, and its Characteristic.

In the shunt dynamo, there are two circuits to be considered ; the main circuit, and the shunt circuit. The symbols used have the following meanings.

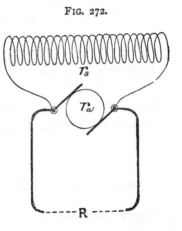

Fig. 272.

R = resistance of external main circuit (leads, lamps, &c.).

r_a = resistance of armature.

r_s = resistance of the shunt circuit (magnet coils).

i = the current in the external main circuit.

i_a = the current in the armature.

i_s = the current in the shunt circuit.

Then, clearly,

$$i_a = i + i_s;$$

because the current generated in the armature splits into these two parts in the main and shunt circuits, and is equal to their sum.

Also, by Ohm's law, we have for e the electromotive-force between terminals,

$$e = R\,i,$$

and also

$$e = r_s i_s;$$

because the terminals for the main circuit are also the terminals for the shunt circuit.

Further, since the nett resistance of a branched circuit is the reciprocal of the sum of the reciprocals of the resistances

2 C

of its parts, the nett external resistance from terminal to terminal is equal to $\dfrac{R\,r_s}{R + r_s}$; and hence it follows that

$$E = \left(r_a + \frac{R\,r_s}{R + r_s}\right)i_a.$$

We may at the same time find an expression for that part of the whole electromotive-force which is being employed solely to overcome the resistance of the armature, and which is, of course, the difference between E the total electromotive-force, and e the effective electromotive-force between terminals.

Ohm's law at once gives us

$$E - e = r_a\,i_a,$$

or

$$E - e = r_a\,(i + i_s).$$

From this we also get

$$e = E - r_a(i + i_s). \qquad\qquad \text{[XIII.]}$$

We will also find an expression for E in terms of e, and the various resistances. Taking as above

$$E = \left(\frac{R\,r_s}{R + r_s} + r_a\right)i_a,$$

and writing for i_a its value as $i + i_s$, and for these $\dfrac{e}{R}$ and $\dfrac{e}{r_s}$, respectively, we get

$$E = e\left\{\frac{R\,r_s + R\,r_a + r_a\,r_s}{R + r_s} \times \frac{R + r_s}{R\,r_s}\right\},$$

or

$$E = e \times r_a\left(\frac{1}{R} + \frac{1}{r_a} + \frac{1}{r_s}\right). \qquad \text{[XIII. *bis.*]}$$

It may be noted that the expression $\left(\dfrac{1}{R} + \dfrac{1}{r_a} + \dfrac{1}{r_s}\right)$ is the sum of three conductivities of three paths, and is therefore equal to the conductivity of these three paths united in parallel with one another; that is to say, the conductivity as measured from brush to brush with the external circuit and shunt circuit joined up. Or, if we write $\mathfrak{R}$ for the resistance of the whole system of machine and circuit, as thus measured from brush to brush, then the equation may be written

$$E = e \times \frac{r_a}{R}$$

Economic Coefficient of Shunt Dynamo.

The economic coefficient η, is the ratio of the useful electric energy available in the external circuit to the total electric energy developed.

By Joule's law there is developed in t seconds in the external circuit

$$\text{useful work} = i^2 R t,$$

and in the same time there is wasted on heating,

$$\text{energy spent in shunt} = i_s^2 r_s t,$$

and

$$\text{energy wasted in armature} = i_a^2 r_a t;$$

whence

$$\eta = \frac{\text{useful work}}{\text{total work}} = \frac{i^2 R}{i^2 R + i_s^2 r_s + i_a^2 r_a}$$

$$= \frac{1}{1 + \dfrac{R}{r_s} + \dfrac{i^2 r_a + 2 i i_s r_a + i_s^2 r_a}{i^2 R}}$$

$$= \frac{1}{1 + \dfrac{R}{r_s} + \dfrac{r_a}{R} + 2 \dfrac{i_s r_a}{i R} + \dfrac{r_a}{R}\left(\dfrac{i_s}{i}\right)^2}$$

$$= \frac{1}{1 + \dfrac{R}{r_s} + \dfrac{r_a}{R} + 2 \dfrac{R}{r_s} \cdot \dfrac{r_a}{R} + \dfrac{r_a}{R}\left(\dfrac{R}{r_s}\right)^2}$$

$$= \frac{1}{1 + \dfrac{R}{r_s} + \dfrac{r_a}{R} + 2 \dfrac{r_a}{r_s} + R \dfrac{r_a}{r_s^2}}$$

$$= \frac{1}{1 + \dfrac{R}{r_s}\left(1 + \dfrac{r_a}{r_s}\right) + \dfrac{r_a}{R} + 2 \dfrac{r_a}{r_s}}$$

Now, for brevity, write for the total internal resistance, $r_a + r_s$, the single symbol r—

$$\eta = \cfrac{1}{1 + \cfrac{R}{r_s} \cdot \cfrac{r}{r_s} + \cfrac{r_a}{R} + 2\cfrac{r_a}{r_s}}$$

For this ratio to be a maximum it is clear that,

$$\frac{d\left(1 + \dfrac{R}{r_s} \cdot \dfrac{r}{r_s} + \dfrac{r_a}{R} + 2\dfrac{r_a}{r_s}\right)}{d\,R} \quad \text{must} = 0,$$

or

$$\frac{r}{r_s^2} - \frac{r_a}{R^2} = 0 ;$$

whence

$$R^2 = \frac{r_a\,r_s^2}{r} = r_a r_s \frac{r_s}{r},$$

and

$$R = \sqrt{r_a r_s}\sqrt{\frac{r_s}{r}}, \qquad\qquad \text{[XIX.]}$$

or

$$R = r_s \sqrt{\frac{r_a}{r}} \qquad\qquad \text{[XIX}a.\text{]}$$

This equation determines what particular resistance of the external main circuit will give the best economy with given internal resistances. Now substitute this value in those terms of the equation for η which contain R, and we get as their values :—

$$\frac{R\,r}{r_s^2} = \frac{r}{r_s}\sqrt{\frac{r_a}{r}} = \frac{\sqrt{r_a r}}{r_s},$$

$$\frac{r_a}{R} = \frac{r_a}{r_s}\sqrt{\frac{r}{r_a}} = \frac{\sqrt{r_a r}}{r_s} ;$$

whence

$$\eta = \frac{\text{useful work}}{\text{total work}} = \cfrac{1}{1 + 2\cfrac{\sqrt{r_a r}}{r_s} + 2\cfrac{r_a}{r_s}}$$

This may be still further simplified, for we know that the resistance of the shunt is very high compared with that of the armature, possibly from 300 to 1000 times as great. If, then, $\frac{r_a}{r_s}$ is so small a term in comparison with the other term as to be negligible, we get

$$\eta = \frac{1}{1 + 2\frac{\sqrt{r_a r}}{r_s}};$$ [XX.]

and, since r_a is small compared with r_s, r is very nearly equal to r_s, so that we may write, as an approximate equality,

$$\eta \doteqdot \frac{1}{1 + 2\frac{\sqrt{r_a r_s}}{r_s}},$$

or

$$\eta \doteqdot \frac{1}{1 + 2\sqrt{\frac{r_a}{r_s}}}.$$ [XXI.]

This latter approximate value is identical with that given by Sir W. Thomson in the report of the British Association for 1881 : the equation No. [XX.] is, however, more correct.

It may be pointed out that it follows from equation No. [XIX.] above, that when the resistance of the armature is small compared with that of the shunt, so that r_s may be taken as equal to the value of r (which would be highly desirable if it could be attained in practice), then we should have

$$R = \sqrt{r_a r_s};$$ [XXII.]

that is to say, when the proportion between r_a and r_s is made as favourable as possible, then the best external resistance to work with from the economic point of view is that resistance which is a geometric mean between the resistances of the armature and of the shunt coils, and any departure from this will diminish the value of the economic coefficient.

Practical Rules for Economic Design.

This affords us some practical information how to apportion the resistances in a shunt dynamo. Let the question be thus stated. Given the resistance of the armature r_a, what must the shunt resistance be so that the dynamo may (under favourable proportions of external resistance R) have an economic coefficient of 90 per cent.? From equation [XXI.] we get

$$\frac{90}{100} = \frac{1}{1 + 2\sqrt{\dfrac{r_a}{r_s}}},$$

$$\frac{100}{90} = 1 + 2\sqrt{\frac{r_a}{r_s}},$$

$$10 = 180\sqrt{\frac{r_a}{r_s}},$$

$$r_s = (18)^2\, r_a,$$

$$r_s = 324\, r_a.$$

No shunt machine can give in the external circuit as much as 90 per cent. of its total electric energy unless its shunt has a resistance at least 324 times as great as that of its armature.

A good practical rule would be the following :—Ascertain what number of lamps will be the usual full load: reckon the resistance of them when connected to the mains. Let the armature resistance be one-twentieth of this ; and let the shunt resistance be twenty times as great as this. In this case about 4 per cent. will be wasted in the armature, and about 4 per cent. in the shunt, leaving a margin of a little over 90 per cent. for the economic coefficient.

In a shunt machine described by Sir C. W. Siemens in the *Philosophical Transactions,* 1880, the results were :—

	Armature.	Shunt Magnet.	r_s/r_a.	η observed per cent.
Siemens	0·234	11·26	48·4	69·3

In an Edison machine (" K," 250 lights) tested at Munich, the values were :—

	Armature.	Shunt Magnet.	r_s/r_a.	η observed.
Edison, "K"	0·0361	13·82	382·8	88·6

An Edison-Hopkinson (200 lights) machine tested by Mr. F. J. Sprague gave :—

(Cold)	0·026	36·5	1403	93·6
(Warm)	0·0325	37·0	1135	

The Edison-Hopkinson machine described on pp. 201 to 203 gave :—

(Cold)	0·009947	16·93	1702	93·66

Characteristic of Shunt Dynamo.

For the shunt dynamo there are two separate characteristics ; the *external characteristic,* in which the quantities plotted are the ampères of current in the external circuit and the volts of potential between terminals ; and the *internal characteristic*, in which the volts and ampères of the shunt circuit are plotted. The internal characteristic of the shunt dynamo is quite similar to the external characteristic of a series dynamo, and shows the saturation of the field-magnets. It is better to plot it with ampère-turns instead of ampères, because the magnetisation depends on the number of turns in the coil as well as the ampères.

The external characteristic of a Siemens shunt dynamo (the same described by the late Sir William Siemens before the Royal Society in 1880 and by Mr. Alexander Siemens in

the *Journ. Soc. Teleg. Eng.*, March 1880) is given in Fig. 273, and
the horse-power lines are shown dotted. The utmost power
of this machine at 630 revolutions was just under 2 horse-
power with a current of 30 ampères, and an electromotive-
force of 47·5 volts.

FIG. 273.

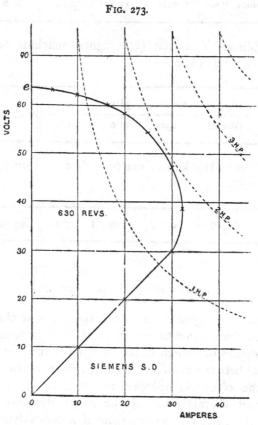

EXTERNAL CHARACTERISTIC OF SHUNT DYNAMO.

The curve of the shunt dynamo is curiously different
from that of the series dynamo. It begins with a straight
or nearly straight portion, which turns up in a curve, and
eventually returns nearly horizontally to the axis of electromotive-force. The straight portion represents the unstable

state when the shunt current is less than its true critical value. The critical external current, if it can be so called, is that current for which the shunt begins to act fully, and in Fig. 273 is about 30 ampères. From this point the shunt current acts with great power and the electromotive-force here rises very rapidly. The slope of the line which constitutes the first portion of the characteristic represents the resistance which for the particular speed may be termed the critical resistance, and in this case is about 1 ohm. If the resistance of the external circuit becomes in the least degree altered, the electromotive-force and current will alter enormously. Any less resistance will cause the

FIG. 274.

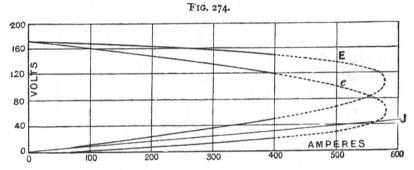

CHARACTERISTIC OF A SHUNT DYNAMO.

magnets to lose their magnetism at once. Any greater resistance will at once run the electromotive-force up above the critical value—in this case about 30 or 31 volts. If the resistance be steadily increased (and the slope of the line from O to the curve be increased in steepness) the electromotive-force will go on steadily augmenting, and become a maximum when the external resistance is infinite, that is to say when the circuit is completely opened and the shunt coils receive the whole of the current flowing in the armature. Fig. 274 depicts the characteristic of a shunt-wound Gramme dynamo capable of giving 400 ampères. In this case the curve e represents the external characteristic, from which the curve E is calculated by adding to the ordinates portions equal to $r_a i_a$.

As the conductors of the armature could not safely carry more than 400 ampères, the dotted portion of the curve represents results not actually observed. It is instructive to contrast the characteristic of the shunt dynamo with that of the series dynamo (Fig. 263). In the series dynamo also, the first part of the characteristic is a sloping line, and the tangent of the angle of its slope is also the critical resistance for the given speed. But the series dynamo will only work if the resistance of the external circuit is *less* than the critical value, and the shunt dynamo will only work if the external resistance is *greater* than the critical value. The contrast is even better shown by drawing a couple of curves in the two cases—not

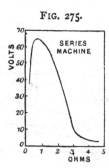

FIG. 275.

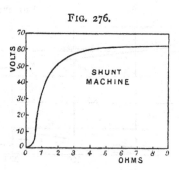

FIG. 276.

characteristics—showing the relation between the potential at terminals and the resistances of the external circuit. Fig. 275 shows this for a series machine, and Fig. 276 for a shunt machine. The electromotive-force of the one drops suddenly when the resistance exceeds 2 ohms; that of the other rises suddenly when the resistance attains 1 ohm.

In the shunt dynamo the characteristic for a doubled speed cannot be obtained as in a series dynamo by doubling the heights of the ordinates. For, even if at double speed we adjust the external resistances so that the external current is the same as before, we do not get a double electromotive-force because we do not get the same current as before round the shunt-magnet circuit. And if, on the other hand, we adjust resistances so that we get the same shunt current as before, and therefore a double electromotive-force, we do not get the same external current as before. If, however, we alter the external resistance, taking a larger current externally, so as to reduce the shunt current to its former value, the magnetisation will remain as before. In that case the double speed will produce very nearly a double electromotive-force ; but the shunt potential may remain as before, the external current being nearly doubled. This is shown in Fig. 277, where *e a* represents the external current in the first case, and *e* A the external current in the second. O A remains a straight line, but at this higher speed the slope is less. From this latter circumstance it may be foreseen that at higher

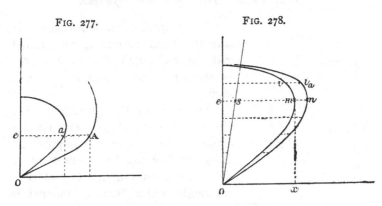

FIG. 277. FIG. 278.

speeds the resistance may be reduced to a lower value before the critical state is reached at which the machine "unbuilds" itself, i. e. discharges the magnetism from its field magnets.

Curve of Total Current in Armature.

In the shunt dynamo the current in the armature is equal to the sum of the currents in the external circuit and in the shunt circuit ; or,

$$i_a = i + i_s.$$

A curve showing the relation between i_a and e is easily obtained. In Fig. 278 let the curve O *m i* be the "external characteristic" at the given speed. Take any point on it *m* ;

at that point the potential between terminals in volts is measured by the length of *m x* or O *e,* and the current in ampères is measured by the length O *x* or *e m.* Now draw the line *s* O at such an angle *s* O *x* that its tangent is equal to the resistance of the shunt. Then *e s* represents the current that will run through the shunt when the potential is O *e* volts. Add on to the end of *e m* a piece *m n* equal to *e s;* then the whole line *e n* represents the armature current i_a when the potential has the value O *e.* A set of similar points may be found giving the new curve O *n* i_a required.

Total Characteristic of Shunt Dynamo.

FIG. 279.

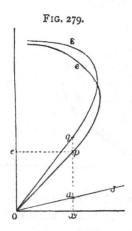

If the total electromotive-force E and the total current i_a be plotted out, we shall obtain the characteristic of the total electrical activity of the dynamo.

Draw, as in the preceding case, the curve for *e* and i_a. Let *p* be any point on the curve where the potential is *p x* or O *e* and the current *e p* or O *x.* Then draw a line O J at such an angle *a* O *x* that its tangent is equal to the resistance of the armature. Call the point where this cuts *p x, a.* Then *a x* represents the number of volts required to drive the current O *x* through the armature resistance. Add a piece *q p* equal to *a x* to the summit of the line *p x.* Then the height *q x* represents the total electromotive-force E when the current i_a has the value represented by O *x.*

Characteristic of Shunt Dynamo, with Permanent Magnetism.

If there is residual magnetism in the field-magnets, there will be an electromotive-force induced, even before the shunt circuit is closed. In this case the characteristic would begin at a point *v*, a short distance along the horizontal axis. In fact the machine behaves as though there were already at work some small electromotive-force (not to be plotted), which had the effect of setting up already a current through the machine, so that the machine excites itself up with currents that are, in the early (and unstable) stages of the magnetisation, proportional to the ampère-turns going round the shunt circuit, *plus* some imaginary ampère-turns causing the permanent magnetism. If there is on the field-magnet a second coil by which an independent magnetisation can be introduced, the same kind of result will follow : the characteristic will begin at some point, such as V, the electromotive-force due to the ampère-turns in the shunt being plotted out above O, whilst the length O *q* below represents the part of the electromotive-force due to the ampère-turns (real or imaginary) of the independent magnetism ; and O V represents the current which the machine will give when short circuited.

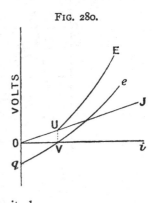

FIG. 280.

There will, in fact, be four curves for a shunt dynamo, namely, those in which the quantities plotted out are respectively *e* and *i*, *e* and i_a, E and *i*, E and i_a. Of these, the first is the *external characteristic*, and the fourth the *total characteristic*.

These four are depicted in Fig. 281, where they are named A, B, C, and D respectively, If D is given, A can be obtained in the following way :—Let the lines O J and O Z represent respectively by their slope the resistance of the armature and of the shunt circuit. Then curve B is got from D by deducting from the ordinates lengths equal to the portions of ordinates intercepted by the line O J, and curve C is got from curve D by deducting from the abscissae lengths equal to the portions of the abscissae intercepted by the line O Z. Then curve A is got by taking ordinates from B and abscissae from C corresponding to any point on D.

It may be noticed that whilst D B represents the lost volts due to internal resistance of armature, C D represents the lost ampères which go round the field-magnets. The lower the resistance of the armature, and the higher the resistance of the shunt, the less will these losses be. In fact, with a well-built modern machine the four curves lie very close together.

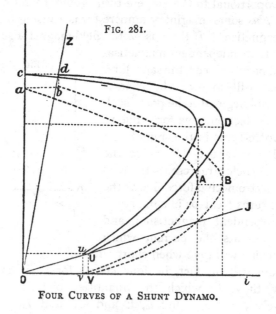

FIG. 281.

FOUR CURVES OF A SHUNT DYNAMO.

Contrast between Series Dynamo and Shunt Dynamo.

The difference between series dynamos and shunt dynamos in their behaviour when the resistance of the current is increased or decreased, has already been touched upon in p. 115. In electric lighting, dynamos are usually required either (*a*) to supply glow-lamps arranged in parallel, in which case the dynamos must maintain a constant potential at the mains, or else (*b*) to supply arc-lamps arranged in series, in which case the dynamo is required to yield a constant current. In the case where the potential is to be constant, the current will vary with the number of lamps in parallel; in the second case, where the current is to be constant, the electromotive-force must vary as the number of lamps in series.

To understand the applicability of series or shunt dynamos to either of these tasks it will be convenient to construct (either from experiment or from theory) comparative curves. In the case of parallel distribution, every additional lamp switched on across the mains adds to the conductance of the circuit an amount equal to its own conductance (i. e. equal to the reciprocal of its own resistance). It is therefore expedient to plot out together the values of e and of $\frac{1}{R}$. This has been done in Fig. 282 for two dynamos, for each of which the

FIG. 282.   FIG. 283.

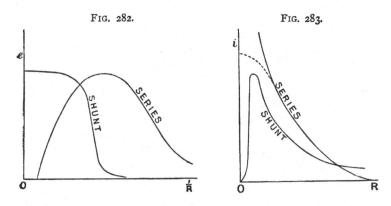

maximum value of e was the same. It will be seen that for neither a series nor for a shunt machine does the value of e remain constant as the number of lamps across the mains is increased. The shunt machine gives the more nearly constant potential, but falls off as the number of lamps is increased.

In the case of single-circuit distribution to lamps in series, every additional lamp adds to the resistance of the circuit, and in this case it is expedient to plot out together the values of i and R. Fig. 283 shows the result for the two kinds of machine. It will be seen that neither machine gives anything like a constant current; but for the shunt machine there is just one brief stage, namely, where its current is at the maximum, where the value is more constant than anything that the series dynamo can give. The dotted part of the curve corresponds to the case of a series dynamo so designed as to have a drooping characteristic (like Fig. 267, p. 379), which gives more nearly (with moderately small resistances) a constant current. But it is abundantly clear that something more than a simple series or simple shunt machine is requisite to give a real self-regulating machine for either purpose.

CHAPTER XIX.

FURTHER EQUATIONS OF THE DYNAMO.

WE now proceed to construct the equations of series-wound and shunt-wound dynamos, applying for that purpose the formula embodying the approximate law of the electromagnet which we discussed on p. 307. It may be worth while, first, to point out that this formula may be arrived at by extremely simple reasoning. In the chapter on magnetic principles it was pointed out that the magneto-motive force in a magnetic circuit is proportional to the ampère-turns (being $= 4\pi S i \div 10$); and the whole enquiry after the true law of magnetisation is merely to find an answer to the question; what will be the value of N, for a given value of $S i$, in a circuit of given form and material. If the number S of turns in the magnetising coil be assumed to be known, then the question resolves itself into finding the value of N (the total number of magnetic lines) for a given value of i (the ampères of current in the coil of S turns). Now it is known that when there is little magnetisation in the iron cores the magnetising effect is nearly proportional to the ampère-turns, but as the magnetisation of the iron cores proceeds and saturation sets in, the addition to N that results from a further addition to i becomes less. The permeability in gross of the magnetic circuit gets less as the permeation becomes more complete. Now make the assumption (see p. 1309), that the permeability is in proportion to the number of magnetic lines that the magnetic circuit can yet receive before becoming practically saturated. Let $\bar{N}$ represent the greatest number of magnetic lines that can possibly be forced through the circuit; we may call $\bar{N}$ the "satural" value of N. Then the number of magnetic lines that the circuit can still receive, when the actual number present is N,

2 D

will be $(\bar{N} - N)$. Our assumption is that the permeability is proportional to this; or in symbols

$$\frac{N}{i} = \frac{\bar{N} - N}{i'}$$

where i' is a quantity the nature of which will be seen later. Multiplying up, transposing, and dividing out by $i + i'$, we get

$$N = \bar{N}\frac{i}{i + i'} \qquad (a)$$

It will be seen on reflection that i' is equal to that particular number of ampères which would excite in the given magnetic circuit exactly half the maximum number of lines, or would half-saturate it; for if we make the variable i' of such a value as to equal i', then $N = \frac{1}{2}\bar{N}$. The number of ampères i' that will thus half-saturate a given magnetic circuit (with coil of given turns) may be called the "*diacritical*" current (see p. 307). Multiplying by the number of turns S the formula becomes

$$N = \bar{N}\frac{Si}{Si + (Si)'} \qquad (b)$$

where $(Si)'$ is the diacritical number of ampère-turns. Or, again, if e volts applied to a coil the resistance of which is r ohms, produces the current i, we may write the formula as

$$N = \bar{N}\frac{e}{e + e'}; \qquad (c)$$

where e' is the diacritical number of volts; that is to say that number of volts which, applied to a magnetising coil of given number of turns and given resistance, will half-saturate the magnetic circuit.

It will be convenient to give here a geometrical illustration of this law.

Upon the horizontal axis lay out, to the left of the origin, $OD = i'$; and at O erect a perpendicular O B representing in height the maximum number of lines $\bar{N}$ on any convenient scale. To find the number of magnetic lines corresponding to any given current i, plot out O A to represent i. At A erect A C equal in height to O B. Join C D, and from O draw, parallel to D C, O P meeting A C in P. The length A P will represent the value of N required. The proof is as follows :

By properties of triangles ;

$$\frac{AP}{AO} = \frac{AC}{AD};$$

or

$$\frac{N}{i} = \frac{\overline{N}}{i + i'}.$$

FIG. 284.

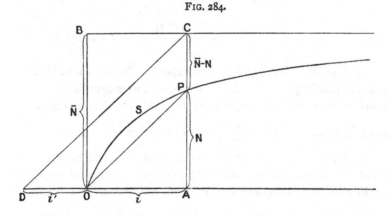

If other values of i are taken, corresponding values of N may be found; and they all lie upon the curve shown in Fig. 284, which may be taken as an approximate curve of magnetisation.[1]

It now remains to introduce this convenient approxima-tion to the law of the electromagnet into the fundamental equations of the dynamo.

---

[1] Of course all this is an extremely simple statement of, and graphic con-struction for, the approximate law of magnetisation explained on p. 307, in the forms given to it by Frölich, and by the author of this work in the former editions.

Equations of the Series Dynamo.

From p. 337 we have the following equations :—

$$E = n\,CN ; \qquad\qquad [\text{I.}]$$

$$E = (R + r_a + r_m)\,i ;$$

and from p. 402, equation (*a*) gives

$$N = \overline{N}\,\frac{i}{i + i'} ;$$

whence

$$(R + r_a + r_m)\,i = n\,C\,\overline{N}\,\frac{i}{i + i'}$$

$$i + i' = \frac{n\,C\,\overline{N}}{R + r_a + r_m}$$

$$i = \frac{n\,C\,\overline{N}}{R + r_a + r_m} - i' \;\ldots\; [\text{XI.}]^1$$

Now $n\,C\overline{N}$ is nothing else than the **maximum** value that E could possibly have if by any means the magnetic circuit could be saturated up to its maximum degree of magnetisa-tion ; and the term $\dfrac{n\,C\,\overline{N}}{R + r_a + r_m}$ is nothing else than the maximum current that this dynamo would give through the particular resistances actually in the circuit, if driven at speed *n*, while the field-magnets were in some way driven up to satu-ration. Hence we may call $\dfrac{n\,C\overline{N}}{R + r_a + r_m}$ the " satural " value of the current under the given conditions of speed and resist-ance, and may indicate it by the symbol $\bar{\imath}$. The equation of the series dynamo then becomes :

$$i = \bar{\imath} - i' \qquad\qquad [\text{XI.}]$$

Multiplying up by $R + r_a + r_m$ we get

$$E = \overline{E} - E' \qquad\qquad [\text{VIII.}]$$

---

[1] The numbers in brackets are the same as those in the earlier editions of this book, in which the notation is more complicated.

Where E' stands for that electromotive-force, which, acting in the circuit with the resistances present, would produce the diacritical current ; and $\overline{\text{E}}$ is the " satural " electromotive-force, namely that electromotive-force which would be generated in the armature at a given speed if the magnetism were saturated. It may be remembered that while i' depends only on the form and materials of the magnetic circuit, and on the number of windings of the coil, and is, therefore, a constant for a given dynamo, E' depends also on the resistances, and is a variable. Neither of these have anything to do with the speed. On the other hand, $\overline{\text{E}}$ and $\bar{i}$ depend on the speed, and on the number of conductors in the armature as well as on the magnetic circuit ; $\bar{i}$ depending also on the resistance.

These equations throw some light on a phenomenon familiar to every electrician, that in every dynamo the current (when resistance of circuit is given) is not proportional to the speed, but is proportional to the speed less a certain number of revolutions per second. The latter number is familiarly known as the " dead-turns." It is also known that (with given resistance) there is a certain speed below which the dynamo does not excite itself. This least speed of excitation (with given resistance) is the same as the " dead-turns " (see p. 101). To investigate the matter take the equation [XI.]

$$i = \frac{n\,\text{C}\,\overline{\text{N}}}{\Sigma\text{R}} - i'$$

and equate it to zero, and call n' the particular value of n, which will give this result, namely :

$$n' = \frac{i'\Sigma\text{R}}{\text{C}\,\overline{\text{N}}}.$$

From this it appears that the least speed of excitation is proportional to the diacritical current, and to the resistance of the circuit. Further, $i'\Sigma\text{R} = \text{E}'$; whence it also follows that

$$\text{E} = (n - n')\,\text{C}\,\overline{\text{N}}.$$

This result shows that n', the least speed of excitation, is the same thing as the "dead-turns"; and indeed the experimental fact expressed in the last equation might have been made a starting point [1] for the theory of the dynamo.

It will be noticed in the expression for n' that the dead-turns of a given series dynamo, though dependent on the resistance in the circuit, are the same for all speeds whatever.

Equation to the characteristic of the series dynamo.

We have as the fundamental equation

$$E = n\,CN\,;$$

and

$$N = \bar{N}\,\frac{i}{i + i'},$$

substituting which gives us

$$E = n\,C\bar{N}\,\frac{i}{i + i'}\,;$$

or, finally,

$$E = \bar{E}\,\frac{i}{i + i'}$$

which is desired equation of the characteristic of the dynamo, giving E in terms of i and of the two quantities $\bar{E}$ and i', which are constants for a given machine driven at constant speed. If these are known it is a simple matter to plot the characteristic; the same construction serving as served for the curve of magnetisation in Fig. 284. Here O B represents the satural value of E at the given speed. Or, instead, the following construction may be used :—At O erect perpendicular O B representing $\bar{E}$ to any convenient scale. Draw B H horizontally to represent i' (Fig. 286). Then, to find the point on the characteristic corresponding to any given current i, lay out O A to represent i, and join H A. This oblique line H A cuts off from O B a length equal to E, which projected across gives the point P on the characteristic.

---

[1] See paper by author in *The Electrician*, xvii. p. 175, July 9, 1886.

Here the line O B represents the number of volts the machine would give, at the given speed, if its magnetic circuit were fully saturated; and O D or B H is the diacritical

FIG. 285.

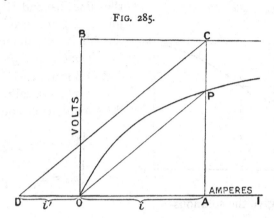

current, or that current which will half saturate the magnetic circuit.

If it is desired to know what are the actual values of $\bar{E}$ and i' for any particular series dynamo, they may be deter-

FIG. 286.

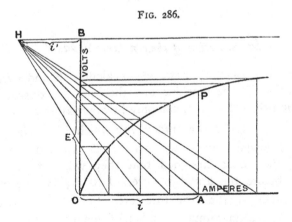

mined either graphically or algebraically, provided two experiments have been made on the machine (at the given speed) to determine two values of E and two corresponding values of i. For example, let E_1 and i_1 be determined when there is much resistance, and $E_2 \, i_2$ with very small resistance in the circuit; this gives P_1 and P_2 as points on the characteristic.

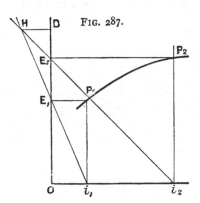

FIG. 287.

Join i_1 to E_1, and produce the line beyond E_1; also join $i_2 \, E_2$, and produce this line till it cuts the former in H. Then H B represents i', and O B represents $\overline{E}$.

Or, algebraically, we get the simultaneous equations

$$\begin{cases} E_1 = \overline{E}\dfrac{i}{i_1 + i'} \, ; \\[2ex] E_2 = \overline{E}\dfrac{i_2}{i_2 + i'} \, ; \end{cases}$$

which give the solutions—

$$\overline{E} = \frac{E_1 \, E_2 \, (i_2 - i_1)}{E_2 \, i_1 - E_1 \, i_2} \, ;$$

$$i' = \frac{i_1 \, i_2 \, (E_1 - E_2)}{E_2 \, i_1 - E_1 \, i_2}.$$

Predetermination of the constants $\overline{E}$ and i'.

It is more satisfactory, however, to calculate $\overline{E}$ and i' from the dimensions of the machine, and the known properties of iron, though the calculation can only be approximate.

Since $\overline{E} = n \, C\overline{N}$, the accuracy with which $\overline{E}$ can be predetermined depends simply upon $\overline{N}$. It is assumed that the field-magnets are so designed that they will not become saturated before the armature-core. Let A_1 be the nett sectional area of the armature-core. Then $\overline{N} = A_1 \overline{B}_1$; where $\overline{B}_1$ is the maximum number of magnetic lines, per square centimetre, that can be forced through the iron. About the actual value there is some uncertainty. For good wrought-iron (Swedish or Low-Moor), the practical maximum appears to be from 16,000 to about 20,000 (see p. 293); though

Ewing, under exceptional conditions (not at all like those that obtain in any real dynamo, where the armature currents always tend more or less to demagnetise the cores), has forced up the induction far beyond that value. For cast iron the practical maximum appears to be about 12,500; but this material is rarely or never used in armature-cores. Subject to the above uncertainty, and provided always that the field-magnet is powerful enough, we may write:

$$\bar{E} = 20{,}000 \times n\,CA_1$$

In the case of the Edison-Hopkinson machine, mentioned on p. 203, the value of $\bar{B}_1$ appears to be 19,800: in the 'Manchester' machine described on p. 165, the value is 23,500.

The determination of i' is more intricate, for i' is that number of ampères, which, given the windings of the magnet-coil, and the magnetic circuit as it stands, will cause the armature-core to become half-saturated; *i.e.* will bring up $N = \frac{1}{2}\bar{N}$; or will make $B_1 = \frac{1}{2}\bar{B}$. Referring to Hopkinson's expressions, p. 320, we may write the present special case as:

$$i' = \frac{10}{4\pi S}\left\{ f\left(\frac{N'}{A_1}\right) l_1 + 2\,\frac{N'}{A_2}\,l_2 + f\left(\frac{\nu N}{A_3}\right) l_3 \right\}$$

now $\dfrac{N'}{A_1} = B'_1 = 10{,}000$ if wrought iron is used and reference to the wrought-iron curve Fig. 234, will show that $f(B'_1)$, or H'_1 as we may call it, will have the value 5 to correspond with $B'_1 = 10{,}000$. Again, in the two gaps in the circuit:

$$\frac{N'}{A_2} = B'_2 = H'_2 = B'_1 \times \frac{A_1}{A_2} = 10{,}000\,\frac{A_1}{A_2}.$$

For the field-magnet part $B'_3 = B'_1\,\dfrac{A_1}{A_3}$; whence

$$\left(\frac{\nu N}{A_3}\right)' = \nu B'_3 = \nu \times 10{,}000\,\frac{A_1}{A_3},$$

where ν is the coefficient of correction for leakage. Then by reference to Fig. 234, or Fig. 235, according to the material of the cores $f\left(\nu \times 10{,}000 \dfrac{A_1}{A_3}\right)$ must be found: its value H'_3 will then be inserted in the formula, which thus becomes

$$i' = \frac{10}{4\pi S}\left\{5 l_1 + 20{,}000 \frac{A_1}{A_2} l_2 + H'_3 l_3\right\}$$

Such a calculation, in which the method of Hopkinson is used to predetermine the constants for the very convenient approximate formulæ of Frölich and of the author of this work, may clear up the difficulties sometimes felt by students in reconciling the two methods of treating theory.

Equations of the Shunt Dynamo.

In this case

$$N = \bar{N}\frac{e}{e + e'}$$

and

$$E = e + r_a i_a = e\frac{r_a}{\mathfrak{R}} = e$$

where (see p. 386) $\mathfrak{R}$ is the resistance of the whole system of machine and its circuits as measured from brush to brush.

From these and the fundamental equation we have

$$e = \frac{n\,C\,\bar{N}\,\mathfrak{R}}{r_a} - e'; \qquad [\text{XIV.}]$$

of which the first term on the left-hand side is nothing else than the satural value of e: that is to say, is that value which e would have with the given resistance and given speed if N were at its maximum. It may therefore be written as $\bar{e}$.

This gives us

$$e = \bar{e} - e'$$

whence also (dividing by r_s)

$$i_s = \bar{i}_s - i'_s,$$

or again,

$$E = \bar{E} - E.$$

Again the dead-turns may be reckoned by equating e to zero in equation [XIV.] above, giving

$$n' = \frac{e' r_a}{C \overline{N} \mathfrak{R}},$$

namely, that number of turns which, given the circuits and the satural magnetism, could produce at the terminals the diacritical difference of potential; so that, once more,

$$e = (n - n') C \overline{N} \frac{\mathfrak{R}}{r_a};$$

a result which might have been obtained direct from experiments on the speed and on the resistance.

Characteristic of the Shunt-Dynamo.

If the curve of magnetisation of the machine is known it is easy to determine the characteristic by a geometrical construction. The curve of magnetisation O P M will show the relation between N and the ampère-turns in the shunt coil, $Z i_s$ in our notation.

Let this curve be set out to the left of the vertical axis, then the line O R may be divided out either in a scale representing ampère-turns or in a scale representing ampères, Z divisions of the former scale corresponding to one of the latter scale. The vertical scale plotted out along O E may in like manner represent either N or E; $n C \times 10^{-8}$ being the ratio of the readings of the scales. Now set out the line O M, making with O R an angle such that the tangent of its slope corresponds, in the units chosen, to the resistance of the shunt-winding; for example, if the shunt-resistance be 16, the line will pass through a point the ordinate of which represents 16 volts and the abscissa 1 ampère. Let this line meet the curve of magnetisation at M. If we consider any point P on the curve, its ordinate P R will represent, according as we please, either the effective magnetism when the magnetising current is O R, or the whole number of volts E induced in the armature; and the part of the ordinate Q R will represent the difference of potential e.   Clearly then PQ will represent E–e, that is to say, the volts lost in the armature, which are equal to $r_a i_a$.   Now if on the right of the diagram we lay out a line OJ at such an angle that the tangent of its slope represents the armature resistance r_a, then if V be taken so far along the horizontal axis that VU = PQ, the length OV will represent i_a.   The most convenient construction is to project points P and Q across to E and e, and from e draw eT, parallel to OU, meeting ET in T; then drop a perpendicular from T giving the points t, U, and V, where Tt = U V.   T will be a point on the curve connecting E and i_a; t a point on the curve connecting e and i_a.   From the latter curve the external characteristic can be got as shown on p. 398.   Of these hyperbolic curves the

FIG. 288.

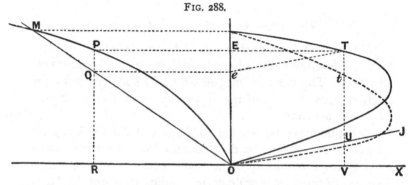

lower limb which returns towards O represents the unstable portion corresponding to the lower part of the magnetic curve. From the existence in the curves of a maximum value of i_a, where the curve turns round on the extreme right, it must not be inferred that the machine can yield this maximum current; on the contrary, the maximum current that the machine can safely give depends on the section of its armature wires, and these are—in the best machines—not intended to carry the current under such conditions. The working part of the curve is usually the top part (Fig. 274), and it will be obvious from the construction that the smaller the internal resistance, the further will the curve extend to the right and the more nearly horizontal will the tops of both curves be; a good shunt-dynamo, with very little internal resistance, being *nearly* self-regulating for constant potential.

If there be no initial or residual magnetisation, the curves will both pass through O ; but neither of them will do so if there is initial or residual magnetisation. In that case the curve of magnetisation will commence above O at some point such as K, Fig. 289, and the lower ends of the two curves for E and *e* will end at points so far to the right that the width U V = K O. With almost every shunt-dynamo it is found that if descending values of *e* are taken (at any given speed) *e* becomes zero whilst i_a has still a definite value.

It will also be noticed that the limiting value of E depends on the slope of M O, that is to say, on the resistance per turn

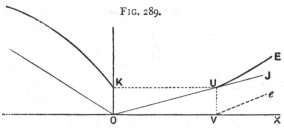

FIG. 289.

of the shunt coil ; any diminution of the resistance per turn will raise E by forcing up to a higher degree the magnetisation corresponding to a given value of *e*.

Equation to the Characteristic of the Shunt-Dynamo

In this case,

$$E = n\, C_x \overline{N}\, \frac{e}{e+e'}.$$

Now,

$$E - r_a i_a = e,$$

and

$$i_a = i + i_s,$$

whence

$$e\,\frac{r_s + r_a}{r_s} + r_a i = n\, C\, \overline{N}\, \frac{e}{e+e'};$$

and putting $b = \dfrac{r_s}{r_s + r_a}$ (a number slightly less than unity),

$$e^2 + e\{ e' + br_a\, i - bn\, C\, \overline{N} \} + be'r_a\, i = 0;$$

whence

$$e = \tfrac{1}{2}\,(bn\, C\, \overline{N} - e' - br_a\, i) \pm \tfrac{1}{2}\,[(bn\, C\, \overline{N} - e' - br_a\, i)^2 - 4\,be'r_a\, i]^{\frac{1}{2}}.$$

This expression, which gives e in terms of i, and is therefore the equation of the external characteristic of the shunt-dynamo, is the equation to a hyperbola (Fig. 290), one branch of which passes through the origin; for if $i = 0$, then the two corresponding values of e are $bn\, C\, \overline{N} - e'$ and zero. The diameter of the hyperbola will be a line, whose equation is

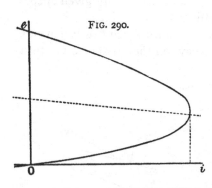

FIG. 290.

$$y = \tfrac{1}{2}\,(bn\, C\, \overline{N} - e') - \tfrac{1}{2}\, br_a\, i,$$

that is, a line cutting the axis of e at half the height of the curve, and inclining gently to the right at an angle proportional to br_a.

The equation to the curve connecting e and i_a (the lower curve in Fig. 288) is

$$e = \tfrac{1}{2}\,(n\, C\, \overline{N} - e' - r_a\, i_a) \pm \tfrac{1}{2}\,[(n\, C\, \overline{N} - e' - r_a\, i_a)^2 - 4\, r_a\, i_a\, e']^{\frac{1}{2}},$$

and the equation to the diametral line is

$$y = \tfrac{1}{2}\,(n\, C\, \overline{N} - e') - \tfrac{1}{2}\, r_a\, i_a.$$

As an example, in the Edison-Hopkinson machine mentioned on pp. 203 and 324, $n = 12 \cdot 5$, $C = 80$, $\overline{N} = 16,040,000$, $b = 0 \cdot 9994$, $r_a = 0 \cdot 009947$, $e' = 52 \times 10^8$, or 52 volts.

For a further discussion of the properties of the shunt-dynamo, the student is referred to papers by MM. Achard, Rechniewski, and Picou, in *La Lumière Électrique,* and to the treatise of Dr. Frölich.

General Equation of the Self-exciting Dynamo.

Assuming still the approximate law of magnetisation, it is possible to put the elementary theory of the simple self-exciting dynamo in a form of still greater generality.

Let ψ be any one of the currents or potentials of the dynamo; $\overline{\psi}$ its "satural" value, that is to say, the value it has when the magnet is separately saturated; and ψ' its "diacritical" value, that is to say, its value such as corresponds to half-saturation.
Then,

$$\overline{\psi} = f(n, C, \overline{N}, [R]),$$

where the form of the function (for a given machine) will be different for each kind of current or potential, and may not, in some cases, depend on the external resistance R. And, by the very nature of the case, whatever the form of the function f,

$$\psi = f(n, C, N, [R]).$$

Now,

$$N = \overline{N}\,\frac{\psi}{\psi + \psi'};$$

whence, since the function so far as N and $\overline{N}$ are concerned is a purely rectilinear one,

$$\psi = \overline{\psi}\,\frac{\psi}{\psi + \psi'};$$

and finally,

$$\psi = \overline{\psi} - \psi'.$$

It will be seen that this includes several of the equations obtained above. The proposition embodied in this most general form was enunciated first by Dr. O. Frölich in 1885. The two terms which appear on the right-hand side of the final equation correspond nearly, but not quite, to his classification of an armature part and a field-magnet part of the expression; for both terms depend on the magnetic circuit: the first on its actual capacity, the second on the current required to magnetise it to half its maximum, and therefore,

unlike the first, dependent on the resistances of the gaps in
the circuit, and on the shape as well as on the quality and
quantity of the iron. The first term depends on the speed
and construction of the armature, whereas the second does
not. The second term depends on the winding of the magnetising coils on the field-magnet, whereas the first does not.
In any case, whether stated for current or for potential, the
second or diacritical term has a minus sign : in every case it
detracts from the output of the machine. It is therefore important to notice that since the one constant diacritical
quantity in a dynamo is the diacritical number of ampère-
turns (namely, that number which will produce semi-saturation), this diacritical current i' may be made as small as
possible by simply increasing the number of windings upon
the field-magnet; and this will always produce economy in
working, provided this increase in the number of windings be
always accomplished by adding to the volume of copper in
the coil.

CHAPTER XX.

CONSTANT POTENTIAL DYNAMOS.

Theory of Self-regulation.

As mentioned in Chapter VI., p. 116, there are various ways of
governing dynamos so as to maintain either a constant potential, or a constant current. Some of these methods involve
hand regulation ; others, automatic switching in or out of
resistances, to vary the excitation of the field-magnets ; others,
automatic adjustment of the brushes ; and others, electrical
governing of the speed. There was, moreover, an important
class of methods, dependent upon the maintenance of a steady
speed of driving, in which a dynamo was made self-regulating
so as to yield either constant potential at its terminals, or constant current in its circuit, by the device of combining in one
machine either a series-winding or else a shunt-winding with
some other source of magnetising power capable of maintaining an independent and fairly constant initial magnetisation.
The most important case—at least for constant potential
machines—is that of the so-called *compound-winding*, the
theory of which is developed in this Chapter.

The two separate ends to be attained in self-regulation
must, however, be kept distinct. For some purposes—as for
feeding a system of incandescent lamps in parallel—the current
must be supplied to the mains at an absolutely *constant potential*, or, as some popularly phrase it, at a *constant pressure ;*
that is to say, the difference of potentials between the main
terminals of the dynamo must be constant. This, of course,
implies that the current delivered by the machine shall vary
exactly in a ratio inverse to that of the resistance of the external

circuit, increasing, as the resistance is diminished by adding to the number of lamps across the mains of the circuit ; and this, again, involves that the whole electromotive-force of the armature shall increase slightly in proportion to the power required to drive the increased current through the fixed resistances in the main circuit of the dynamo. For some other purposes, as for supplying a set of arc lamps connected in a simple series, or for charging a number of sets of accumulators in different houses, or for running a number of motors in different places on one line, it is necessary to maintain in the line an absolutely *constant current*, no matter how many or how few lamps or motors may be at work. This, of course, means that when the resistance of the main circuit is increased by the switching-in of more lamps, the dynamo must of itself put forth a proportionate increase of electromotive-force.

The two ends to be attained by self-regulation are therefore not only distinct, but incompatible with one another ; a dynamo cannot possibly keep its electromotive-force constant, and at the same time vary it in proportion to the varying resistance of the external circuit. The two systems must therefore be kept absolutely apart. They are adapted to entirely different cases of electric distribution. Their theory is different.

But, having said this, it may be pointed out that though the conditions of *constant-potential* distribution are distinct from those of *constant-current* distribution, the combinations for attaining either result in the generating machinery are closely similar. The general method of arranging such a self-regulating system consists in causing the machine to augment its own magnetism in the proportion desired. In the first case, since the increased magnetism is required to be in proportion to the current in the main circuit of the dynamo, a *series coil* on the field-magnet will, if properly proportioned, provide the required regulation. In the second case, a *shunt* coil should theoretically regulate, since the additional magnetism is required to be proportional to the electromotive-force that is applied to drive the current through the resistances of the external circuit.

To obtain a *constant-potential* distribution, one of the following combinations must be made :—

(*i.*) Series regulating coils + permanent magnets to partly excite the field, with an independent constant magnetisation.

(*ii.*) Series regulating coils + an independent current circulating in separate coils round the field-magnets, to produce an independent constant magnetisation.

(*iii.*) Series regulating coils + an independent current circulating in the main circuit (and generated either by a battery or by an independent magneto-dynamo) having the effect of partly exciting the field-magnets, with an independent constant magnetisation.

(*iv.*) Series regulating coils + shunt-magnet coils supplied by a portion of the current of the machine itself, thereby partly exciting the field-magnets, with an independent and nearly constant magnetisation.

(*v.*) For alternate-current dynamos, coils in series with the main circuit cannot be used, but they may be compounded by providing them with regulating coils supplied with a current derived (by a suitable transformer) from the main-circuit currents, and proportional to them, the derived currents being first sent through a suitable commutator to rectify them. The independent magnetisation may be derived either from an auxiliary exciter, or from a separate coil or group of coils in the armature, or, in fact, by another transformer, the primary of which is placed across the mains as a shunt ; in either of the latter cases the current being properly commuted.

To obtain a *constant-current* distribution, one of the following analogous combinations must be adopted :—

(*i.*) Shunt regulating coils + permanent magnets, to partly excite the field with an independent constant magnetisation.

(*ii.*) Shunt regulating coils + independent current circulating in separate coils round field-magnets, to produce an independent constant magnetisation.

(*iii.*) Shunt regulating coils + independent current circulating in the shunt coils along with the shunt current (and generated either by a battery or by an independent magneto-dynamo) having the effect of partly exciting the field-magnets, with an independent constant magnetisation.

(*iv.*) Shunt regulating coils + series coils on field-magnet supplied by the main-circuit current of the machine itself, thereby partly exciting the field-magnets, with an independent and constant magnetisation.

The two methods marked (*iv.*) in each set are often spoken of as methods of "*compounding*" the dynamo; the term "compound dynamo" having been introduced by Messrs. Crompton and Kapp to signify a dynamo with mixed series and shunt-winding, by analogy with the engineers' term "compound engine" for a steam-engine working with both high- and low-pressure cylinders. Theoretically, several other self-regulating combinations are possible; for example, a series machine with unsaturated magnets combined with a (quasi-independent) series machine with over-saturated magnets on the same shaft; a series machine having two sets of field-magnet poles at different leads, one of the sets of poles being the series-excited set, the other excited independently, or in shunt circuit; &c.

In considering both kinds of distribution we shall proceed as follows :—First find an expression for the potential (or, in the other case, for the current). This will, in general, consist of three terms. Secondly, we shall consider these three terms as to whether their factors are constants or variables. Then, having ascertained which of the terms contain variable factors, we must consider what conditions must be laid down (such as prescribing a particular speed or a particular number of windings) in order that the terms containing variable factors shall disappear. These conditions will be embodied in an "equation of condition," which will be then discussed. In general it will be found that if the number of regulating coils is prescribed beforehand, there will be a particular or "critical" speed at which self-regulation holds good ; or on the other hand, if the speed is prescribed beforehand, there will be a certain "critical" number of regulating coils to be deduced. The case of constant-current regulation is reserved for the next Chapter.

DISTRIBUTION AT CONSTANT POTENTIAL.

It is possible to treat the theory either algebraically or geometrically. Both methods will be here used.

Case (i.). *Series Regulating Coils + Permanent Magnets* (Fig. 76, p. 121).—If the field-magnets are partly permanently magnetised, or if there are permanent steel magnets in addition to the electromagnets, giving a partial permanent field, independent of that due to the current in the circuit, we may denominate the number of magnetic lines in this independent field as N_1.

Now the fundamental equation of the series dynamo is

$$E = n \, C \, N,$$

and the difference of potential between the terminals, otherwise called the pressure, is shown on p. 367 to be

$$e = E - (r_a + r_m) \, i.$$

But N, the number of magnetic lines that pass through the armature at any instant, is made up of two parts, the permanent independent part N_1, and a part depending upon the current i, and equal to

$$\frac{4 \pi \, S \, i \div 10}{\Sigma \dfrac{l}{\mu \, A}} \, ;$$

where S is the number of turns in the regulating coil, l the length of the magnetic circuit, A its cross-section, and μ the *average* value of the permeability (see p. 318) between the two extreme values that it has when i is zero and when i is at its maximum. If for brevity we write

$$\frac{4\pi}{10}\bigg/ \Sigma \frac{l}{\mu \, A} = q,$$

we may then write the variable part of N as $q\,S\,i$; and therefore,

$$N = N_1 + q\,S\,i;$$

and we get, as the complete expression for e,

$$e = n\,C\,(N_1 + q\,S\,i) - (r_a + r_m)\,i,$$

or

$$e = n\,C\,N_1 + n\,C\,q\,S\,i - (r_a + r_m)\,i.$$

The expression on the right-hand side of this equation consists of three terms, of which the first contains the speed and two constants as factors. The last two contain a variable, the current, and one of them also contains as factors the speed n and the number of regulating coils S. If S is prescribed beforehand, then the particular speed at which the dynamo is self-regulating will be clearly that speed at which the expression for e will contain nothing but constants. If n is prescribed beforehand, then we must vary S so as to eliminate the terms that contain the variable factor. Since the two last terms are of opposite sign, it is clear that by varying S or n, or both, the value of $n\,C\,q\,S$ may be made numerically equal to $r_a + r_m$. Then at the constant speed, which we will call n_1, the last two terms will cancel one another out, or,

$$n_1\,C\,q\,S\,i - (r_a + r_m)\,i = 0.$$

That is to say, S and n_1 must be such that

$$n_1\,C\,q\,S = r_a + r_m. \qquad \text{[XXIII.]}$$

This is the equation of condition.

If the condition laid down in this equation is observed, then the last two terms for e disappear, and we have simply,

$$e = n_1\,C\,N_1 = a \; constant.$$

Having thus proved that, at the given speed, e is a constant, it is worth while to enquire what it is that determines the value of e. Clearly e is directly proportional to N_1, the independent and permanent field magnetism. Therefore, we can arrange that the dynamo, still driven at the given speed, shall give any potential we please, within limits, provided we alter N_1 in the requisite proportion.

Returning to the equation of condition, we will write it in the second form—

$$n_1 \, C \, q \, S = r_a + r_m,$$

which gives us as the value of the critical speed,

$$n_1 = \frac{r_a + r_m}{S} \cdot \frac{1}{C \, q}.$$

This shows us that there is another way of getting a higher value of e. If we increase n_1 we know e will also increase proportionally, and we may increase n_1, provided we decrease S at the same time in the inverse proportion. If this is done the critical condition still holds good, as $n_1 \, C \, q \, S$ will still equal $r_a + r_m$.

Lastly, we may write the last equation in the following way,

$$\text{critical speed} = \frac{\text{total internal resistance}}{\text{number of turns of magnet coil}} \times \text{a quantity}$$

depending only on the armature windings and on the magnetic circuit and its working permeability within the range for which regulation is required.

Suppose, on the other hand, that the speed is prescribed by mechanical considerations, then the proper or critical number of regulating coils is given by the expression

$$S = \frac{r_a + r_m}{n \, C \, q}.$$

This is instructive also. The higher the internal resistance of a dynamo, the higher must be the driving speed, or the greater must be the number of the regulating coils in series, if it is to be self-regulating.

Case (ii.).   Series Dynamo + Separately - exciting Coils (see " *Series and Separate,*" Fig. 75, p. 119).—In this case there

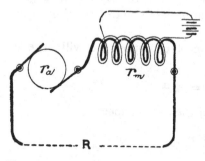

FIG. 291.

is an independent mag-netism due to a current carried round the field-magnets in separate coils, and providing a part of the field magnetism. The connexions are shown in Fig. 291.

If we call the number of magnetic lines due to the independent excitation N_1, the same argument holds good as in the preceding case, and the same conclusions. N_1 will not, however, be really a constant, for the effect of the introduction of a constant amount of magnetising force will vary with the degree of saturation resulting from the *whole* magnetising force.   If, however, the average working permeability through-out the range of regulation is taken into account in the calcula-tion of N_1 as well as in that of S, then any falling-off in the effect of the independent exciting current is implicitly pro-vided for.

Messrs. Wright & Kapp have accomplished the same end as Case ii., using one coil only, in the main circuit, to which coil, however, a battery, or the circuit of a separate dynamo, is applied as a shunt.   M. Picard has made a somewhat similar suggestion.

Geometrical Demonstration of Cases (i.) and (ii.).

On p. 355 it was shown how the values of the potential at terminals fall off in magneto and separately-excited dynamos as the current increases, *e* always being less[1] than E by an amount equal to $r_a i$.

To represent the facts, let O X and O P be taken as the

---

[1] This is illustrated in Fig. 261, p. 359, where the curve *e* shows the drop due to this cause, and the curve B the actual drop due chiefly to this cause and partly to the demagnetising reactions going on in the armature.

axes for plotting out ampères and volts, and let O P represent the electromotive-force ($E = n_1 C N_1$) due to the permanent or independent magnetism, as measured when no
current is running through the armature. Now, assuming
that the armature reactions are small enough to be neglected,
E will at constant speed remain the same with all currents,
but e will drop. From O set off the line O J at an angle
such that its tangent represents the internal resistance of the
machine. Now consider the case when the current i has the
particular value corresponding to the length O V. The height
U V will be the drop in the external electromotive-force; for
$U V = \tan U O V \times O V = r_a i$. Cut off from t V a portion,
$t Q = U V$, and Q V will represent e. While the curve for

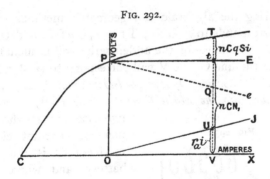

FIG. 292.

E and i is approximately a horizontal line, the curve for e and
i (the external characteristic) drops, as shown by the dotted
line. Any point on the e curve can be got from the E curve
by simply deducting from the height a piece equal to the
corresponding width across the triangle J O X. Now it must
be obvious that if when the E curve is horizontal the e curve
slopes downward, we must use an E curve that slopes upward
by a precisely equal amount, if we want to get a horizontal
e curve; that is to say, if we want to get constant potential.
How are we to get an upward sloping E curve? Remember
that at a given speed, n_1, the value of E is $n_1 C N_1$, where N_1
means that the magnetic circuit has somehow (either permanently or by a separate current) been excited up to such a

degree that N_1 lines go through the armature. The same plotting that serves for volts will serve for values of N by choosing the appropriate scale; or, O P may represent N_1. Therefore P is a point on a curve of magnetisation, which will rise still higher if only we put on more ampère-turns of excitation. Therefore all that is required to be done is to put on the magnets a coil in series, having such a number of turns S that the ampère-turns S i will have the effect of raising the magnetism in the right proportion; in fact, so that T t shall be equal to U V. We have now got an E curve which slopes up—not quite a straight line, to be sure, but such that when we subtract the volts required to drive the current through the armature resistance, we get an e curve that is approximately level.

Comparing the algebraic and geometric methods, we see that t V corresponds to n_1 C N_1; T t to n_1 C q S i; and U V to $r_a i$, or if the resistance of the added series coil is included in the slope of the line O J, U V will correspond to $(r_a + r_m) i$.

Case (iii.).　Series Dynamo + Independent Electromotive-force thrown into the Main Circuit.—This really comprises two cases: where the independent constant electromotive-force is due to a battery, and where it is due to a separate magneto machine driven at a constant speed ("Series and Magneto," see p. 121). The argument is the same, however, for both cases. Fig. 293 will represent either case.

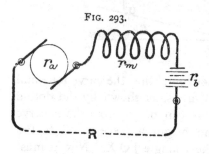

FIG. 293.

We have here as the whole electromotive-force of the combination E, the electromotive-force of the armature, plus E_b the independent electromotive-force thrown in from the battery or magneto machine. The difference of potential between the terminals, which we have always denominated as e, will be got by subtracting from $E_b + E$ that part of the electromotive-force which is devoted to sending current i

through the internal resistances, which are now r_a, r_m, and r_b; so that we have

$$e = E_b + E - (r_a + r_m + r_b)\,i.$$

Now $E = n\,C\,q\,S\,i$; therefore, in order to make the last two terms cancel one another and leave e a constant, we must give the speed the value n_1 such that

$$n_1\,C\,q\,S = r_a + r_m + r_b, \qquad \text{[XXV.]}$$

which is the equation of condition. In this case,

$$e = E_b.$$

This proves that in this case, too, the constant potential at the terminals is identical with that due to the independent excitation. Of course this does *not* mean that the dynamo does no work. On the contrary, it means this : that when the resistance of the external circuit is infinitely great, so that the dynamo does no work, then the only electromotive-force in the circuit is that due to the independent source.

Case (iv.). Series Regulating Coils + Shunt Exciting Coils: "Compound" Dynamo.—The compound-wound dynamo may be regarded as either a series dynamo to which some shunt windings have been added, so as to provide an initial magnetisation, or as a shunt dynamo to which some series windings have been added to compensate for the drop of potential at the terminals. There are two possible methods of connecting the shunt coils to the dynamo, and the proportions differ slightly in the two cases. The shunt coils may be joined as a shunt to the armature part of the dynamo only, being connected across from brush to brush. This case is denominated in the earlier part of this work as "Series and Shunt" (see Fig. 77, p. 121). In the second method the shunt coils are connected across the terminals of the machine, and may, therefore, be regarded either as a shunt to the external circuit, or as a shunt to the armature and series coils together. This arrangement I have termed "Series and Long-Shunt" (see Fig. 78, p. 122). In the former arrangement the current through the shunt is not constant, because the potential at the brushes ϵ is not the same as e; and though e may remain fairly constant, ϵ does not, but increases when the external circuit's resistance decreases. In the latter arrangement ("long shunt") the current through the shunt is constant if e is constant, and the case becomes one analogous to those already discussed, of an independent constant excitation. The connexions of the series and shunt method are indicated in Fig. 294. Since in a well-designed dynamo r_a is very small, r_m will also be very small, for few regulating coils in series are required. Moreover, as the shunt resistance r_s is relatively very great, the shunt current will be relatively small, and it makes very little difference therefore whether the shunt is connected across the brushes or across the terminals of the main circuit. The connexions of the series and long-shunt arrangement are as shown in Fig. 295.

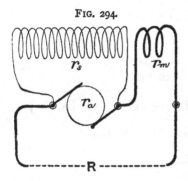

FIG. 294.

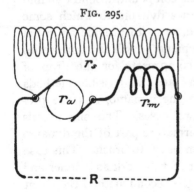

FIG. 295.

The calculations for the two cases are practically identical, and involve the same kind of arguments. That for long shunt is a little more simple, and is accordingly given.

We have then

$$E = n\,C\,N\;;$$
$$e = E - (r_a + r_m)\,i_a\;;$$

and as the magnetism depends on the total number of ampère-turns circulating around the field-magnet, we shall write,

$$N = q\,(Z\,i_s + S\,i_a)\;;$$

where q has the same signification as before [(p. 422), namely,

$$= \frac{4\pi}{10} \Big/ \Sigma \frac{l}{\mu\,\mathrm{A}}\,;$$

or more strictly is the variable number representing, at the various stages of magnetisation, the numerical ratio between N and the total number of ampère-turns for the magnetic circuit of the particular dynamo. It is of course best obtained by reference to such a diagram as Fig. 239. For the present purpose it is necessary to consider (1) the value which q has when the external current is zero, and when the only excitation is that due to the shunt, namely, $Z\,i_s$ ampère-turns : this may be called q_o; and (2) the value which q has when the current in the armature is at the maximum for which the dynamo is intended to be used. If the maximum current is called x, this value may be called q_x. Then, as the current varies from o to x, the corresponding values of N will vary from

$$\mathrm{N}_o = q_o Z\,i_s$$

to

$$\mathrm{N}_x = q_x (Z\,i_s + \mathrm{S}\,x).$$

Now between these two limits, which we may call the range for which the dynamo is required to be made self-regulating, q will have intermediate values, and one of these must be selected for use in the formula. The value $\frac{1}{2}\,(q_o + q_x)$ will not be far from a fair average. Let this average value be denominated by q_1 in the equations which follow. Then from the three preceding equations we have

$$e = n\,\mathrm{C}\,q_1\,Z\,i_s + s\,\mathrm{C}\,q_1\,\mathrm{S}\,i_a - (r_a + r_m)\,i_a.$$

Now here we have three terms, the first containing, as factors, the speed (which may be kept constant), and the shunt current i_s which will be made constant if e can be made constant; the second contains the speed also as a factor; the second and third both contain the variable current i_a. The two variable terms are of opposite sign. Now e cannot possibly be a constant, when two of its terms contain a variable as factor, unless the coefficients of that variable factor are such that they make those two terms cancel one another ; e cannot be constant unless either the speed n or the windings S, or both, are so adjusted as to fulfil this. But these can be adjusted, and even with a given value of S a particular value of the speed n_1 can be found, such that

$$n_1 \, C \, q_1 \, S = r_a + r_m \, .$$

This is then one of the two equations of condition ; and then the critical speed will be

$$n_1 = \frac{r_a + r_m}{S} \cdot \frac{1}{C \, q_1} \, .$$

Or, if the speed be given, then the critical number of series turns will be

$$S = \frac{r_a + r_m}{n_1} \cdot \frac{1}{C \, q_1} \, .$$

If this condition be observed, then e will be constant and have the value

$$e = n_1 \, C \, q_1 \, Z \, i_s \, .$$

But this would leave e indeterminate. But we may reflect that though this equation might not give us the value of e, there will nevertheless be a determinate value for it, namely, the same value that e would have when there is no current taken from the dynamo at all, but when it is running on open circuit only.   Under these conditions the value of e will be

$$e = n_1 \, C \, N_o - (r_a + r_m) \, i_s \, ;$$

or, since here q has the value q_o,

$$e = n_1 \, C \, q_o \, Z \, i_s \, .$$

But $e = r_s \, i_s$, whence we get

$$n_1 = \frac{r_s}{Z} \cdot \frac{1}{C \, q_o} \, .$$

Comparing this value of n_1 with that obtained in the first equation of condition, we get

$$\frac{r_s}{Z \, q_o} = \frac{r_a + r_m}{S \, q_1} \, ;$$

whence, finally, as the *second* equation of condition,

$$\frac{Z}{S} = \frac{r_s}{r_a + r_m} \cdot \frac{q_1}{q_0}.$$

As q_0 is proportional to μ_0 the permeability when there is no external current, and q_1 proportional to the average permeability μ_1 for the range of working between zero current and maximum current, it follows that if there were no alteration of saturation, $q_1 \div q_0$ would equal 1. In the former editions of this work, wherein the theory of compounding was expressly based upon the supposition that there was no saturation—or in other words, that the permeability was constant—the formulæ obtained were admittedly incorrect for this reason. Dr. Frölich found for a certain Siemens series and shunt dynamo,

$$\frac{Z}{S} = 17\cdot7, \text{whereas } \frac{r_s + r_a}{r_a + r_m} = 61\cdot9.$$

From which it is clear that μ_0 must have been about $3\cdot5$ times as great as μ_1; in other words, this machine had an insufficient quantity of iron in its magnets or armature core, or in both. This dynamo must have been both badly designed and of low efficiency; r_s ought to have been not $61\cdot9$, but at least 300 times as great as $r_a + r_m$. There is also another cause which tends to make S greater than the theoretical value, namely, the demagnetising tendency (p. 92) of the currents in the armature coils, which must be compensated for by putting on more turns in series. It will be noticed that the amount of excitation to be provided for by the series coils is always proportional to the resistances that are in the main circuit and internal to the points for which the constant difference of potential is desired; this renders it possible in a case where the mains leading from the dynamo to the lamps are long, so to compound the dynamo by adding more coils in series as to give a constant potential, not at its terminals, but at the distant point of the circuit where the lamps are to be used.

Practical Process for Compounding.

It is clear from the foregoing argument that the compound machine, when run on open circuit, with only the shunt-current flowing, must give the same potentials at its terminals as it is to give as a compound machine. Hence this leads to the following practical process for compounding. Let the armature of the machine be run at the proper speed dictated by mechanical considerations, and let a voltmeter be applied at the terminals. Two experiments are then necessary. First, by means of temporary coils, having a known number of turns wound on the field-magnets, and furnished with measured currents from some accumulators or another dynamo, ascertain the number of ampère-turns that will suffice to excite up the magnets to this point. From this Z can be determined; for it is known beforehand that r_s must be at least 300 or 400 times (or sometimes as much as 1000 times) r_a; and therefore i_s is really known beforehand. Secondly, put into the main circuit some resistance to represent the maximum load of lamps, and while the machine is running at its proper speed, ascertain, using still the temporary coil and accumulators, how many ampère-turns of excitation are needed in total when the machine is doing full work, subtracting from this the value of $Z i_s$ obtained in the first experiment, the remainder gives the number of ampère-turns which the series coil must furnish, and as the maximum current is known, S can at once be calculated.

Design of Constant Potential Machines.

It is obviously of importance in such machines that the iron parts should be so designed that (1) the characteristic should be as nearly straight as possible in that portion of it corresponding to the working range of currents for which regulation is desired; (2) that it should not turn down. Consequently it is of importance in such machines that there should be plenty of iron in the armature, and that the reactions due to the armature currents should be small.

There should, in fact, be so much iron in the armature that it does not get more saturated than the iron of the field-magnets ; and the number of turns of wire on the armature should be small, and the field-magnets relatively powerful.   Also, of course, the resistance of the armature should be kept as small as possible.

Characteristic of Compound Dynamo.

In the original theory of constant potential machines, devised by Marcel Deprez, the argument was based upon the absence of saturation, and the presence of an initial independent magnetisation.   The following was the argument of Deprez.   If there is a permanent excitement of magnetism quite independent of that due to the main-circuit coils of the dynamo, the characteristic (Fig. 296) will not start from O, but from some point above it depending on the amount of independent magnetisation and on the speed.   Let the starting point be P.   O P is the electromotive-force between terminals when the main circuit is open, but there is no external current until the circuit is closed, and then the characteristic rises in the usual fashion from P to Q. Draw O J at the proper slope to represent the resistance of the armature and series coils together.   Now consider a line O E drawn at such an

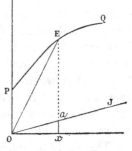

FIG. 296.

angle that the tangent of its slope represents the total resistance of the circuit at any particular moment.   Then E x is the total electromotive-force at that moment, and a part of this equal to $a x$ will be employed in driving the current O x through the resistance of armature and series coils.   The remaining part E a represents the difference of potentials between the terminals of the external circuit.   So the problem resolves itself into this : how to arrange matters so that E a shall always be of the same length as O P, no matter how much or

2 F

how little the line O E may slope. Clearly the only way to do this is so to arrange the speed of the dynamo that the part from P to Q shall be parallel to O J. If the speed is reduced exactly to the right amount the inclination of the character-istic will be equal to that of the line O J. Then, as shown in Fig. 297, the potential between the terminals will be constant. It will be seen that this agrees with the deductions arrived at in the algebraic treatment of the question : namely, that the critical speed is proportional to the internal resistance ; and that the constant difference of potential E a is equal to that due to the independent magnetisation O P at the critical speed.

FIG. 297.

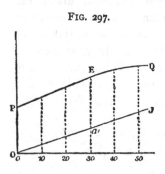

It should also be noticed that if the part of the characteristic be not straight, that is to say, if the field-magnet cores are becoming saturated, the regulation cannot be perfect. If the line P Q be curved, then the potential for large currents will not be equal to that for small currents. If, in the practical process for winding the magnets, the coils have been wound so as to make e the requisite number of volts, both on open circuit (*i.e.* at O P), and at another point (say at Q J), when the dynamo is feeding its maximum load, then there will in general be a slightly greater potential for intermediate loads, owing to the slight convexity of the curve between P and Q.

The above argument holds good whether the independent excitation be due to permanent magnetism or to a combina-tion with separately-exciting coils (see pp. 118 and 419), or to the machine being "compounded" by the addition of some shunt-exciting coils. In the latter case O P represents the potential at terminals due to shunt circuit alone.

The case of the "compound" dynamo is worth looking at from another point of view also. On p. 394 two curves— not characteristics—are given, showing the relation of electro-motive-force to external resistance in a series machine and in a shunt machine. One begins at a certain height and falls when the resistance has attained a certain value; the other begins low and rises when the resistance has attained a certain value. It is conceivable that if a dynamo were wound with both shunt and series coils so that each worked up to the same potential at the same speed, and so proportioned that the number of ohms at which one fell should be the same as that at which the other rose, then the compound machine should, as indicated in Fig. 298, give as a result of the double-winding, a constant potential. It remains to be seen how far this is attained in practice.

FIG. 298.

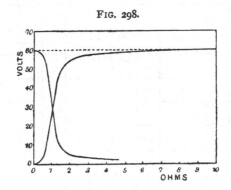

External Characteristics of Self-regulating Dynamos.

Simultaneous observations of the external current i and the external potential e enable us to plot the external characteristic; which in a perfectly self-regulating dynamo would be a horizontal line. The curves given in Fig. 299 relate to a Siemens dynamo,[1] a Schuckert-Mordey " Victoria " dynamo[2] (see p. 172), and a Gülcher machine (see p. 179).

If the number of regulating coils in series is too small, the characteristic will fall as the current rises; if too large, will droop slightly at the end near the origin (see Fig. 299). This latter case, however, is not always a disadvantage, for with machines worked singly on an engine the speed often rises

[1] See Richter in *Elektrotechnische Zeitschrift*, April 1883.
[2] See *Journal Soc. of Arts*, Mar. 7, 1884.

in consequence of imperfect governing when the load on the dynamo is small.

Esson's Observations.—Some observations by Mr. W. B. Esson in the *Electrician* of June 1885 are worthy of consideration in connexion with recent theory. Mr. Esson asks why is it that compound dynamos wound so as to be self-regulating for a given speed, regulate fairly well at any speed within considerably wide limits? To explain this peculiarity he observes that in no dynamo is the quantity or quality of the iron such that the saturation effect can be neglected.

FIG. 299.

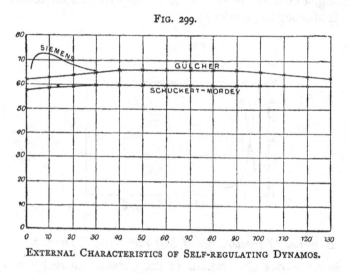

EXTERNAL CHARACTERISTICS OF SELF-REGULATING DYNAMOS.

If the magnetism were strictly proportional to the ampère-turns of excitation, there would be literally a critical speed. The approximate rule $\frac{S}{Z} = \frac{r_a + r_m}{r_s}$ gives the number of series coils much too low, for when the shunt coil has already excited a certain degree of magnetisation, the series coil cannot produce its proportionate increase. In a series machine (designed to give a current of 20 ampères), the electromotive-force added to the machine by increasing the exciting current from 5 to 10 ampères is much greater than the electromotive-force added by increasing the current from 10 to 15 ampères. Again, a 100-volt machine (self-regulating), in which therefore the shunt gave excitement enough for 100 volts on open circuit, had series coils upon it which were able, when the shunt was removed and the full current on, to give 60 volts between terminals. The value of the series coil to excite magnetism is diminished as the excitation due to the shunt is increased. All this is, of course, due to the diminution in permeability of the iron of the machine as the degree of saturation increases. From this it follows that a certain relation must subsist between the speed of the machine and the degree to which the magnets are excited by the shunt coil. But the magnetism furnished by the shunt coil also depends on the speed and increases with it. If, therefore, at one speed this relation is such as to produce self-regulation, the relation will be almost equally true at other speeds. At the high speed the relative value of the series coils is less, and at the low speeds it is greater; but the sum of the two effects may be constant. At speeds lower than the normal speed, the potential is lower when there is a large resistance in circuit than where there is a small resistance in circuit. At speeds higher than the normal, the potential falls as the resistance is diminished. Mr. Esson deduces from the foregoing considerations certain practical hints as to how to improve the regulation of a dynamo whose potential rises either when many or when few lamps are in circuit.

Students desiring further information on compound winding of dynamos, are referred to a series of articles in the *Electrician*, in 1883, by Mr. Gisbert Kapp, and another in the *English Mechanic*, in 1884, by Mr. W. B. Esson, also to two articles by Mr. Esson in the *Electrician* of June 1885. Articles by M. Hospitalier in *L'Électricien*, and by Herr Uppenborn in the *Centralblatt für Elektrotechnik*, should also be consulted; and the student should above all read the series of papers published by Dr. Frölich in the *Elektrotechnische Zeitschrift* for 1885, and a still more remarkable paper by Professor Rücker in the *Philosophical Magazine* of June 1885. Some account of these is given in Appendices VII. and IX. The latest contributions to this question are by C. Zickler in the *Centralblatt für Elektrotechnik*, ix. 264, 1887, and by M. Baumgardt, *Ibid.*, x. 281, 1888.

CHAPTER XXI.

Constant-current Dynamos.

As mentioned in the previous chapter, the methods similar to those adopted for giving constant potential can be applied to produce machines for giving constant currents.

For reasons which will be discussed, machines of this kind are not yet practical, hence the arguments may be treated the more briefly. We must find expressions for i, the current in the external circuit; and having found them, deduce such equations of condition as will make their values constant.

Case (i.). *Shunt Regulating Coils + Independent Initial Magnetisation.*—We have at the outset

$$\mathrm{E} = n\,\mathrm{C}\,\mathrm{N};$$
$$\mathrm{N} = \mathrm{N}_1 + q_1\,\mathrm{Z}\,i_s;$$

where N_1 is the initial and independent value of N, and q_1 has the same meaning as in the preceding chapter. Now, further,

$$\mathrm{E} = e + r_a i_a = r_s i_s + r_a i_a;$$
and
$$i_a = i + i_s;$$
whence
$$r_a i = n\,\mathrm{C}\,\mathrm{N}_1 + n\,\mathrm{C}\,q_1\,\mathrm{Z}\,i_s - (r_s + r_a)\,i_s.$$

Now, if i is to be constant, the coefficients of the last two terms (which contain the variable i_s) must be such as to cancel one another; in other words, if n_1 be the prescribed speed, the number Z of regulating shunt coils must be such that we may write as the equation of condition

$$n_1\,\mathrm{C}\,q_1\,\mathrm{Z} = r_s + r_a;$$

giving, as the number of coils,

$$Z = \frac{r_s + r_a}{n_1} \cdot \frac{1}{C\,q_1},$$

and then

$$i = \frac{n_1\,C\,N_1}{r_a}.$$

This means that N_1 provides magnetism enough to yield, at the particular speed, the required constant current when the machine is running on short circuit, when the only resistance is that in the armature. The shunt current provides all the rest of the magnetism required when other resistances are in circuit.

Comparing together the above values of n_1, it further follows that Z must be such that

$$\frac{Z\,i}{N_1} = \frac{r_s + r_a}{r_a} \cdot \frac{1}{q_1}.$$

Cases (*ii.*) *and* (*iii.*), where there is an independent auxiliary current, were treated in the former editions of this work, but are unimportant.

Case (*iv.*). *Shunt Regulating Coils + Series Exciting Coils.* —Here, again, are two cases—shunt and long shunt: we take the former. The connexions are the same as Fig. 294, p. 428. The equations are :—

$$E = n\,C\,q\,(Z\,i_s + S\,i);$$
$$E = r_a\,i_a + r_s\,i_s = r_a\,i + (r_s + r_a)\,i_s;$$

whence

$$\{n\,C\,q\,Z - (r_s + r_a)\}\,i_s = \{r_a - n\,C\,q\,S\}i.$$

Now as i_s is a variable and i is to be a constant, the equation cannot be true unless at the prescribed speed n_1 such values are given to Z and S that the quantities in brackets shall be zero. This gives

$$n_1 = \frac{r_s + r_a}{Z} \cdot \frac{1}{C\,q_1};$$
$$n_1 = \frac{r_a}{S} \cdot \frac{1}{C\,q_0}.$$

We write q_0 in the latter case, because it corresponds to the case where, under short circuit, the series coils alone are in action. This then gives

$$\frac{Z}{S} = \frac{r_s + r_a}{r_a} \cdot \frac{q_o}{q_1}.$$

The graphic method well illustrates the principles of the foregoing theory. Fig. 281, B, p. 398, gives the curve connecting e and i_a for a shunt dynamo which has initial magnetisation. In Fig. 300 the characteristic begins at a point U on the axis of currents. At any point, such as W, the length Y W gives the value of i_a. By drawing from O a line O Z such that the tangent of its inclination represents the shunt resistance, we cut off from Y W a length Y Z representing the current through the shunt; hence the remainder Z W represents the external current. Now Z W cannot be always equal to O U unless the ascending portion of the characteristic be approximately straight, and unless the speed be such that U W is approximately parallel to O Z. Also it is clear that the constant current will be equal to the current due to the initial magnetisation alone when the dynamo is short-circuited, and the armature's resistance is the only resistance in the circuit. The practical objection to the theory is that at this lower part of the characteristic the magnetism is very unstable just so long as U W is approximately straight.

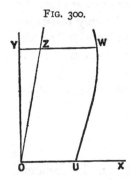

FIG. 300.

Another way of viewing the same problem is to follow backwards a construction similar to that given in Fig. 288. Let the vertical line U W, in Fig. 301, represent the desired constant current, being the external characteristic of the supposed constant-current dynamo. From this we determine the required curve of magnetisation, as follows :—

Lay off below O U the line O J to represent the armature resistance. Set out O M at a slope representing the resistance per turn of the shunt coils. Take any point W in the constant-current line, and draw W e parallel to O U and W E parallel to O J. The height O E will represent the corresponding value of E, whilst O e represents that of e. Produce W e to meet O M in Q, and through Q draw the perpendicular R P = O E. Then in the proper units R P represents (see p. 411 for argument) the magnetic lines cut by the armature when O R represents the ampère-turns in the shunt. Hence P is a point on the curve of magnetisation. Following the same construction for other points, it is obvious that the required curve of magnetisation will start from K (O K being the initial magnetisation, or that due to the series coil in the case of the compound machine), and must run parallel to O M

FIG. 301.

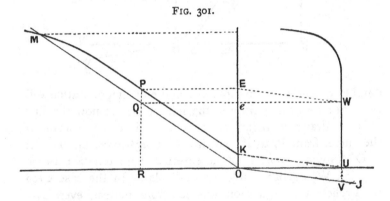

just so far as U W is carried straight upwards. But all curves of magnetisation sooner or later begin to bend over, as in the diagram to the left of P. Beyond this point the line U W cannot be straight; it also will bend over as shown. Now it is just this portion R P of the curves of magnetisation that corresponds to the most unstable stages. A very small raising or lowering of the point P, such as might correspond simply to the difference between a curve of ascending and a curve of descending magnetisation (see p. 300, and Figs. 234 and 235) will make an immense difference in the distance of W to the right of the vertical axis. In other words, the supposed constant current will follow the instability of the magnetisation.

Messrs. Siemens and Halske have proposed a method of compounding, to obtain an approximately constant current in which series coils are so connected as to act differentially *against* the shunt windings.[1] The graphic theory of this device is as follows :—It will be seen from Fig. 301 that U W is parallel to O E only so far as K P is parallel to O M. Now in Siemens and Halske's plan there is a constant negative

FIG. 302.

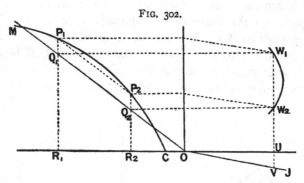

number of ampère-turns, and the curve of magnetisation will begin at a point C a little to the left of O. If now the line O M be drawn as before, there will be a certain portion of the curve, from P_1 to P_2, which will be *approximately* parallel to O M, giving a portion of the external characteristic between W_1 and W_2 *approximately* perpendicular. In the case cited by Frölich the regulation was far from perfect, even over a restricted range, the values of e and i and of the external resistance being as follows :—

R	e	i
0·98	76	77·3
1·00	82·8	82·5
1·04	90	86·2
1·11	98	88·5
1·19	106	89·3
1·29	111	86·1
1·44	118	82·2
1·67	124	74·4

Arc Lighting Dynamos.—A method which, though not in itself securing constancy of current, is much followed in the

[1] *Elektrotechnische Zeitschrift*, vol. vi. p. 136, 1885.   See also Frölich's *Dynamo-elektrische Maschine*, p. 97.

construction of arc-lighting dynamos, should here be explained. Attention was drawn on p. 379 to the drooping
form of the characteristics of certain series-wound machines.
It is obvious that if this effect is sufficiently exaggerated, the
drooping portion of the characteristic will correspond to the
case of an approximately constant current. The drooping
characteristic is important in promoting the steady working of
arc lamps in the circuit. Suppose an arc lamp to be running
on a series-wound dynamo under such conditions of working
that the characteristic is ascending, any shortening of the
arc will be followed by a reduction of resistance and a large
increase of current. Whereas if the conditions of working
are such as to fit to a point on the drooping part of the characteristic, any decrease in resistance in the circuit will result
in a comparatively small increase of current.

The various influences which cause the characteristic of
the dynamo to turn down after reaching a maximum
height are : (1) the demagnetising effect of the armature
current when there is a positive lead at the brushes; (2) the
saturating of the iron of the
armature core before that of the
field-magnets ; (3) the leakage
of magnetic lines from the field-
magnet; (4) the peculiar commuting arrangements in certain
machines, for example, the open-
coil dynamos mentioned in
Chapter X., which make their
effective electromotive - force
vary greatly with the position
given to the brushes. As the
demagnetising effect of the
armature current is nearly pro-

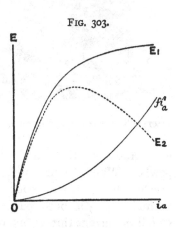

FIG. 303.

portional to the strength of the current and to the sine of the
angle of lead, and as the sine of the angle of lead is itself
nearly proportional to the armature current, it follows that
the whole demagnetising effect is nearly proportional to the
square of the armature current. In Fig. 303, let the curve E_1

represent the electromotive-force (at a given speed) when the field-magnets are separately excited, the armature circuit being left open ; this includes the effect of (2) and partially (3) above. On the same diagram a curve having ordinates proportional to i^2_a and of such a magnitude as to represent the demagnetising action of the armature current, may be plotted. Deducting the ordinates of this curve from those of curve E_1 we get curve E_2, the drooping characteristic. The trouble with all machines of this class is the sparking at the brushes consequent on the variability of the angle of lead.

Statter's Regulator.—Mr. J. G. Statter has devised a constant-current dynamo based upon the principle of auto-matically varying the posi-tion of the brushes (see p. 129). The regulator shown in Fig. 304 consists of a main-circuit solenoid, the core of which is connected by a rod to a double pawl which is continually rocked by means of a vibrating lever, driven by an eccentric pin on the end of the shaft. Two ratchet wheels, mounted upon a common spindle, and having their teeth facing in reverse directions, are set so that one or other of them is operated by the pawls according to the position of the pawls. When these ratchet wheels turn, they shift the brushes forward and backward. When the current is of normal strength, both pawls are out of gear, but any rise or fall of current throws one or other ratchet wheel into gear. Sparking is minimised by peculiar shaping of the pole-pieces, in such a way that through a considerable range of angular positions the induced electromotive-force in any one of the sections of the armature is of equal amount, and sufficient (see p. 98) to balance the induction effects in the coils that

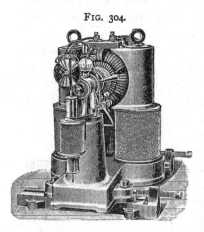

FIG. 304.

STATTER'S DYNAMO, WITH REGULATOR.

are momentarily short-circuited as they pass under the brush. With this device the brushes may be shifted forward through a considerable angle, in order to diminish the electromotive-force, without causing greater sparking than occurs at the neutral point.

OTHER METHODS OF CONSTANT-CURRENT REGULATION.

The other methods of Constant-current Regulation have been briefly enumerated in Chapter VI. We can here deal only with one other class of methods, namely, that depending

FIG. 305.

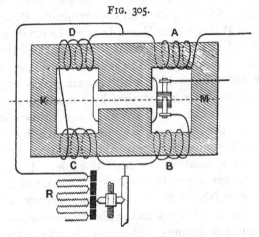

on the application of a magnetic shunt, the plan originally suggested by Mr. J. W. Langley, but much more practically carried out by Messrs. Ravenshaw & Trotter. In the arrangement described by Trotter the number of magnetic lines through the armature core is reduced to any desired degree, without producing instability, while the field-magnet is saturated to its usual degree, by diverting the magnetic lines out of their usual path, into a path of lower magnetic resistance. At first they proposed to do this as Langley had suggested by employing a movable keeper of iron ; but this plan was superseded by Trotter's suggestion to fix the iron keeper and to vary its effect by surrounding it with a counter-magnetising coil. The double-circuit type of field-magnet is suited for this purpose, one half of the machine being kept saturated as a field-magnet of constant power, whilst the other half acts as keeper. In Fig. 305 the acting field-magnet is the right-hand half M, with the magnetising coils A and B; the keeper is the left-hand half K, with the coils C and D. These four coils are connected up in the usual way to give consequent poles at the pole-pieces, but there is a regulating set of resistances R arranged as a shunt to the coils C and D. When the resistance R is small, the coils C and D scarcely receive any current, and consequently nearly the whole of the magnetic lines generated in the magnet pass round the keeper instead of flowing through the armature. By introducing resistance at R the coils C and D begin to excite a counter magnetism and drive back some of the lines out of the keeper and into the armature. The resistances at R are arranged in a graduated set provided with contact pieces with which contact is made by a slider worked by a mechanical regulator driven from the shaft of the dynamo, and thrown into gear so as to move the slider one way or the other by means of an electro-magnetic arrangement according as the current exceeds or falls short of its normal value. It was found necessary[1] to cut a gap in the magnetic circuit of the keeper to prevent sluggishness in the descending magnetisations.

The author of this work has made the suggestion that the same end would be attained without any regulating mechanism at all by winding the "keeper" part of the magnetic circuit with coils connected as a shunt to the armature. This is virtually a new method of compound winding.

M. Reignier[2] has drawn attention to a solution of the problem of exact governing to procure a constant current by automatically varying the number of exciting coils through which the current is permitted to pass.

An account of some of the various sorts of electrical governors in use for controlling the power transmitted to the dynamo will be found in Appendix VI.

[1] See paper by A. P. Trotter, in *Electrician*, vol. xix. p. 374, 1887. A drawing of the regulator itself is given in the *Electrical Review*, xix. p. 289, Sept. 17, 1886.

[2] *La Lumière Electrique*, vol. xxvi. p. 420, 1887.

CHAPTER XXII.

FURTHER USE OF GRAPHIC CONSTRUCTIONS.

Applications of Characteristics.

The following examples of the further use of characteristics are taken from Dr. Hopkinson's paper in the *Proc. Inst. Mech. Engineers* for April 1880 :—

To Determine Lowest Possible Speed of Dynamo running an Arc Lamp.

It appears that with the ordinary carbons, and at ordinary atmospheric pressure, no arc can exist with a less difference of potential than about 20 volts; and that in ordinary work, with an arc about ¼ inch long, the

FIG. 306.

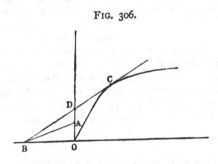

difference of potential is from 30 to 50 volts. Assuming the former result, about 20 volts, for the difference of potential, the use of the curve of electromotive-forces may be illustrated by determining the lowest speed at which a given machine can run and yet be capable of producing a short arc. Taking O as the origin of co-ordinates (Fig. 306), set off upon the axis of ordinates the distance O A equal to 20 volts; draw A B to intersect at B the negative prolongation of the axis of abscissæ, so that the ratio $\dfrac{OA}{OB}$ may represent the necessary metallic resistance of the circuit. Through the point B thus obtained draw a tangent to the curve touching it at C, and cutting O A in D.   Then the speed of the machine, corresponding to the particular curve employed, must be diminished in the ratio $\frac{OD}{OA}$, in order that an exceedingly small arc may be just possible.

Use of Characteristic to Explain Instability of Arc Light.

The curve may also be employed to put into a somewhat different form the explanation given by Dr. Siemens at the Royal Society respecting the occasional instability of the electric light as produced by ordinary dynamo-electric machines.   The operation of all ordinary regulators is to part the carbons when the current is greater than a certain amount and to close them when it is less ; initially the carbons are in contact.   Through the origin O (Fig. 307), draw the straight line O A, inclined at the angle representing the resistances of the circuit other than the arc, and meeting the curve at A.   The abscissa of the point A represents the current which will pass if the lamp be prevented from operating. Let O N represent the current to which the lamp is adjusted; then if the abscissa of A be greater than O N, the carbons will part. Through N draw the ordinate B N, meeting the curve in the point B ; and parallel to O A draw a tangent C D, touching the curve at D.   If the point B is to the right of D, or further from the origin, the arc will persist ; but if B is to the left of D, or nearer to the origin, the carbons will go on parting till the current suddenly fails and the light goes out. If B, although to the right of D, is very near to it, a very small reduction in the speed of the machine will suffice to extinguish the light.

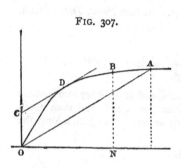

FIG. 307.

Relation of Characteristic to Size of Machine.

Suppose that a certain dynamo of a given construction has for its characteristic the curve O *a* (Fig. 308).   What will be the characteristic of a dynamo built of precisely the same type, but with all its linear dimensions doubled?   The surfaces will be four times as great, the volume and weight eight times as great.   There will be the same number of turns of wire, but the length will be doubled and the cross-section quadrupled, and therefore the internal resistances will be halved.   If the resistances were adjusted so as to give the same current as before, the new machine would have only half the intensity of field of the small one.

But if adjusted to give the same intensity of field as before, the current will be doubled.

Now as the area of the rotating coils is increased fourfold, there will be four times as many lines of force cut (at the same speed), and therefore the electromotive-force will be four times as great. But we only wanted the current doubled. That is to say, the resistance will have to be doubled if the field is to be of same intensity. To represent this state of things, take the point *a* on the characteristic of the small machine, and draw the ordinate *a m*. Draw O M, double O *m*, and at M erect an ordinate A M four times the length of *a m*. The new characteristic will pass through A. Also the resistance—the slope of O A—will be double

FIG. 308.

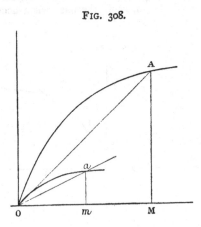

that of O *a*. The points *a* and A are similar points with respect to the saturation of the iron of the magnets; and it is this which determines the practical limits to the economic working of a dynamo of given type at a given speed. Hence we see, with quadrupled electromotive-force and doubled current, the electric energy evolved per second will be eight times as great as with the smaller machine when worked up to an equal saturation limit. These points may be compared with the discussion of the relation of size to efficiency on p. 363.

Application of Characteristics to Dynamos used in Charging Accumulators.

The following problem is of great practical importance :—*Suppose a dynamo is used for charging an accumulator, and is driven at a given speed, what current will pass through it ?*

Dr. Hopkinson has given a solution of this problem for the case of a series dynamo. Draw the total characteristic of the dynamo (Fig. 309) for

2 G

the given speed. Along O Y set off O E to represent the electromotive-force of the accumulator, and through E draw the line C E A, making an angle with O X such that its tangent represents the resistance of the whole circuit, including the accumulators. This line will cut the characteristic in the points B and A; and, if the characteristic be repeated backwards, in C also. This negative branch of the characteristic is simply the characteristic of the dynamo when the current through it is reversed, and its electromotive-force therefore also inverted. Then O L represents the actual current in the circuit; O M represents an unstable current which might exist for a moment; and O N represents the current which would traverse the circuit were the accumulators to overpower the dynamo and reverse it, as indeed frequently happens when series dynamos are so used. For it will be observed that if, as is the case when accumulators are

FIG. 309.

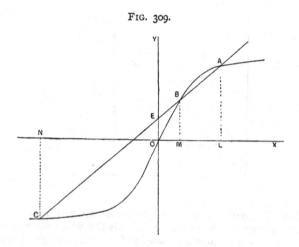

reaching their full charge, their electromotive-force were to rise, or the resistance of the circuit to increase in consequence of heating, the inevitable result would be to diminish A L, the effective electromotive-force, and to diminish the current O L, so that the magnetism of the field-magnets will also drop, and the point A will be brought nearer to the position of instability at the bow of the curve.

With the shunt dynamo the case is different. Let Fig. 310 represent the characteristic of the shunt dynamo, the external current being plotted along O X, and the total electromotive-force along O Y. Draw the line C E A as before. Then it cuts the positive branch at A, and O L is the current in the main circuit. If, now, either the counter electromotive-force of the accumulators, or the resistance of the circuit increases, the effect will be to move the point A to a higher point on the curve. The charging current O L may diminish, but the shunt current will increase, for the effective electromotive-force A L will be increased. Therefore with the shunt dynamo there will be no likelihood of the accumulators over-powering and reversing the dynamo.

FIG. 310.

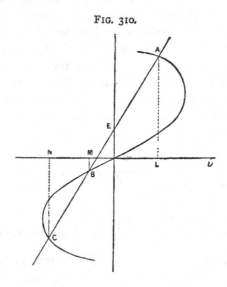

Curves of Torque.

The torque [1] or turning-moment is, in a series dynamo, both when used as a generator and when used as a motor, very nearly proportional to the current. On p. 130 it was shown that the work per second of the dynamo or motor may be expressed mechanically as the product of the angular velocity and of the torque, or

$$\omega\, T = \text{mechanical work per second};$$

and electrically as the product of volts and ampères, or

$$E\, i = \text{electrical work per second}.$$

---

[1] Sometimes also called "the couple," the "moment of couple," the "angular force," the "axial force"; also called in Frölich's memoirs the "Zugkraft"; and in those of Deprez the "effort statique," or the "couple mécanique."

And since in the series dynamo E is very nearly proportional to ω, it followed that T was proportional to *i*. Frölich has given[1] curves showing these relations, and has also argued from the law of magnetic saturation that these curves should for small speeds be curved, and for large speeds become nearly straight lines. He has also shown that in a motor, where the armature current helps to magnetise the field-magnet, the torque is less nearly proportional to the current than in a generator. The following tables summarise the results of his experiments on a series-wound Siemens dynamo used in both functions :—

Generator—

Current	..	2·83	9·56	14·3	19·8	24·3	36·6	ampères.
Torque	..	5·1	10·61	14·8	21·3	29·6	44·0	kilos.

Motor—

Current	..	13·3	21·0	28·1	36·8	ampères.
Torque	..	10	20	30	40	kilos. at circumference.

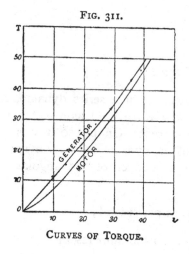

FIG. 311.

CURVES OF TORQUE.

These results are plotted out in Fig. 311 for the two cases.

Similar curves have been given by Deprez[2] for the Gramme machine, and by Ayrton and Perry[3] for a De Meritens motor. It can be shown that the torque is proportional to the square root of the heat-waste in the motor or dynamo. As moreover the current in a motor cannot be maintained without the continual expenditure of energy equal to $i^2 r$ watts, it follows that the continuous torque or turning-moment in a motor costs a certain expenditure, which

[1] *Elektrotechnische Zeitschrift*, vol. iv. p. 61, Feb. 1883.
[2] *La Lumière Électrique*, t. xi. p. 42, Jan. 5, 1884.
[3] *Journal Soc. Teleg.-Eng. and Electricians*, vol. xii. No. 49, May 1883.

will not only vary with the actual load on the motor, but is different in different types of motor. In a badly-designed motor a strong current running through a high internal resistance (and therefore expending much energy as heat) will produce but a feeble torque. For economy, it is therefore important to know at what cost in heat the torque is attained. The ratio may be expressed algebraically as

$$\frac{T}{\text{heat-waste}} = \frac{E\,i}{2\,\pi\,n\,i^2\,r} = \frac{E}{2\,\pi\,n\,i\,r};$$

where r is the internal resistance, E the total electromotive-force of the dynamo, and n the number of revolutions per second. It is, however, preferable to measure T by a direct dynamometric process. Marcel Deprez, who has given to this important ratio the rather awkward name of the "price of the statical effort," has also given curves showing the variation of this ratio with the speed at which the machine is run. Professors Ayrton and Perry have shown in their memoir on electro-motors, that as the speed increases it requires a greater and greater current through the motor to produce a given torque. Herr Hummel[1] has more recently investigated the same point.

Curve of Variation of Economic Coefficient.

The economic coefficient η (see p. 361) varies when the external resistance and external currents change. In a series dynamo we have

$$\eta = \frac{R}{R + r_a + r_m} = i\,\frac{R}{E} = \frac{e}{E}.$$

The second of the three expressions for η shows us how to obtain the curve of efficiency from the characteristic. Consider any point on the curve corresponding to a particular value of i. Divide the corresponding value of the external resistance by the whole electromotive-force, and the quotient

---

[1] *Elektrotechnische Zeitschrift,* vol. viii. p. 426, Oct. 1887.

multiplied by i is the value of η. Or η may be obtained by directly measuring e and i and calculating E. Lastly η may be calculated beforehand from the resistance alone. The curve will be of the following kind:—Let O B (Fig. 312) be the characteristic of the series dynamo, and O J the line representing by its slope the armature resistance. Then, with any particular current P, B P = E, and A B = e, and $\eta = \dfrac{A B}{B P}$. Taking as unity any convenient height, say the height O E, set off P C equal to the fraction $\dfrac{A B}{B P} \times$ O E, giving C as a point on

FIG. 312.

the required curve. It is clear that this curve will descend from E, where at first it is nearly horizontal, and will terminate at a point on O i opposite J. In a shunt dynamo the calculation is much less easy, but it shows a curve which, for small values of i, has smaller values of η than for large values of i. If, instead of plotting out the relation between η and i, we plot the values for η and R, we shall find that with small values of R, η is small, and as R increases η increases, until R has the value

$$R = \sqrt{r_a r_s} \sqrt{\frac{r_s}{r_s + r_a}},$$ when η is a maximum, after which the values of η diminish and become zero when R is infinitely great. In a compound dynamo wound for constant potential the value of η is almost constant, whatever the value of R or i, within the limits of working.

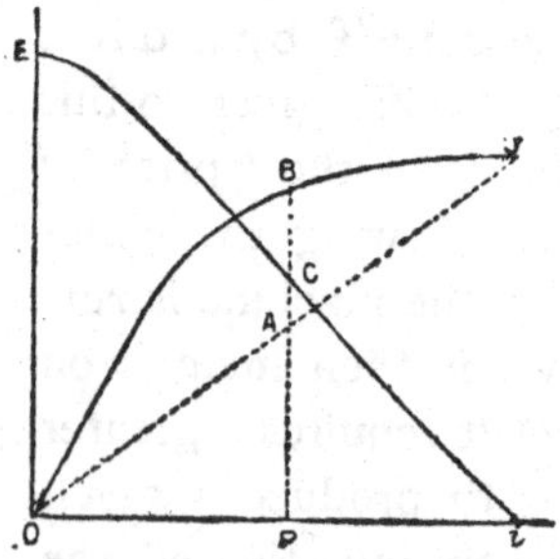

Curve of Horse-power Expended on Maintaining the Field-magnetism.

The energy spent per second in maintaining the magnetism of any magnet may be readily calculated as the product of the square of the magnetising current into the resistance of the coil. Thus, for a shunt machine the energy spent per second is $i_a^2 r'$ or $e\,i_a$ watts, and the electric horse-power is $i^2 r_a \div 746$. It is convenient to exhibit the relation between this expenditure of energy and the current in a curve. Such a curve, taken from tests on an Edison-Hopkinson dynamo, is shown in Fig. 313. The curve would be a parabola if the resistance were constant. This is not so, however, on account of the greater heating effect of the stronger current.

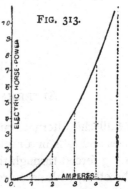

FIG. 313.

For further use of graphic methods in the study of the dynamo, the reader is referred to papers in *La Lumière Électrique*, by MM. Rechniewski, xviii. 481, 1885 : xix. 99, 1886 ; Picou, xxiii. 13, 56, 88 : xxiv. 169, 1887 ; and Reignier, xxiii. 468, 618, 1887.

CHAPTER XXIII.

ALTERNATE-CURRENT DYNAMOS.

IN all the alternate-current dynamos the electromotive-force rises and falls in a rapid periodic fashion, a wave of electricity being forced through the circuit first in one direction, then in the other, with very great rapidity. To calculate this rise and fall we must remember that the motions of the rotating coils are supposed to take place with a uniform speed, and that the induced electromotive-force will be proportional to the rate of change in the number of lines of force induced through the circuit. To get a complete account of the action we must take into consideration the number of lines of force *induced by the circuit on itself.*

Consider a simple loop of wire traversed by a current. Every portion of the loop will be surrounded by a whirl of lines of force similar to those of Fig. 5, and those belonging to the current in one-half of the loop tend to influence the current in the other half. Such influence or tendency of the current's lines of force to influence the other parts of the circuit is, however, only manifested when the strength of the current or the shape of the circuit is changing. We know that any increase in the number of lines of force that thread through a circuit (as, for example, by poking a magnet pole into it) tends to set up a current that will oppose the motion. Any increase in the strength of the current in the loop will increase the number of lines of force through the loop, and that increase will of itself tend to set up an opposing current. On the other hand, any decrease in the current in a circuit, by reducing the number of lines of force which thread through the circuit, tends to oppose the reduction of the current. A current, in fact, acts as if it had inertia and tends to keep the number of its lines of force constant. This inertia of the current in a circuit is also known as the induction of the circuit on itself, or, briefly its self-induction. The self-induction in a circuit, which, as we have seen, is the number of lines of force threaded through the circuit itself by the current flowing in it, is always made up of two factors. When discussing the earlier case of the induction of a current in a loop by rotating it in a uniform magnetic field we neglected self-induction and considered the number of lines of force which were cut by the loop as being the product of two factors, viz. the total number of magnetic lines N that actually traverse the iron core of the armature, and C, the number of conductors at its periphery. In the case of self-induction the two factors will be different. The number of lines of force induced on itself by the current in a loop will be proportional to the strength of the current i. The number will, for a simple circular loop, be also proportional to the area of the loop. But for coils that are not circular, and consist of many turns, and have iron in them, a much more complicated investigation would be required than could be undertaken here. It is found that where coils of many turns, furnished with iron cores, are in question, the self-inductive effect depends on the properties of the iron, as well as on the number of coils and on the current. The magnetising force of the current is proportional to the number of turns in the coil ; hence, supposing the number of magnetic lines thus created in the core to be proportional to the magnetising force, the self-induced effect (due to a given current) which is proportional to the number of times that the magnetic lines thread their way through the turns of the coil, will be proportional to the *square* of the number of turns of the coil. Hence, the two factors chosen are current i, and a quantity symbolised by the letter L, and called " *the coefficient of self-induction,*" which represents the number of lines of force which the circuit would enclose or induce on itself if the current flowing in it were one "absolute unit." It follows at once that if a current of i units flow through a circuit whose coefficient of self-induction is L, the whole self-induction of the circuit will be equal to L times i; and the product L i will represent the total number of lines of force belonging to the circuit itself. It will also be evident that if a current begins from strength o and grows until its strength is i, the *average* self-induction in the circuit will be $\frac{1}{2} L i$.

Returning now to the case of a loop having S_2 turns, placed at such an angle θ (measured from the initial position as in Fig. 245, where it stands right across the field), we see that it no longer encloses the whole number of lines of force which are present in the magnetic circuit. When we omit all account of self-induction, we may write

$$N_1 = S_2 N \cos \theta,$$

where N_1 is the virtual amount of "cutting" of lines by the circuit.

We now know that if there is a current i in the circuit, we ought to write the equation in full—

$$N_1 = S_2 N \cos \theta + L i. \qquad \text{[XLII.]}$$

Our omission of the self-induction term in all the previous equations was only justifiable on the assumptions—firstly, that the field-magnets so overpowered the armature as to make the second term negligible; secondly, that the equations we obtained were the equations for *steady* currents.

Now we know that any variation in N_1 will set up induced electromotive-force, and that at any moment the electromotive-force will have the value

$$E = -\frac{d N_1}{d t};$$

where we use the negative sign to show that an increase in N_1 will produce an inverse or negative electromotive-force. We are obliged to take note henceforth of the signs of the various quantities. Any change in N_1, from whatever source arising, will set up electromotive-force. We cannot well alter S_2, the number of coils of our armature. N can be altered; and in most modern alternate-current machines it is arranged that N, the whole number of magnetic lines in the field, can be controlled by hand or otherwise, the field-magnets of the alternate-current machine being usually separately excited by a constant current from a smaller dynamo, called the "exciter" (see Fig. 217, p. 267). While the dynamo is at work, the coefficient of self-induction L cannot be changed at will, as that depends on the size, shape, coiling, and coring of the armature, and on magnetic permeability. The only quantities that are really important in respect of the variations they will undergo — the only quantities whose variations contribute to the variations of N_1—are, then, θ and i. The angle of position θ varies from o to 2π (radians); that is to say, from o° right round to 360°, and then recurs; and its cosine therefore fluctuates between 1 and − 1. The current i varies also from a certain maximum value $+ i_{max.}$ to an equal negative value $- i_{max.}$ We will neglect all the variations of the other quantities, not because these variations would not be instructive—for that would be quite untrue—but because of their lesser practical importance. Then we have

$$E = -\frac{dN_1}{dt} = -\frac{d\,(S_2N \cos\theta + L\,i)}{dt}.$$

Now suppose that while the armature loop has turned through the angle θ, the time occupied—a small fraction of a second—is t. Also take T to represent the time taken for one revolution; so that if there were n revolutions per second, T will be $\frac{1}{n}$ of a second. Then obviously θ will be the $\frac{t}{T}$ part of a whole revolution, and as there are 2π radians in a circle, the angle expressed in radians will be

$$\theta = 2\pi\,\frac{t}{T} = 2\pi nt = \omega t.$$

Inserting this value, and performing the differentiation, we get

$$E = \frac{2\pi S_2N}{T} \sin\frac{2\pi t}{T} - L\frac{di}{dt}; \qquad [XLIII.]$$

$$= \omega S_2N \sin \omega t - L\frac{di}{dt}.$$

Consider this equation carefully. It shows us that if there

were an open circuit, so that there could be no i, then self-induction would not come in at all. Also if the motion were so slow that the rate of change of i were inappreciable, then the second term might be neglected. The negative sign also indicates that that part of the electromotive-force which is due to self-induction opposes the other part. Suppose we pause for a moment to consider the case of slow motion, and neglect the self-induction term; remember that $\frac{1}{T}$ is the same thing as n, and we get

$$E = 2 \pi n S_2 N \sin \theta,$$

as we did on p. 339. Also, since the average value of the sine between 0° and 90° is $\frac{2}{\pi}$,[1] we get, as the average value of E,

$$E \text{ (average)} = 4 n S_2 N.$$

Also it is important to notice, that with slow rotation, if the resistance of the circuit be R, the current will be at any moment

$$i = 2 \pi \frac{S_2 N}{R T} \sin \theta,$$

and the *average* value of the current

$$i \text{ (average)} = 4 \frac{S_2 N}{R T}.$$

This average value of i being found, we are now able to entertain the case where the rotation is so quick that we must take in the self-induction term. Remember that $E = R i$; and we may write

$$R i = \frac{2 \pi S_2 N}{T} \sin \frac{2 \pi t}{T} - L \frac{d i}{d t}.$$

---

[1] Or more strictly

$$\frac{1}{\theta} \int_0^\theta \sin \theta \, d\theta = \frac{1 - \cos \theta}{\theta},$$

whence, if $\theta = \frac{\pi}{2}$, the average is $\frac{2}{\pi}$.

Here is a differential equation requiring to be solved. Its solution is

$$i = \frac{2\pi S_2 N}{T} \cdot \frac{\sin\left(\dfrac{2\pi t}{T} - \dfrac{2\pi \tau}{T}\right)}{\sqrt{\left(\dfrac{2\pi L}{T}\right)^2 + R^2}} + C\,\epsilon^{-\frac{R}{L}t} \, ;$$

or

$$i = \frac{2\pi S_2 N}{T} \cdot \frac{\sin(\theta - \phi)}{\sqrt{\left(\dfrac{2\pi L}{T}\right)^2 + R^2}} + C\,\epsilon^{-\frac{R}{L}t}, \quad \text{[XLIV.]}$$

where ϕ is called the retardation or angle of lag, and is determined by the condition,

$$\frac{2\pi L}{R T} = \tan\phi = \tan\frac{2\pi \tau}{T}.$$

Here the symbol τ stands for the short interval corresponding to the time-retardation. The symbol ϵ is used here in its common mathematical sense to represent the number $2 \cdot 7182$ which is the base of the Neperian logarithms; it has nothing to do with the potential at the brushes, for which we happen to have used the same symbol elsewhere ; and C is a constant of integration. It may be pointed out that of the two terms, the second may be neglected, because it relates only to the momentary growth of the current and dies out after a very short interval. The first of the two terms may also be written more simply. Remembering that $\frac{1}{T} = n$, we have

$$i = \frac{2\pi n S_2 N}{\sqrt{\left(\dfrac{2\pi L}{T}\right)^2 + R^2}} \sin(\theta - \phi). \quad \text{[XLV.]}$$

This shows us that our current is virtually due to an electromotive-force of the value $2\pi n\, S_2 N \sin(\theta - \phi)$, acting through a resistance of the value $\sqrt{\left(\dfrac{2\pi L}{T}\right)^2 + R^2}$. Now when there was no self-induction to take into account, the electromotive-force was (see p. 339) $2\pi n\, S_2 N \sin\theta$, and the resistance simply R. The effect of self-induction is then to *retard* the rise and fall of the current, so that it attains its maximum not when $\theta = 90°$, but when $\theta = 90° + \phi$, the latter symbol standing for the angle of lag. Looking at the value of ϕ stated above, we see that its tangent is $\dfrac{2 \pi L}{R T}$, or $2 \pi n \dfrac{L}{R}$. From this we learn that the retardation will increase with increased speed, and that it depends on the ratio between the self-induction and the resistance. There will be less lag therefore if the machine is so designed that it can be driven at a slow speed, and if the coefficient of self-induction is small compared with the resistance of the circuit. This indicates that the number of turns of the coils in the armature part should be kept as small as possible, and the magnetic field made enormously powerful: a rule which applies equally to continuous-current dynamos. It may also be noticed that the resistance is apparently increased from R to $\sqrt{(2 \pi n L)^2 + R^2}$ the apparent increase depending both on speed and self-induction. Both speed and self-induction tend, therefore, to retard the oscillations of the electromotive-force, and to diminish the available amount of current. High speeds are, therefore, both mechanically and electrically bad.

To find the amount of work-per-second W done by the alternate-current machine, we may remember that

$$W = i^2 R \text{ (watts)},$$

and that for i^2 we must take the *average square*[1] of the values of i. Since the average value of the *square* of the sine of any angle between 0° and 90° is $\frac{1}{2}$, we thus obtain

$$W = \frac{2 R \pi^2 S_2^2 N^2}{R^2 T^2 + 4 \pi^2 L^2},$$

and this expression, by a well-known algebraic formula, is a

---

[1] The student will not forget that the average of the square of any variable quantity is different from (and greater than) the square of the average value of that same quantity.

maximum for variations of R, when the two terms in its denominator are equal, that is, when

$$R = \frac{2\pi L}{T} = 2n\pi L,$$

or when the apparent resistance is such that its square is double the square of the real resistance. If the speed be such that this relation between resistance and self-induction is observed, then also tan ϕ will = 1, and the retardation is 45°. These calculations are taken from the investigations of M. Joubert. In the preceding argument we have supposed the machine to consist of a simple armature rotating in a simple field. But in the majority of alternate-current machines (see pp. 256 to 278), the field is complex, and the coils of the armature pass a series of poles set symmetrically round a circle. In this case all the alternations of induction will recur several times in a revolution. If, as in the Siemens alternate-current dynamo, there are sixteen sets of poles, of which eight are N. poles, and the intermediate eight S. poles, eight times in every revolution all the periodic fluctuations in the induction will recur. The electromotive-force is, how-ever, found to follow the sine law fairly well, provided we take *n* to represent the number of *alternations* per second, or eight times the number of revolutions per second. Also θ must be understood to be not the actual angle of position of the coil at any moment, but equal to the $\frac{t}{T}$ part of 2π; T being the period, not of one revolution, but of *one alternation*.

Two of the most important of M. Joubert's results may be summed up in a diagram, Fig. 314. Let the curve A represent the rise and fall of the induced current, as it would be if there were no self-induction. Then, since the effect of self-induction in the armature circuit is both to retard the rise and fall of the effective electromotive-force, and to increase the apparent resistance, it will have the effect of causing the current to rise and fall like the curve B, which has a smaller amplitude and is shifted along.

It may be remarked that mere retardation does not waste

any of the power, nor does the mere introduction into the
circuit of the opposing electromotive-forces of self-induction.
If induction could be limited to these two effects, it would
not be very prejudicial, it would simply make the machine
act as a smaller machine. But, unfortunately, induction
cannot be so limited. It occurs in every moving mass of
metal in the armature. It occurs in the iron cores and even
in the driving shaft. Also, in the continuous-current machines
it operates most prejudicially in the separate sections at the
moment when they are short-circuited by passing the brushes,
as explained in Chapter V., and at that moment the section
that is short-circuited does not electrically belong to the

FIG. 314.

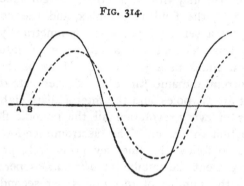

circuit. All these inductive effects are attended with waste
of power, because the currents generated in any conductor
that does not form an actual part of the circuit simply
degenerate into heat. A curious case occurred with Lontin's
alternate-current machine, which had solid iron cores. When
the coils of the armature were actually supplying the circuit,
the currents were generated in them, and also to some extent
in the cores. But when the circuit was open, so that no
current could be generated in the coils, then currents were
generated instead in the cores. Two results followed. Whenever the machine ran on an open circuit, the cores heated up
to a destructive point, *and it required more power to drive the
machine when it was doing no work than when it was lighting
its maximum number of lamps.*

It has been shown by Dr. Hopkinson[1] that two alternate-current machines, independently driven, so as to have the same periodic alternation and the same electromotive-force, cannot well be connected in series, otherwise they will tend to annul one another's currents: but that they may be coupled in parallel to one another (see Chapter XXIV.).

The facts of alternate-current machines are more complex in yet another respect than in the simple case taken for calculation. The question how far the gradual rise and fall of a curve of sines corresponds to the rise and fall of the electro-motive-force in the armature, depends on the form of the armature coils and on the form of the polar surfaces between which the coils pass. Consider the case of a machine in which the field-magnets consist of a double crown of opposing poles (as in the machines of Wilde, Siemens, Ferranti, Mordey, &c.). If the armature coils and magnet cores are both of circular form, and equal in diameter, as the coils approach the polar ends of the cores they will, it is true, gradually enter the field, and the number of lines cut by the coil during equal displacements will gradually increase and become a maximum when the axis of coil and core coincide, and from that point it will again decrease, almost in a sine law; the greatest rate of cutting being when the edge of the coil is opposite the centre of the core : but if coil and core be rectangular in section, the greatest rate of cutting will not be when the axes coincide, but just when one edge of the coil is passing the edge of the pole. In this case the sine law cannot be true for the electromotive-force. In order to test whether in any given dynamo the rise and fall of electromotive-force and of current in the armature coils conforms to the law of sines, experiments are necessary. Joubert, in order to measure the currents of a Siemens' dynamo, employed an electrometer method, and took off the current at any desired phase by a special commutator, and found an approximate curve of sines. Another method, analogous to that used in investigations on direct-current machines, has been used by Mr. Mordey. It

---

[1] *Proceedings of Inst. Civil Engineers,* April 1883, " Some Points in Electric Lighting " ; and *Journ. Soc. Teleg.-Eng. and Electr.,* vol. xiii. p. 496.

consists in connecting the armature coils, or one of them, to a
suitable (ballistic) galvanometer, exciting the field-magnet
with a current from some constant source, placing the armature in any desired position, and then suddenly, by a quick-
acting switch, turning off the exciting current. The throw of
the galvanometer measures the effective induction in the
particular position: the armature is then moved on through a
small angle, and another throw obtained. A curve is then
plotted showing the variations in the induction of magnetism
through the armature. The steepness of the curve measures
the electromotive-force at every point.

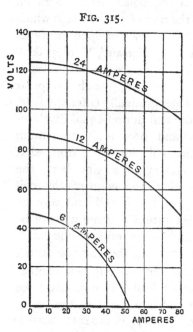

FIG. 315.

This method does not
require the machine to be
driven: and it does not take
into account the possible reactions of the armature current
in a running machine, which
will doubtless displace the
distribution of the magnetism.

The demagnetising in-
fluence of the armature current has been studied by Mr.
Esson,[1] who has determined
(Fig. 315) the external characteristic of an alternate-current
dynamo when separately excited with different amounts of
current. The three curves correspond to the three cases
when the separate exciting
current measured respectively
24, 12, and 6 ampères. The
effect of the armature current
is most marked when the field-magnets are weakly excited,
and apparently roughly in proportion to the square of the
armature current. In the Mordey alternator (p. 273) the
field-magnet is so powerful that the diminution of the electro-

[1] See *Electrical Review*, xviii. p. 248, March 1886.

motive-force from this cause with the full current is less than 3 per cent. of the whole, the resulting droop in the characteristic being extremely slight.

It may be pointed out that owing to the great self-induction in the coils of any electro-dynamometer constructed of fine wire, it is generally impracticable to use such an instrument as a voltmeter for measuring the average difference of potential between two points of an alternate-current circuit, and, even if practicable, its indications would be false, as they would depend, not alone on the average potential, resistance, and self-induction, but also on the frequency of the alternations. In such cases it is better either to use a voltmeter depending on the stretching of a long thin wire heated by the current, completely enclosed in a tube having the same coefficient of thermal expansion, as suggested by Cardew, or else to apply a quadrant electrometer in the manner suggested in 1880 by M. Joubert. In this latter case the difficulty is to calibrate the scale-readings. This is best done by taking readings first with a continuous current on some portion of the circuit in which there is little or no self-induction, such as a straight thin wire, and comparing these readings with others obtained *on the same wire* when the alternate current is passed through it. Another method is to measure in a delicate calorimeter the heat evolved from a thin wire of high resistance used as a shunt. If a Cardew voltmeter is employed, the *maximum* electromotive-force will be equal to that shown by the instrument multiplied by $\sqrt{2}$.

The student is referred for further reading to Joubert, *Annales de l'École Normale Supérieure*, x., 1881, and *Journal de Physique*, s. ii. t. ii. p. 293, 1883 ; and particularly to the paper of Dr. J. Hopkinson in Vol. XIII. of the *Journal of the Society of Telegraph-Engineers and Electricians* (with the discussion which followed). More recently Dr. J. Hopkinson has published in the *Proceedings of the Royal Society* (February 1887) a theory of the alternate-current dynamo in which the periodic variations in the self-induction in the machine are taken into account. Dr. E. Hopkinson has also touched the subject in his paper read before the British Association, 1887, reprinted in the *Electrician*, xix. p. 379, 1887.

CHAPTER XXIV.

On Coupling Two or More Dynamos in One Circuit.

It is sometimes needful to couple two or more dynamos together, so that they may supply to a circuit a larger quantity of electric energy than either could· do singly. Thus it may occur that two dynamos, neither of which can safely carry a greater current than 10 ampères, are required to supply jointly a 20-ampère current: or two machines, each of which can run at 60 volts, are required to furnish an electromotive-force of 120 volts. Simple as these cases may seem, it is not so easy to carry them out, because it depends upon the construction of the machine, and especially upon the mode of excitation of the field-magnets, whether they can be coupled together without interfering with each other's running. For it may, and does, occur that if not rightly arranged, one machine will absorb energy from the other and be driven as a motor instead of adding anything to the energy of the circuit.

Coupling Machines in Series.—Series-wound dynamos may be united in series with one another for the purpose of doubling the electromotive-force. Thus two Brush machines, each working at 10 ampères, and each capable of working 6 arc-lamps, may be joined in one circuit with 12 arc-lamps in series. The only needful precaution is to see that the + terminal of one machine is joined to the − terminal of the other, precisely as with cells of a battery. Shunt-wound dynamos may also be coupled in series, though the arrangement is not good unless the two shunt coils are also put in series with one another, so as to form one long shunt across the circuit. Compound-wound dynamos may be connected

in series with one another, provided, the shunt parts of the
two are connected as a single shunt, which may extend
simply across the two armatures (double short-shunt), or
may be a shunt to the external circuit (double long-shunt),
or may be a mixture of long and short shunt. The same
considerations apply to more than two machines. The
coupling of alternate-current dynamos is considered later.

Coupling Dynamos in Parallel. — Two series dynamos
cannot be coupled in parallel in a circuit without a slight
re-arrangement, otherwise they interfere. For, suppose one
of them to fall a little in speed, so that the electromotive-force
of one machine is higher than that of the other machine with
which it is in parallel. The machine having the higher
electromotive-force will then drive a current in the wrong
direction through the other machine, reversing the polarity
of its field-magnets and driving it as a motor. To obviate
this, Gramme made the suggestion that the machines should
be coupled in parallel at the brushes
as well as at the terminals. This is
shown in Fig. 316. The terminals
$T_1 T_1$ of one machine are respectively
joined to $T_2 T_2$ of the second machine,
and a third wire joins B_1 with B_2. If
both machines are doing precisely
equal work, there will be no current
through the wire $B_1 B_2$. If either
machine falls behind, part of the
current from the other machine will
flow through $B_1 B_2$ and help to main-
tain the excitement of the magnets of the weaker machine.
This effectually prevents reversals.

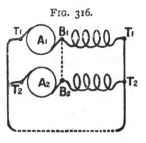

FIG. 316.

COUPLING OF TWO SERIES
DYNAMOS IN PARALLEL.

In the case of shunt machines there is no great difficulty in
running them in parallel, as indeed is done on a large scale
with the eight dynamos of Edison's New York lighting
station. The chief precaution to be taken is that, whenever
an additional dynamo has to be switched into circuit, its field
must be turned on, and it must be run at full speed before its
armature is switched into connexion with the mains, otherwise

the current from the mains will flow back through it and
overpower the driving force.[1]   Another method of coupling
two series machines is to cause each to excite the other's
field-magnetism.   This equalises the work between the two
machines.

Coupling of Compound Dynamos in Parallel Circuit.—It
has been stated on high authority that there is difficulty in
working compound dynamo machines in parallel circuit,
chiefly on account of the tendency possessed by any particular dynamo, which from any cause may temporarily be
doing more work than the other, or others, in parallel circuit
with it, to do still more work.   It is stated that the exaltation
of the field strength in such a case, due to the increase of the
current traversing the series circuit, raises the electromotive-
force of the machine, and causes it not only to do more than
its due share of the work, but even to send current through
the machine which is connected with it, which on account of
its reduced electromotive-force becomes merely a part of
the external circuit of the more powerful generator.   There
is no doubt that such an action may, and does, take place ;
and if unavoidable would render compound dynamos un-
suitable for employment in many situations where their self-
regulating powers must render them very useful.   Mr. W. M.
Mordey, to whom the author is indebted for the paragraphs
which follow, has, however, pointed out that the difficulty
may be overcome by connecting the parallel machines in
such a way that not only are the shunt portions of the field-
magnets in parallel circuit, but the series circuits of the field-
magnets are also a shunt on one another ; in other words,
by connecting the brushes, as well as the terminals, in parallel
circuit, precisely as Gramme has done for series-wound machines.   This mode of connexion is shown by the accompanying diagram (Fig. 317).

$A_1 A_2$ are the armatures of two compound dynamos, $T_1 T_1$

[1] See Burstyn, in the *Zeitschrift für angewandte Elektricitätslehren*, 1881,
p. 339, also Schellen (2nd Edition), p. 717 ; Ledeboer, in *La Lumière Électrique*,
xxvi. 210, 1887 ; Meylan, in *La Lumière Électrique*, xxvi. 379, 1887 ; and
Feussner, in *Zeitschrift für Elektrotechnik* (1887), 108.

and $T_2 T_2$ are the terminals; the wire $B_1 B_2$ acting in conjunction with the lead $T_1 T_2$ on the left, puts the armatures in parallel. When compound dynamos are connected in this way,[1] they work quite satisfactorily, and exercise a considerable power of mutual adjustment. No necessity exists for driving by clutch or other "positive" connexion, ordinary belt driving being quite admissible, even when the belts have

FIG. 317.

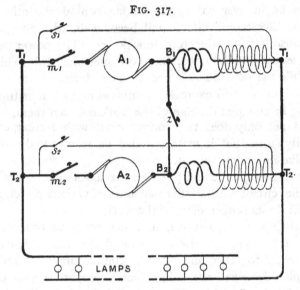

COUPLING OF TWO COMPOUND DYNAMOS IN PARALLEL.

different percentages of slip—as may happen when they are not alike in tightness or character.

This mutual adjustment extends also to the case of slight differences in the sizes of the driving or driven pulleys where a single steam-engine or other motor is driving both or all of the machines, as well as to the case of separate engines being employed to drive separate machines. Of course the power of mutual adjustment must not be unduly strained by trusting to it to remedy inequalities of a serious nature.

[1]. The method proposed by M. Ledeboer in *La Lumière Électrique*, xxvi. 210, 1887, is practically identical with the above.

The *rationale* of this adjustment is very simple. Taking the case of two exactly similar compound dynamos, connected as described above, it will be evident that as the shunt fields as well as the series fields are similar, and are respectively in parallel circuit, the strength of the magnetic fields in the two dynamos will be alike. Then at the same speed their electromotive-forces will be equal, and they will absorb power equally, and will do equal work. But if from any cause one of them begin ever so slightly to lag behind the other in speed, its electromotive-force will become slightly lower, and it will absorb proportionately less power. The power being thus unequally distributed, the slow machine will tend to race, while the fast one will tend to slow down. In this way the two dynamos will exercise a continual mutual adjustment, resulting in an equal division of the work between them.

And not only does this control exist with similar compound dynamos, but it may be relied on when the dynamos are unlike in size, power, and speed.

For instance, large and powerful machines may be worked in parallel circuit with smaller machines of various power, and each will do its proper share of the work.

In such a case, however, it is necessary to observe an additional precaution. Not only should the various dynamos be connected together, and to the external circuit, according to the plan described above, but such a proportion should be observed between the resistances of the series coils of the various connected machines that with the varying resistance of the external circuit the fall of potential in all the series coils may be similar. This is the case when the respective resistances of the series coils are inversely proportional to the full (or any equal proportion of the full) current intended to be generated by each dynamo.

When the resistances of the series fields of the parallel dynamos are thus inversely proportional to their currents, they will work satisfactorily in parallel circuit, and will possess the desired power of adjustment under any circumstances likely to arise in practice where ordinary care and skill are exercised.

The examination of the subject does not, of course, cover all the details of the actions connected with the working of compound dynamos in parallel circuit. A fuller inquiry reveals a theoretical necessity for giving an exactly similar formation to the characteristic curves of all the connected machines. For practical purposes, however, the foregoing precautions will generally be found sufficient.

A method of connecting the machines, differing from the above, has been suggested. It is very similar to that which has been used, as mentioned above, with Gramme dynamos, consisting in employing the current of one machine A to excite the fields of a second machine B, while the current of B in turn is made to magnetise the fields of A. This is a perfectly practicable plan. With compound dynamos the series coils will alone require to be operated in this way. But there are some objections to such an arrangement. It can only be used when the dynamos are exact copies of each other, and is therefore out of the question when it is desired to utilise machines of various sizes and speeds to operate one circuit. Another objection is that with such a method it is always necessary to have at least two machines working, even when one is sufficient or more than sufficient for the requirements of the moment. In such a case when it may be desired to use one machine only, an arrangement of switches, always more or less unsatisfactory, must be adopted; while the making of the involved change could scarcely be effected, while the machines were working, without causing some interruption to the external current—an event, however momentary, to be carefully avoided in practical work. Again, an accident to one machine would incapacitate not only that machine, but also the second one which relied on the former for its field excitation.

The plan suggested by Mr. Mordey appears the more satisfactory one, and may be used in a lighting station, or in any situation where the varying requirements of the circuit render it desirable to bring additional machines into operation as the work increases, and to disconnect them from the mains as the demand for current falls off. To accomplish this the following arrangements and order of operations should be observed. The dynamos should each be furnished (see Fig. 317) with a switch s in the shunt circuit ; they should each have also a switch m in their main circuit between the armature part and the point where the shunt circuit joins on, so that the armature part may be interrupted without interrupting the shunt circuit. The connecting wire from brush to brush, which should be at least as thick as the mains, should also be furnished with a switch z. Suppose dynamo No. 1 is at work alone, its two switches, s_1 and m_1, will be closed. If, now, dynamo No. 2 is to be thrown in, the following order must be observed. First get up the speed of No. 2 to its full value, then close s_2, then z ; this will fully excite its magnetism ; lastly, close m_2. When No. 2 has to be thrown out of circuit the order must be exactly reversed : first open m_2 ; then z ; then s_2 ; lastly, slow down the machine. A special combination-switch, which will perform these successive operations in their proper order, is desirable.

Coupling of Alternate-current Dynamos. — The chief principles governing the working of two or more alternate-current machines on the same circuit were experimentally discovered by Wilde, and described by him in a paper published in the *Philosophical Magazine* as far back as January 1869. In the midst of the labour devoted during succeeding years to the development of direct-current machines, Wilde's paper appears to have been quite forgotten, and it was not until Dr. Hopkinson independently took up the question and treated it in his lecture on "Electric Lighting" before the Institution of Civil Engineers in 1883, that attention was recalled to the subject. Dr. Hopkinson's method of procedure differed essentially from Wilde's in that he first deduced the behaviour of certain alternate-current dynamos under given conditions from theoretical considerations, and afterwards when opportunity occurred practically tested and verified his theoretical conclusions. In the following remarks we shall chiefly follow the line of argument used by Dr. Hopkinson.

In order that two or more alternate-current machines may work usefully together, it is easily seen that the periodic time of the alternations of one machine must be exactly equal to or at least some very small multiple of the periodic time of the other. In practice only the first case has been carried out hitherto. Let us consider then the case of two exactly similar and equal machines A and B, running at the same speed, so that the periodic time of the alternations of electromotive-force is the same in each machine. If the phase[1] of these two machines happens to be exactly the same, and they are joined in series, then evidently the two electromotive-forces will be added together and the two machines will behave as

FIG. 318.

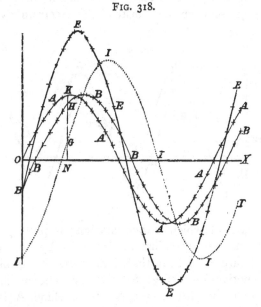

one. But such a condition of affairs will be unstable ; and if anything happens, such as a slip of one of the driving belts, to slightly alter the exact agreement of phase, the mutual electrical action will *tend* to increase the difference of phase instead of counteracting it. In Fig. 318 let the abscissæ measured along O X represent time and the vertical

[1] By the phase being the same we mean that the maximum of positive electromotive-force occurs at *exactly the same instant* in each machine. In any other case the phases differ.

distances electromotive-force; then the curves A A A and
B B B will represent the march of the alternations of electromotive-force in the two machines, the curve B B B, which
lies to the right of A A A, being the one corresponding to
the machine which lags behind the other in phase. The
curve E E E, which is obtained by adding the ordinates of
the other two curves, gives the resultant electromotive-force in
the circuit at any instant. As shown on p. 461, the current
will have the same periodic time as this last electromotive-
force, but will lag behind it in phase. It may therefore
be represented by the dotted curve I I I, in which the
ordinates represent current instead of electromotive-force.

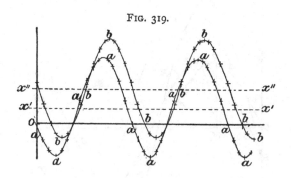

FIG. 319.

Now the rate at which either machine is putting energy
into the circuit at any instant is given by the product
of the ordinate of I I at that instant by the ordinate of
the electromotive-force curve for that machine at the same
instant. Thus at the instant N the machine A is doing
work at the rate NK×NG, and the machine B at the
rate NA×NG. The meaning of the product being sometimes negative is that the machine is at that moment *absorbing*
energy from the circuit. The *total work* done by either
machine during a complete alternation of current is obtained
by summing up for the whole alternation the above products,
each multiplied by the small interval of time $d\,t$ during
which it can be assumed to be constant. This is most readily
summed analytically, but may be done graphically as follows.

Plot a new set of curves in which the abscissæ are the same as before, but the ordinates are the products of electromotive-

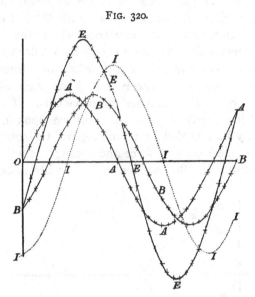

FIG. 320.

force and current for each machine for each instant of time. The result will somewhat resemble Fig. 319, which has been obtained thus from Fig. 318. Here the curve *a a a*

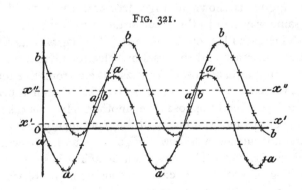

FIG. 321.

refers to machine A, and *b b b* to machine B. The total work is obtained by measuring the area included between each curve and the axis of abscissæ, remembering that areas *below* the axis are to be reckoned as *negative* and arithmetically deducted. A moment's inspection of these curves will show that machine B is doing more work than machine A. In fact, the curve *a a a* is symmetrical about the horizontal axis *x′ x′*, whereas the curve *b b b* is exactly similar to it, but is symmetrical about the higher axis *x″ x″*, and therefore the positive area included between it and the axis of time is necessarily greater than that for the curve *a a a*. The *lagging* machine B has therefore most work thrown upon it, and will consequently be retarded ; thus the lag will be increased, and the resultant electromotive-force and consequently the current thereby diminished. But that the tendency will still be towards

FIG. 322.

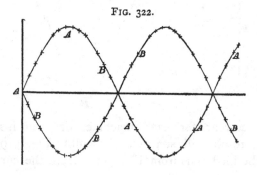

further lagging is shown in Figs. 320, 321, which are drawn in the same way and to the same scale as Figs. 318 and 319, but with increased lag of machine B. The lag will therefore continue to increase, and the resultant electromotive-force and current to diminish, until such time as the electromotive-forces of the two machines differ in phase by exactly half a period and therefore directly oppose one another. In this case the resultant electromotive-force will be continually zero and therefore no current will flow. Figs. 319 and 321 will be reduced to Fig. 322. This condition of affairs is stable, since if anything disturbs the exact opposition of phase the electrical action will tend to re-establish it.

Another deduction from the above proof is that the machines will theoretically work perfectly well in parallel. For let A *a* (Fig. 323) represent the collectors of machine A, and B *b* those of machine B; then, as shown above, if these are joined, A to B, and *a* to *b*, as in the dotted lines of the figure,

FIG. 323.

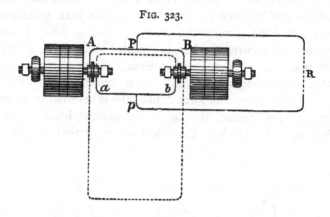

no current will flow, and if an arc-lamp be placed in *b a* or B A it will not light up. In fact, A and B are both at their maximum positive potential simultaneously, and at the same instant

FIG. 324.

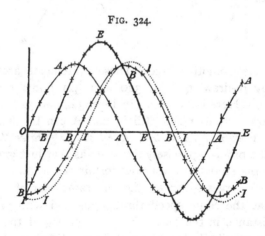

a and *b* are at their maximum negative potential. But this is exactly the state of affairs which will enable us to obtain a current through the circuit P R *p* joined on to the wires

A *a* and B *b* at the points P and *p*. With this arrangement the machines are working in parallel through the circuit P R *p*.

A still further deduction is that an alternate-current machine can be used as a motor. In this case machine B (the *lagging* machine) is the generator and is doing positive work upon the current, whereas machine A is doing negative work upon the current, *i. e.* is receiving energy therefrom, and is therefore acting as a motor. The conditions are that the lag of one electromotive-force behind the other shall be greater than a quarter period, the lag of the current being as usual either equal to or less than a quarter period behind the

FIG. 325.

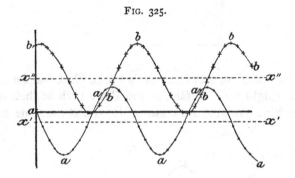

resultant electromotive-force. These statements are readily proved by re-drawing Figs. 320 and 321 under the above conditions. Here is the result in Figs. 324 and 325.

A more curious result still is that A can be driven as a motor by B *even if its electromotive-force is greater than that of* B. The proof is precisely the same as that just given, and is left for the student to work out for himself as an exercise.

All the curves given are for the same pair of machines running at the same speed throughout and with the same total resistance in the outer circuit. The lag of the current behind the resultant electromotive-force is therefore the same in each case (see p. 462).

It is scarcely necessary to say that the analytical proofs are in perfect accord with the graphic ones. A further result of the analytical method which is not so amenable to graphic treatment is that the energy wasted in Foucault currents in the iron when the machine is *short-circuited* is *less* than when it is running on *open circuit,* and therefore the machine will be cooler when used to generate a current than when allowed to run without doing so. This fact has long been well known in connection with alternating-current machines, and is well illustrated by the following experiments made by Dr. Hopkinson on a De Meritens machine:—

Power given to machine as measured on belt ..	3·1	4·8	5·6	6·5	5·4
Electric power developed	0·7	3·4	4·3	5·7	3·4
Power lost	2·4	1·4	1·3	0·8	2·0
Mean current in ampères	7·7	38·6	51·7	73·6	151

It will be noticed that the loss of power is *least* with maximum load.

The above conclusions have been brought by Dr. Hopkinson and Prof. Adams to the test of experiment with the three large De Meritens machines used for the investigation on "Lighthouse Illuminants" at the South Foreland.

Two of the machines were connected in parallel and clutched together until they had attained their usual speed, when they were unclutched and each was driven by its own belt. The electromotive-force on open circuit remained steady, the machines continuing to rotate in unison, and was the same as that of one of the machines when tested by itself. No current passed along the connecting wires. The circuit P R *p* (Fig. 323) was now closed through an arc-lamp ; the machines continued to run as steadily as before, although a large current of 221 ampères was passing through the arc. Lastly, the lamp circuit was broken, the machines were short-circuited on one another, and the belt was thrown off one of them ; it continued to run at the same steady speed, being driven as a motor by the current from the other machine. Other experiments were made, all confirming the theoretical conclusions. These results are summarised in the following table,[1] some photo-

---

[1] *Journal of Soc. Telegr.-Eng. and Electricians,* vol. xiii. p. 524.

metric measurements of the arc light being included, as they
are interesting in themselves :—

	E. M. F. at the Arc.	Current.	Watts in the Arc.	Illumination.	
				By red light.	By blue light.
With one machine	33	175	5,775	8,000	16,000
With two machines .. ..{	35·5	275	9,750}	13,500	23,200
	36	278	10,000}		
With three machines.. ..{	37	300	11,100}		
	42 to 32	300	11,000}	16,000	31,000
	41 to 37	310	11,600}		
With two machines driving the third as a motor ..}	52 to 41	240 to 285	12,000	17,300	—

The governing of the speed of a motor running thus is
perfect, but is accompanied by the serious disadvantages
(*a*) that it can only run at one speed which depends on the
speed of the generator, (*b*) that it has to be brought up to this
speed by some extraneous means before it can be run as a
motor at all, (*c*) that if any of the conditions (such for instance
as the load being too great) are unfavourable it pulls up
altogether.   One way of bringing the motor up to the required
speed is to drive it by a belt until this speed is attained and
then throw the belt off.   Another method is to start the
generator slowly and turn the motor by hand until it falls
into step with the generator ; the speed of the latter may then
be increased, when it will be found that the motor also
increases its speed, keeping pace exactly with the generator.
This of course involves the serious disadvantage of having
to stop or at least slow down the generator every time the
motor has to be started.

[For further information on this subject the reader is referred to
an article by Dr. G. Schmidt, in *Centralblatt für Elektrotechnik*, ix.
440, 1887, dealing with methods of indicating difference of phase in
alternate-current dynamos; and by Elihu Thomson, in *Electrical
World*, vol. ix. p. 258, May 1887 : and vol. xi. p. 39, Jan. 1888.]

CHAPTER XXV.

TRANSFORMERS.

DURING recent years methods of distributing electric energy have come into vogue, in which, for the sake of economising the cost of the metallic conductors, distribution is effected at a high electromotive-force or electric pressure from the central generators, and is received at different points by apparatus known as *transformers*, which transform the electric energy supplied to them, and give it out again at a lower electro-motive-force or electric pressure.

To comprehend fully the bearing of the matter, it must be remembered that the energy supplied per second is the product of two factors, the current and the electromotive-force at which that current is supplied ; or, in our notation,

$$e\,i = \text{electric energy per second (in watts).}$$

The magnitudes of the two factors may vary, but the value of the power supplied depends only on the product of the two : for example, the energy furnished per second by a current of 10 ampères supplied at a pressure of 2000 volts is exactly the same in amount as that furnished per second by a current of 400 ampères supplied at a pressure of 50 volts : in each case the product is 20,000 watts. Now the loss of energy that occurs in transmission through a well-insulated wire depends also on two factors, the current and the resistance of the wire, and in a given wire is proportional to the square of the current. In the above example the current of 400 ampères, if transmitted through the same wire as the 10-ampère current, would, because it is forty times as great, waste sixteen hundred times as much energy in heating the wire. Or, to put it the other way round, for the same loss of energy one may use, to carry the 10-ampère current at 2000 volts, a wire having only $\frac{1}{1600}$th of the sectional area of the wire used for the 400-ampère current at 50 volts. The cost of copper conductors for the distributing lines is therefore very greatly economised by employing high pressures for distribution of small currents.

For transforming from high pressures to low, several kinds of apparatus are known, namely :—

1. *Storage Batteries.*—A large number of these to be charged in series at a high potential ; the series afterwards divided up or rearranged so as to discharge larger currents at lower pressure. This system is applicable to direct-current working only, not to alternate currents, and has the advantage of storing the consumer's supply.

2. *Induction-coils*, also called for this purpose *Secondary Generators*, or *Transformers*, or *Converters.*—This system will only answer with alternating currents, which being transmitted through the distributing mains at high pressure, and traversing the primary wires of the induction-coils, set up in the secondary wires currents which feed the separate circuits of lamps at the desired low pressure.

3. *Motor-generators.*—These are either two separate machines : a motor, adapted to receive small currents at high potential, and be driven by them, and a dynamo, driven by the motor, and capable of generating large currents at low potential ; or they consist of single machines having a double-wound armature, one set of windings of fine wire to receive the incoming currents at high potential, and another set, of thick wire, to furnish the outgoing currents at low potential.

4. *Commuting Transformers.*—These are a variety of the last, but neither armature nor field-magnet revolves, the polarity of the magnetic circuit being caused to vary by special commutators.

Storage batteries we cannot here discuss : motor-generators will be very briefly touched upon. Induction-coil transformers for alternate currents being a species of dynamo-electric machine, will be dealt with only as such : space does not admit of discussion of the various systems of distribution in which they are employed.

HISTORICAL NOTES ON TRANSFORMERS.

The first induction-coil was used by Faraday,[1] and consisted of a solid iron ring six inches in diameter and seven-eighth of an inch thick, having a primary wire seventy-two feet in length, and a secondary sixty feet in length, coiled around opposite parts (Fig. 326, No. 1). Small modifications were made by Masson,[2] who introduced a bundle of iron wires as core, Pohl,[3] Wright, Henley,[4] Dove,[5] who examined the properties of various kinds of iron cores and the damping effects of solid metal conductors and tubes, Sinsteden,[6] Magnus,[7] who also investigated the effects of the form of the iron core, Stöhrer,[8] who constructed the vertical pattern of coil, Ritchie,[9] who suggested the use of cloisons or partitions in winding the secondary, where high insulation was wanted, Ruhmkorff,[10] who gave the spark-coil its classical proportions, and Varley,[11] who in 1856 described a form of induction-coil having a laminated and closed magnetic circuit, closely resembling some of the modern forms of transformers. Various suggestions for use of induction-coils in distributing current for electric lighting were made by Jablochkoff,[12] who proposed to place in the circuit of an alternate-current machine, at different points of the circuit, the primary coils of a number of transformers, the secondary coils of which were used to supply lamps of the "electric-candle" type, and by Sir C. Bright,[13] who proposed similarly to distribute currents to vacuum-tube lamps. Jablochkoff, and later Edwards and Normandy,[14] suggested a peculiar construction of copper-strip conductors, set edge-on, in place of round wires in the coils. More practical forms of transformers were suggested by Fuller[15] and by De Meritens.[16] In 1881 Hopkinson[17] proposed to utilise the

[1] *Experimental Researches*, i. 7, 1831.
[2] *Ann. Chim. Phys.*, lxvi. 5, 1837 ; and iv. 129, 1842.
[3] *Pogg. Ann.*, xxxiv. 185, 500, 1835.
[4] *Annals of Electricity*, v. 349, 1840 ; vii. 322, 1841.
[5] *Pogg. Ann.*, xlix. 72, 1840 ; lvi. 251, 1842.
[6] *Ibid.*, lxxxv. 465, 1851 ; and xcvi. 366, 1855.
[7] *Ibid.*, xlviii. 95, 1839.
[8] See Wiedemann's *Lehre von der Elektricität*, iv. 339.
[9] *Phil. Mag.*, xiv. 239, 480, 1857.
[10] *Comptes Rendus*, xxxvi. 649 ; and xxxvii. 801.
[11] Specification of Patent 3059 of 1856.
[12] *Ibid.*, 1996 of 1877.
[13] *Ibid.*, 4212 of 1878.
[14] *Ibid.*, 4611 of 1878.
[15] *Ibid.*, 5183 of 1878.
[16] *Ibid.*, 5257 of 1878.
[17] *Ibid.*, 3362 of 1881.

self-induction of a coil, with iron core, to "choke" the currents of an alternate-current system instead of introducing wasteful resistance. In the same year a very important patent was taken out by MM. Marcel Deprez and J. Carpentier[1] for a system for the economical transport of electric energy by means of transformers. At the generating station the alternating currents of low potential were to be transformed by means of an induction-coil to currents of high potential, which could then be economically conveyed to a distance through long thin conducting-wires, and there, entering the fine-wire primary coil of another transformer could be transformed down to low potential suitable for the lamps or motors. In 1882, Gravier[2] suggested the motor-dynamo as a transformer for continuous currents, a suggestion which had been partially anticipated by Sir W. Thomson[3] and by Gramme.[4] This suggestion was generalised in 1883 by Cabanellas,[5] and systems of distribution on the same plan were independently proposed by Edison,[6] and by Lane-Fox.[7] In 1882, Gaulard and Gibbs[8] revived the use of alternate-current distribution by induction-coils; their first patent proposes a coil of Ruhmkorff type. Their second patent[9] states that in their invention the alternate current in the primary circuit was to be maintained of constant strength, while its electromotive-force was varied, according to the demand on the secondary circuits. This implies distribution in series; in the same year they revived the construction of induction-coils with sheet-copper conductors set edge-on to the core under the name of "secondary generators." In 1883, Kennedy used a modified Gramme ring as a transformer; and in 1885, Deri and Zipernowsky[10] proposed the use of transformers for affecting the self-regulation of the alternate-current dynamo, and revived the use of induction-coils having laminated and closed magnetic circuits. In the same year Ferranti[11] brought out transformers constructed of iron strips. Ken-

[1] Specification of Patent 4128 of 1881.
[2] *Ibid.*, 1211 of 1882.
[3] *Rep. Brit. Assoc.*, 1881.
[4] *Comptes Rendus*, Nov. 23, 1874.
[5] *La Lumière Électrique*, iii. 44, and iv. 206, 1883; and Specification of Patent 2880, 1885.
[6] Specifications of Patents 3752 and 3949 of 1882; and U.S. Patents, Nos. 266,793, Oct. 31, 1882, 278,418, May 29, 1883, and 287,516, Oct. 30, 1883.
[7] Specification of Patent 3692 of 1883.
[8] *Ibid.*, 4362 of 1882; and *La Lumière Électrique*, xiv. 40, 156, 1884.
[9] *Ibid.*, 2858 and 3173 of 1884.
[10] *Ibid.*, 3379 and 5201 of 1885.
[11] *Ibid.*, 15,251 of 1885.

nedy [1] had proposed to make the distribution at a constant potential, a method which has been adopted by Zipernowsky, Ferranti, and subsequently by Gaulard and Gibbs, and by Westinghouse in the United States, and which is nearly self-regulating. Messrs. Sharp and Kent, Snell, and Kapp have proposed the compounding of transformers so as to make them automatically self-regulating. Improvements in details of construction have been made by Kapp and Snell, Mordey, Westinghouse, and Statter. Recent improvements in motor-generators are due to Scott and Paris, [3] R. P. and J. S. Sellon, [4] and J. Swinburne, [5] the latter of whom suggests machines adapted for transforming from a circuit supplied at constant current to a secondary circuit fed at constant potential.

SYSTEMS OF DISTRIBUTION BY ALTERNATING TRANSFORMERS.

These are briefly two in number :—(1) *At constant potential*, the distributing mains branching to the various local points where the transformers supply the lights at low potential. In this case the primary wires of the local transformers are coils of fine wire of many turns, and are all in parallel across the mains ; the secondary coils being of thick wire and of few turns. (2) *With constant current*, the single main circuit going from the primary of one transformer to that of the next. In this case, as the primary wires are in series, they are thicker and of few turns. The first of these systems is suitable for incandescent lighting, the second for arc lighting.

GENERAL PRINCIPLES OF ALTERNATING TRANSFORMERS.

An alternating transformer may be regarded as a species of dynamo, in which neither armature nor field-magnet revolve, but in which the magnetism of the iron circuit is made to vary through rapidly repeated cycles of alternation, by separately

---

[1] *Electrical Review*, xii. 486, June 9, 1883.
[2] *Journal Soc. Teleg.-Engineers*, xvii. 96, 1888.
[3] Specification of Patent 6260 of 1884.
[4] *Ibid.*, 3525 of 1885.
[5] *Ibid.*, 6682 of 1887.

exciting it with an alternating current. The primary coil of the transformer corresponds to the field-magnet coil of the dynamo; the secondary of the transformer to the armature coil of the dynamo.

Many of the rules for construction of dynamos apply with equal force to the construction of transformers: for example, the rules concerning insulation, lamination of iron cores, and the like. In all cases where transformers are used with very high potentials, the utmost care must be taken about the complete and efficient insulation of all the wire (and con-nexions) of the high-potential coils: it should on no account be wound in alternate layers between the windings of the low-potential circuit, nor stranded with it into a cable; it should, while properly enclosing the iron circuit, not only be as per-fectly insulated, but as completely isolated as may be from the low-potential wire and from all other metal work in the transformer.

In the alternating transformer, by whatever name called, the function of the iron core is to carry the magnetic lines of force (that are created by the current in the primary coil) through the convolutions of the secondary coil. The rate at which the magnetic lines due to the primary current are cut by the secondary circuit is the measure of the electromotive-force given to the secondary circuit. In order to be able to calculate the amount of cutting of magnetic lines that goes on with various strengths of primary current and various rates of alternation, it is convenient to know the amount of cutting of magnetic lines that takes place when unit current is made to flow, or is stopped in the primary coils. Let M be used as a symbol for this quantity. It will be proportional to the number of turns in the secondary coil, because each turn encircles the iron core and cuts the magnetic lines; it will also be proportional to the number of turns in the primary coil, because, *cæteris paribus*, the magnetism evoked in the iron core is proportional to the ampère-turns that excite it; it will also be proportional at every stage to the permeability of the iron core. We may, in fact, calculate M by the magnetic principles laid down in Chapter XIV. Suppose the iron core to form a closed circuit of length l, section A, permeability μ; and that S_1 and S_2 are the respective numbers of turns in primary and secondary. Then, if the primary current is unity, the magnetomotive-force due to it will be $4\pi S_1$, and the magnetic resistance will be $l/A\mu$. Dividing the former by the latter, we shall have an expression for the number of lines in the core; this multiplied by S_2 gives the amount of cutting of lines by the secondary circuit; or in symbols

$$M = 4\pi S_1 S_2 A \mu/l.$$

The name given to this quantity is the *coefficient of mutual induction*. If the current in the primary have the value i_1 (absolute C.G.S. units), then the amount of cutting by the secondary on turning this current on or off will be $M i_1$. And if the rate of increase or decrease of the primary current at any instant is known, this, multiplied by M, will give the electromotive-force impressed at that instant on the secondary circuit. Now it is obviously advantageous in a transformer that this quantity should be as great as possible, for it is desirable to attain the requisite electromotive-force in the secondary with as little primary current as possible. The conditions are, however, a little conflicting: S_2 must not be made large, on account of internal resistance and cost of copper; but it is clear that a compact magnetic circuit, having l small, A large, and the iron of good quality, will be of advantage. In the particular case (for series working), where the primary and secondary wires are to carry equal currents, it is clear that for a given total weight of copper, the maximum product $S_1 S_2$ will be afforded when there are equal numbers of turns in the coils, and therefore equal weights of copper. In other cases, provided the wires are proportioned to the currents they have to carry, and that equal heat can be developed in each coil, equal weights of copper is still a good rule, though the numbers of turns in the two coils will be widely different. It has been assumed that there is no magnetic leakage; that all the lines of force created in the core pass through the secondary coils. This assumption is very nearly fulfilled in those transformers in which the magnetic circuit is effectively closed upon itself.

The next important point is that the magnetic lines of force created by the primary current in the core pass through the convolutions of the primary coil itself, and that therefore there will be self-induced electromotive-forces in the primary coil, which will tend to oppose the impressed variations of current. Considerations precisely analogous to those above will show that there will be a *coefficient of self-induction*, which we will call L_1, which represents the amount of cutting, by the primary coils, of the magnetic lines created in the coil when the primary coil carries unit current; and we may calculate, on magnetic principles, as before, the value of this coefficient to be

$$L_1 = 4\pi S_1^2 A \mu/l.$$

That is to say, L_1 will be large in proportion to the goodness of the magnetic circuit; and (as in the case in transformers for distribution at constant potential) as S_1 is itself large, L_1 will be enormous. As will presently be seen, this has important results in the automatic action of transformers.

The electromotive-forces induced (by the variations of magnetisation of the core) in the secondary circuit will, of course, produce no currents if the secondary circuit is open; but if the secondary circuit is closed there will be currents, the strength of which will depend also upon the resistances and upon the counter electromotive-forces (if any) in this circuit. But these secondary currents, as they circulate around the convolutions of the secondary coil, exercise a magnetising (or rather a demagnetising) action, the effect of which on the core will depend, not simply upon their strength, but also upon their phase. When the transformer is doing full work, the secondary current is rising in the positive sense, almost exactly as the primary is rising in the negative direction, their phases being almost exactly in opposition. That is to say, at full work the secondary current will exercise a great demagnetising action. The total magnetising force[1] at work is equal to

[1] See p. 317.

$4 \pi (S_1 i_1 + S_2 i_2) \div 10$, where the proper signs $+$ and $-$ must be assigned to i_1 and i_2 according to the sense in which they circulate around the core. Further, there will be a coefficient of self-induction L_2 in the secondary circuit, such that

$$L_2 = 4 \pi S_2^2 A \mu / l.$$

As will be seen later, there is an inter-action between the primary and secondary circuits, owing to mutual induction, tending to neutralise the effects of self-induction in both circuits.

In a well-built transformer it is clear that

$$M = \sqrt{L_1 L_2}.$$

If, however, all the magnetic lines due to one circuit are not enclosed by the other, M will have a less value than is indicated by the above relation.

The relation between the two electromotive-forces and the two sets of windings is readily stated in a way sufficiently accurate for ordinary purposes, by saying that the ratio of the two electromotive-forces is equal to that of the windings, or

$$\frac{E_1}{E_2} = \frac{S_1}{S_2} = p.$$

The number p which represents the ratio of the windings we shall call the *coefficient of transformation.*

If it is assumed that there are equal weights of copper used in the primary and secondary coils, then the following relations will hold good:—

	Primary.	Secondary.	Ratio.
Windings	S_1	S_2	p
Resistance	r_1	r_2	p^2
Self-induction	L_1	L_2	p^2
Electromotive-force	E_1	E_2	p
Current	i_1	i_2	p^{-1}
Heat-waste	$i_1^2 r_1$	$i_2^2 r_2$	1

Also $$M = \frac{L_1}{p} = p L_2.$$

It must be always remembered that the symbol E is here

being used for the square-root of mean square of the alternating electromotive-force: the actual value of E at any moment being represented by the expression $E = D \sin 2\pi n t$; where D is the maximum value to which E rises, and is equal to $2\pi n S_2 N$ (see p. 461); *n* being the number of alternations per second. Further, for alternate currents (see p. 461) Ohm's law must be written as

$$i = \frac{D \sin (2\pi n t - \phi)}{\sqrt{R^2 + 4\pi^2 n^2 L^2}}.$$

We are now able to find a more exact expression for the induced electromotive-force in the secondary circuit; for the number of magnetic lines induced by the primary circuit into the core (when there is no current in the secondary) is M i; and therefore

$$E_2 = -M\frac{di}{dt} = M'\frac{2\pi n D \sin \left(2\pi n t - \phi - \dfrac{\pi}{2}\right)}{\sqrt{r_1^2 + 4\pi^2 n^2 L_1^2}};$$

or, writing $2\pi n = \omega$, we have (neglecting retardation of phase)

$$E_2 = \frac{\omega M}{\sqrt{r_1^2 + \omega^2 L_1^2}} E_1;$$

or, putting

$$\frac{\omega M}{\sqrt{r_1^2 + \omega^2 L_1^2}} = k,$$

$$E_2 = k E_1.$$

Examining the expression k, it will be seen that if r_1 is small compared with ωL, as will be the case where there are very rapid alternations and plenty of iron in the core, k will be equal to M/L_1, or will be the reciprocal of p, and equal to S_2/S_1.

By leaving out of consideration differences of phase in the above argument, it must not be supposed that these differences of phase are small. On the contrary, they may be very large. If, however, assuming E_1 to be a sine-function, we remember that the average value of the square of the sine (between 0° and 360°) is equal to ½, then we shall have the average value of

E_1 (the square-root of mean square) equal to D divided by $\sqrt{2}$. It is in this sense (namely, taking the square-root of mean square as the basis for comparison) that we may say that E_2 is to E_1 in the same ratio as S_2 to S_1, or that the ratio of the primary and secondary volts is equal to the ratio of the primary and secondary windings. In any case, the work given out (per second) by the secondary will be very nearly equal to the work imparted to the primary ; the difference being the loss due to the production of eddy-currents, and the waste work spent in carrying the residual magnetisation of the iron through repeated cycles of alternation. If $E_1 i_1$ be the watts imparted to the primary, and $E_2 i_2$ those given out by the secondary, it is found that $E_2 i_2$ may be as much as 94 per cent. of $E_1 i_1$. A current of 10 ampères at 2000 volts imparted to a good transformer, having twenty times as many turns on its primary as it has on its secondary coil, will be transformed by the secondary into a current of 200 ampères [1] at very nearly 100 volts.

One of the most difficult points for beginners to comprehend in the action of transformers is the way in which, when used in parallel distribution, they of themselves adjust the amount of primary current that flows through them from the mains. Suppose a transformer, having a primary coil of many turns of fine wire to receive a current of 1 ampère from mains at, say, 2000 volts, and having a secondary coil adapted to give out a current of 20 ampères at 100 volts. This would supply 40 lamps, each taking a $\frac{1}{2}$ ampère of current. When all the lamps are on, the 20-ampère flows through the secondary, and the 1-ampère flows through the primary. If now half the lamps are turned out, only 10 ampères will be supplied by the secondary, and now only half an ampère is wanted from the mains to flow through the primary. As a matter of

---

[1] It must be remembered by beginners that the gain in quantity of current is balanced by the diminution in electromotive-force ; otherwise the work would not be equal. It is a curious fact in the history of invention that in 1883 the United States Patent Office refused to grant a transformer patent to Mr. Bernstein on the ground that he could not possibly get a larger quantity of current out of the secondary than was supplied to the primary. Nevertheless, the same Patent Office in 1886 issued a patent to Gaulard and Gibbs for this very thing !

fact, under these circumstances, the current supplied through the primary automatically diminishes in the proper proportion. As lamp after lamp is turned out in the secondary circuit, so, little by little, the primary current also diminishes; when all lamps are out, and the secondary circuit is completely opened, the primary current is found to have of itself all but stopped, although the mains are still supplied at full pressure. Nothing has been done to the primary circuit: there is still a perfect path through which the current might flow, no resistance has been inserted in the primary coil, yet the current does not flow to any extent through it. The clue to the matter lies in the inductive reaction of the iron core. As the waves of primary current circulate around the core, they set up periodic magnetisations in the iron; but the periodic fluctuations of the magnetism of the iron necessarily set up electromotive-forces in the surrounding coils. If, in a short time dt, the change that takes place in the number of magnetic lines in the core is dN, then during that short time the electromotive-force induced in a wire which is coiled S times round the core will be equal to $S \dfrac{dN}{dt}$ (this number should be divided by 10^8 to bring to volts). The electromotive-force in the secondary coil will be $S_2 \dfrac{dN}{dt}$; and there will of course be also another electromotive-force induced in the primary equal to $S_1 \dfrac{dN}{dt}$. Now it is this self-induced counter electromotive-force in the primary which dams back the currents from the mains when the secondary circuit is open. Consider what happens when there are no lamps on: the secondary circuit is entirely idle, and takes no part in the action. The alternating electromotive-force of the mains impresses alternating currents in the primary coil, which in turn impress an alternating magnetisation on the core, and this finally induces alternating electro-motive-forces into the primary circuit, tending to stop the primary currents. The greater the degree of magnetisation of the core, the greater the induced electromotive-force. Now consider what happens when the lamps are on. There will be considerable currents in the secondary : these are always in almost exactly opposite phase to the primary currents. When the primary current is increasing, the secondary current is increasing negatively, that is to say, is increasing whilst circulating around the core in the opposite sense to the primary. The magnetising tendency is of course proportional to the difference between the ampère-turns of the primary and those of the secondary. [It is, as previously stated, equal to $4 \pi (S_1 i_1 + S_2 i_2) \div 10$, the proper signs being assigned to the currents according to their sense.] As a result, the larger the number of lamps switched on, the less do the total magnetising forces acting on the core become, and the back electromotive-forces in the primary coil which in the first case dammed back the current from the mains, are less ; hence more current flows in the primary.

Graphic constructions illustrating the variations of phase and magnitude of the electromotive-forces and currents have been given by Kapp and by Blakesley. The magnetisation is of course carried to the greatest degree when the transformer is working on open circuit. It is expedient to arrange that there shall be so much iron in the core that the induction in it need never be carried above 6000 or 7000 lines to the square centimetre. It is possible to work above these values, but the transformer emits a singing noise and the iron grows hot. Work is, of course, spent in carrying the iron to a high degree of magnetisation. Of this work, the greater part is returned to the secondary circuit during the next part of the cycle as the magnetisation of the iron descends ; but part, namely, that spent in producing the so-called permanent portion of the magnetisation, is lost, being frittered down into heat. As to the phases of the various waves, as mentioned above, the secondary current is always in almost exactly opposite phase to the primary current; and the secondary electromotive-force is in almost exactly opposite phase to the primary electromotive-force. But the difference of phase between E_1 and i_1 varies greatly with the work done. When there are no lamps on, and no secondary current, i_1 is small, and its phase differs by very nearly 90° from that of E_1 : as lamps are switched on and i_2 grows, i_1 grows also, and the angle of lag between E_1 and i_1 grows less. The phase of the magnetisation of the core (which is always necessarily 90° in advance of the phase of E_2), is always [1] very nearly 90° behind the phase of E_1.

CONSTRUCTION OF ALTERNATING TRANS-FORMERS.

All transformers consist of a core of iron, around which the primary and secondary coils are so wound that the core carries the magnetic lines from the one set of coils through the other. Faraday's first induction-coil, No. 1 in Fig. 326, is thoroughly typical of the principle common to all transformers. In Faraday's actual ring the iron was not laminated. No. 2 of Fig. 326 is the cylindrical type of induction-coil introduced by Masson and Ritchie, and perfected for spark purposes by Ruhmkorff. No. 3 depicts the form given to the transformer by Varley for telegraphic purposes. The core of iron wires is made long, and the ends are bent over so as to constitute practically a closed magnetic circuit. The Ferranti transformer resembles this, but has the core of narrow strips of sheet iron. A similar form has been used recently by Gaulard and Gibbs. No. 4 is a form introduced by Zipernowsky; in this the primary and secondary coils are laid upon one another, and the iron core is then wound through and over them by a shuttle, so that the whole of the copper is enclosed within the iron. In the drawing of Nos. 3 and 4 the front portion of the iron winding is represented as removed to show the interior. Mr. Kapp has proposed the name of "shell-transformers" for this type of apparatus as distinguished from those with a mere straight or a non-expanded core. No. 5, which is also a form due to Zipernowsky, is constructed somewhat like a Gramme ring. There is an iron wire ring core, upon which the copper coils are wound in sections; but with alternate sections connected together, one set as primary, the other as secondary. In the drawing, one of the ten sections

---

[1] See experiments by Captain Cardew, in *Electrical Engineer*, July 1887, p. 358.

is represented as removed to show the core. Nos. 6, 7, and 8 represent three closely allied methods of procuring the necessary structure of laminated iron, the parts in these instances being stamped from sheet iron. The copper coils are not shown in these three, but in each they consist of two sets of

FIG. 326.

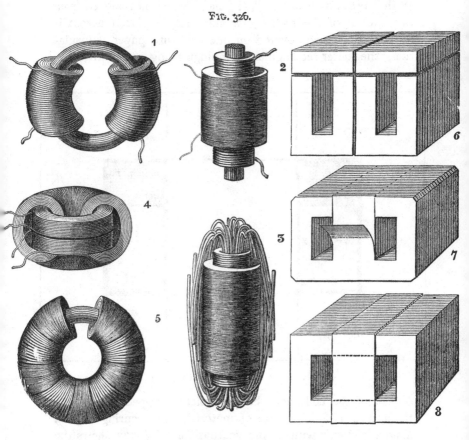

VARIOUS TYPES OF ALTERNATING TRANSFORMERS.

windings, previously wound on rectangular formers of such a form as to admit of insertion into the iron stampings. In No. 6, which shows the method of Kapp and Snell, there are two sets of U-shaped stampings, each 3½ inches broad, and 5 inches high, set side by side, forming two parallel channels

2 K

to receive the coils. The stampings that are removed between
the limits of the **U** pieces serve to complete the magnetic
circuits above. No. 7 is the form adopted by Westinghouse.
The stamping in this form must be bent up in order to slip it
on over the coils; the core being put together over the coils.
With Mordey's form, No. 8, the piece which is removed from
the interior of the rectangular stamping is placed across it
(after having been covered on one side with paper as insulation), the outer rectangular stampings being passed over the

FIG. 327.

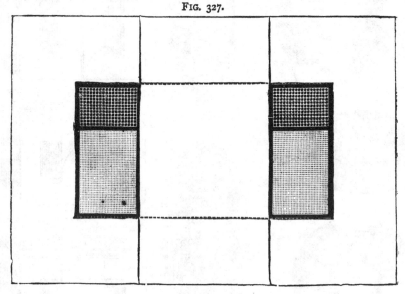

MORDEY'S TRANSFORMER (Section).

outside of the coils, and the cross-pieces slipped in between
them, thus avoiding waste of material, and securing a good
magnetic circuit with ample ventilation. Fig. 327 shows the
exact relative proportions of the stampings of Mordey's transformer, including a section of the coils. This apparatus, which
is constructed by the Brush (Anglo-American) Company, is
further depicted in Fig. 328.

As an example, the following dimensions are given for a transformer
capable of transforming a current of 1·5 ampère at 1000 volts down to a
current of 37·5 ampères at 40 volts. Total external size, 20 × 6 × 4 inches ;

$S_1 = 300$; $S_2 = 12$; $r_1 = 10$ ohms; $r_2 = 0.014$ ohms; gauge of primary wire, 0.035 inch; secondary wire 25 in parallel, each 0.12 inch gauge. Weight of copper in coils, about 5 lbs. in primary and 5½ lbs. in secondary. Weight of iron used, about 50 lbs. Efficiency, 97.2 per cent.

FIG. 328.

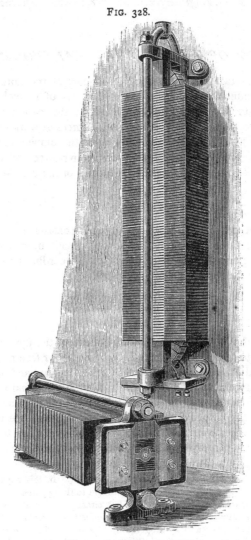

MORDEY'S TRANSFORMER.

For further information on the constructional details of transformers the reader is referred to an excellent series of articles by R. Kennedy, in the *Electrical Review,* xx. 247 and seqq., 1887, since republished in separate form under the title *Electrical Distribution by Alternating Currents and Transformers;* to an article by Rechniewski, in *La Lumière Électrique,* xxvi. 95, 1887; and to the paper by G. Kapp, in *Journal Soc. Teleg. Engineers,* xvii. 96, 1888, to which reference is made above.

THEORY OF ALTERNATE TRANSFORMERS.

There are two ways of treating the theory of transformers—the first by introducing the notion of coefficients of mutual and self-induction into the differential equations for the two circuits; the second by considering the magnetomotive-forces at work in the iron core as the result of the algebraic sum of the ampère-turns in the two circuits, and deducing the electromotive-forces which result from the variations in the magnetic induction of the core—

First Method.

This method, introduced by Maxwell,[1] consists in finding the electromotive-force induced in the second circuit by the variations of current impressed upon the first circuit. Accordingly we write as the differential equation of the first circuit—

$$\mathrm{E}_1 - \mathrm{M}\frac{d\,i_2}{d\,t} - \mathrm{L}_1\frac{d\,i_1}{d\,t} - \mathrm{R}_1\,i_1 = \mathrm{o}\,; \qquad (1)$$

where E_1 is the impressed electromotive-force of the dynamo, which is supposed to fulfil the condition $\mathrm{E}_1 = \mathrm{D}\sin 2\,\pi\,n\,t$ (see p. 492), and where L_1 and R_1 are the coefficient of self-induction and resistance of the primary circuit. If the supposition is admitted that a constant (alternating) potential can be maintained at the terminals of the primary coil (by proper compounding of the dynamo, see p. 277, or otherwise), then the letters E_1, L_1, and R_1, may be taken to apply to that part of the primary circuit only which lies between the terminals

---

[1] *Philosophical Transactions,* clv. pt. i. p. 459, 1865. In this paper Maxwell shows that the effect of the second circuit is to add to the apparent resistance and diminish the apparent self-induction of the first circuit. The student will find the equations more fully treated by Mascart and Joubert, *Électricité et Magnétisme,* i. p. 593, and ii. p. 834; also by Hopkinson, *Journ. Soc. Teleg. Engineers,* xiii. p. 511, 1884; Ferraris, *Mem. Accad. Sci.* (Turin), xxxvii. 1885; and by Vaschy, *Annales Télégraphiques,* 1885-6, or *Théorie des Machines Magnéto et Dynamo-électriques,* p. 31.

of the primary coil. From this differential equation we have to deduce a value for $M \frac{d i_1}{d t}$. For brevity we will write ω for $2 \pi n$; and $- \omega^2 i$ for $\frac{d^2 i}{d t^2}$, because i is a sine-function. Then differentiating equation (1) we get—

$$\frac{d E_1}{d t} + M \omega^2 i_2 + L_1 \omega^2 i_1 - R_1 \frac{d i_1}{d t} = 0. \qquad (2)$$

Now multiply this by R_1 to get equation (3), and multiply equation (1) by $L_1 \omega^2$, to get equation (4); and add (3) and (4) to get (5).

$$R_1 \frac{d E_1}{d t} + M \omega^2 R_1 i_2 + L_1 \omega^2 R_1 i_1 - R_1^2 \frac{d i_1}{d t} = 0, \qquad (3)$$

$$L_1 \omega^2 E_1 - L_1 \omega^2 M \frac{d i_2}{d t} - L_1^2 \omega^2 \frac{d i_1}{d t} - L_1 \omega^2 R_1 i_1 = 0, \qquad (4)$$

$$\left(R_1^2 + L_1^2 \omega^2 \right) \frac{d i_1}{d t} = R_1 \frac{d E_1}{d t} + L_1 \omega^2 E_1 + M \omega^2 \left(R_1 i_2 - L_1 \frac{d i_2}{d t} \right). \qquad (5)$$

Now multiply every term by $\frac{M}{R_1^2 + L_1^2 \omega^2}$, and write the following abbreviations:—

$$\frac{M \omega}{\sqrt{R_1^2 + L_1^2 \omega^2}} = k,$$

$$k^2 R_1 = \rho,$$

$$k^2 L_1 = \lambda,$$

$$- \frac{k^2}{M} \left(\frac{R_1}{\omega^2} . \frac{d E_1}{d t} + L_1 E_1 \right) = E_2 = k E_1 \sin (\omega t - \phi), \text{ where } \phi \text{ relates to}$$

the phase of the electromotive-force; and we may write equation (5) as—

$$M \frac{d i_1}{d t} = \rho i_2 - \lambda \frac{d i_2}{d t} - E_2. \qquad (6)$$

Now the differential equation for the second circuit is—

$$M \frac{d i_1}{d t} + L_2 \frac{d i_2}{d t} + R_2 i_2 = 0; \qquad (7)$$

there being in this circuit no other electromotive-forces than those due to mutual and self-induction. Inserting in (7) the value obtained in (6), we get as the final equation—

$$(R + \rho) i_2 + (L_2 - \lambda) \frac{d i_2}{d t} - E_2 = 0. \qquad (8)$$

This shows us that the whole effect is equivalent to that which would happen if, the primary circuit being absent, there were introduced into the secondary circuit an electromotive-force equal to $k\,E_1$, and at the same time the resistance were increased by a quantity equal to $k^2\,R_1$, and the self-induction were diminished by a quantity equal to $k^2\,L_1$. Further, examination of k shows us that if the frequency of alternation and the permeability of the iron core be great enough, k becomes equal to $M \div L_1$, or equal to the reciprocal of p, the ratio of the primary and secondary windings. But if there are equal weights of copper in the two coils, $L_1 = p^2\,L_2$ and $R_1 = p^2\,R_2$. So that the effect, when the transformer is fully at work, is to make $\lambda = L_2$, and $\rho =$ the resistance of the secondary coil; or self-induction is wiped out, and the internal resistance of the secondary is virtually doubled.

Second Method.

In this method, due to Hopkinson,[1] we begin by considering the magnetomotive-force requisite to force N magnetic lines through a core having length l and permeability μ. This magnetomotive-force is due to the algebraic sum of the ampère-turns in the two coils. Following the argument on p. 319, Chapter XIV., we may therefore write :—

$$\frac{4\,\pi}{10}\left(S_1\,i_1 + S_2\,i_2\right) \div \frac{l}{A\,\mu} = N; \tag{1}$$

or if the core has area of cross-section A, and there are B lines to the square centimetre, we may write $N = A\,B$; and then, the magneto-motive-force will be—

$$\frac{4\,\pi}{10}\left(S_1\,i_1 + S_2\,i_2\right) = A\,B\frac{l}{A\,\mu} = \frac{B}{\mu}\,l = H\,l. \tag{2}$$

We then write two equations for the electromotive-forces impressed in the two circuits as follows :—

$$\begin{cases} E_1 = R_1\,i_1 - S_1\dfrac{d\,N}{d\,t} = R_1\,i_1 - S_1\,A\dfrac{d\,B}{d\,t}, & (3)\\[2ex] o = \left(r_2 + R_2\right)i_2 - S_2\dfrac{d\,N}{d\,t} = \left(r_2 + R_2\right)i_2 - S_2\,A\dfrac{d\,B}{d\,t}; & (4) \end{cases}$$

where R_2 is the resistance of the lamp circuit, and r_2 the internal resistance of the secondary coil. Multiplying (3) by S_2 and (4) by S_1, we deduce—

$$S_2\,E_1 = S_2\,R_1\,i_1 - S_1\,(r_2 + R_2)\,i_2;$$

[1] *Proc. Roy. Soc.*, February 1887.

which with equation (2) gives—

$$i_1\{S_2^2 R_1 + S_1^2 (r_2 + R_2)\} = S_2^2 E_1 + S_1 (r_2 + R_2) (10\,H\,l \div 4\,\pi),$$
$$i_2\{S_2^2 R_1 + S_1^2 (r_2 + R_2)\} = - S_1 S_2 E_1 + S_2 R_1 (10\,H\,l \div 4\,\pi);$$

and

$$A \cdot \frac{dB}{dt} = - \frac{(r_2 + R_2) S_1 E_1}{S_2^2 R_1 + S_1^2 (r_2 + R_2)} + \frac{10\,H\,l}{4\,\pi} \cdot \frac{R_1 (r_2 + R_2)}{S_2^2 R_1 + S_1^2 (r_2 + R_2)}. \quad (5)$$

The second term may be neglected simply because $H\,l$ is small compared with $\frac{4\pi}{10} S_1 i_1$; in other words, because it is only the difference between $S_1 i_1$ and $S_2 i_2$ (both of which are very large), that is needed as a magnetomotive-force, and this is small if the permeability of the iron is great as it is when, as in actual practice, B is not carried above 8000 or 10,000 lines to the square centimetre. We have then—

$$A \frac{dB}{dt} = \frac{dN}{dt} = - \frac{(r_2 + R_2) S_1 E_1}{S_2^2 R_1 + S_1^2 (r_2 + R_2)}.$$

And remembering that $E_2 = S_2 \frac{dN}{dt}$, we get—

$$E_2 = - \frac{(r_2 + R_2) E_1 \frac{S_2}{S_1}}{\frac{S_2^2}{S_1^2} R_1 + r_2 + R_2}.$$

But E_2, the effective electromotive-force in the secondary circuit, is equal to $(r_2 + R_2) i_2$; hence it follows that—

$$i_2 = - \frac{E_1 \frac{S_2}{S_1}}{\frac{S_2^2}{S_1^2} R_1 + r_2 + R_2}. \quad (6)$$

Now if the condition of supply in the mains is such as to give constant-potential at the terminals of the primary circuit, we may take E_1 and R_1 as referring to that which is internal to the primary coil; and it is evident that the effect of the transformer is to reduce the volts in proportion to the ratio of the windings, and to add to the resistance of the secondary circuit a term equal to the resistance of the primary circuit reduced in proportion to the square of the ratio of the windings. It also follows, from the negative sign in (6), that the two currents are in exactly opposite phases.

For further discussions on alternating transformers see—

Ferraris: *La Lumière Électrique*, xvi. 399, 1885 ; xxvii. 518, 1888 ; and *Electrical Review*, xvi. 256, 343 seqq. 1885; xxii. 221, 252, 1888.
G. Kapp: *Industries*, April 1887 ; and *Electrician*, 1887. Also *Journal Soc. Teleg. Engineers*, xvii. 96, 1888, with discussion thereon.
G. Forbes: *Journ. Soc. Teleg. Engineers*, xvii. 153, 1888.
Rechniewski: *La Lumière Électrique*, xxv. 613, 1887.
Peukert and Zickler: *Zeitschrift für Elektrotechnik*, 1886 ; and *La Lumière Électrique*, xxi. 276, 1886. Also *Electrical Review*, xix. 80, 1886.

CONTINUOUS-CURRENT TRANSFORMERS.

Gramme, in 1874, constructed a machine with a ring-armature wound with two circuits—one of coarse wire, the other with fine wire, having eight times as many turns. Two separate commutators were connected with the two windings. This machine could be used for transforming either from high to low potential or *vice versâ*. The same end can be less conveniently attained by uniting on one shaft the armatures of two dynamos, one to be used as a motor driving the other as a generator ; and these may have separate field-magnets, or a common field-magnet. When it is desired to transform from an alternate current to a continuous one, the armatures and field-magnets must be kept separate. In England, continuous-current transformers have been introduced with success by Messrs. Paris and Scott,[1] who employ a two-pole machine with cast-iron frame and an armature wound with double circuits. There is very little sparking with such machines, as the reactions in the two sets of coils tend to correct each other. The field-magnet is excited as a shunt to the low-potential armature coil. Swinburne has discussed many possible combinations, including one for transforming from a constant-current to a constant-potential condition of distribution.

A somewhat different system of continuous-current transformation has been suggested by Cabanellas,[2] and patented by Edison,[3] in which neither armature nor field-magnet

[1] See *Electrician*, xix. 517, October 1887 ; and *Electrical Review*, xxii. 4, 1888.
[2] See *La Nature*, p. 43, 1882.
[3] Specification of Patent 3949 of 1882 ; and *Electrician*, xix. 479, 1887.

revolves, but in which, by means of a revolving commutator, the magnetic polarity of a double-wound armature is continually caused to rotate. In a further modification of this idea, due to Jehl and Rupp, a mass of iron, which completes the magnetic circuit, rotates within the double-wound ring.[1]

For further notices of the methods of continuous-current transformation, the reader is referred to articles by Elihu Thomson, in *Electrical World*, x. 108, 1887; by R. P. Sellon, in *Electrician*, xx. 633, 1888; and .by Rechniewski, in *La Lumière Électrique*, xxv. 416, 1887.

THEORY OF MOTOR-GENERATORS.

Let $\mathscr{E}$ be the potential at terminals of the primary or motor part, and e that at terminals of the secondary or generator part. Let the i_1, r_1, and C_1 stand respectively for the armature current, armature resistance, and number of armature conductors of the primary part; and i_2, r_2, and C_2 for the corresponding quantities of the secondary part. Then the two induced electromotive-forces will be—

$$E_1 = n\,C_1\,N, \text{ and } E_2 = n\,C_2\,N \text{ ; and}$$
$$E_1 = \mathscr{E} - r_1\,i_1, \text{ and } E_2 = e + r_2\,i_2.$$

Now write p for $C_1 \div C_2$ (the reciprocal of the *coefficient of transformation*), and we have—

$$p\,e = \mathscr{E} - r_1\,i_1 - p\,r_2\,i_2.$$

But the electric work done on and by the armature is equal, or $E_1\,i_1 = E_2\,i_2$; whence $i_2 = p\,i_1$, so that the last equation becomes—

$$e = \frac{\mathscr{E}}{p} - \left(r_2 + \frac{r_1}{p^2}\right)i_2.$$

This shows that everything goes on in the secondary circuit as though the electromotive-force were reduced from that supplied by the primary mains in proportion to the respective numbers of windings on the armature; and as though there were added to the internal resistance of the secondary circuit a resistance equal to that of the primary winding multiplied by the square of the coefficient of transformation.

[1] See *Electrician*, xix. 514, 1887: xx. 7, 1887; and Specification of Patent 2130 of 1887.

CHAPTER XXVI.

THE DYNAMO AS A MOTOR.

IN the first chapter, the definition was laid down that dynamo-electric machinery meant "machinery for converting energy in the form of mechanical power into energy in the form of electric currents, or *vice versâ.*" Up to the present point we have treated the dynamo solely in its functions as a generator of electric currents. We now come to the converse function of the dynamo, namely, that of converting the energy of electric currents into the energy of mechanical motion.

An electric motor, or, as it was formerly called, an electro-magnetic engine, is one which does mechanical work at the expense of electric energy ; and this is true, no matter whether the magnets which form the fixed part of the machine be permanent magnets of steel or electromagnets. In fact, any kind of dynamo independently excited, or self-exciting, can be used conversely as a motor, though, as we shall see, some more appropriately than others. But whether the field-magnets be of permanently magnetised steel or of tempo-rarily magnetised iron, all these motors are electro-magnetic in principle ; that is to say, there is some part either fixed or moving which is an electromagnet, and which as such attracts and is attracted magnetically.

Every one knows that a magnet will attract the opposite pole of another magnet, and will pull it round. We know also that every magnet placed in a magnetic field tends to turn round and set itself along the lines of force. As a first illustration of the nature of the forces at work in the magnetic field, let us take the case of Fig. 329. Here there is, in the first place, a simple magnetic field produced between the

poles of two strong magnets, one on the right, the other on
the left. Between the two, confined forcibly at right angles
to the lines of force, is placed a small magnetic needle
Iron filings sprinkled in the field reveal the actions at work
in a most instructive way. Faraday, who first taught us the
significance of these mysterious lines of force, has told us
that we may reason about them as if they tended to contract
or grow shorter. Now a simple inspection of Fig. 329 will
show that the shortening of the lines of force must have the
effect of rotating the magnetic needle upon its centre, through

FIG. 329.

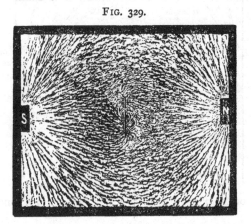

ACTION OF MAGNETIC FIELD ON A MAGNETIC NEEDLE.

an angle of 90°, for the lines stream away on the right hand
above, and on the left hand below, in a most suggestive
fashion. It is not, therefore, difficult to understand that
very soon after the invention of the electromagnet, which
gave us for the first time a magnet whose power was under
control, a number of ingenious persons perceived that it would
be possible to construct an electro-magnetic engine, in which
an electromagnet, placed in a magnetic field, should be pulled
round ; and, further, that the rotation should be kept up continuously, by reversing the current at an appropriate moment.
As a matter of fact, a mere coil of wire, carrying a current, is
acted upon when placed in the magnetic field, and is pulled

round as a magnet is. Fig. 330 shows how, in this case, the
lines of force reveal the action. The magnetic field is, as before,
produced between the ends of two large magnets. The two
round spots are two holes drilled in the sheet of glass, where
the wire which carried the current came up through the glass
and descended again. We shall notice how the lines of iron
filings, which would, if there were no current, run simply across
from right to left, are bent out of their course. If these lines

FIG. 330.

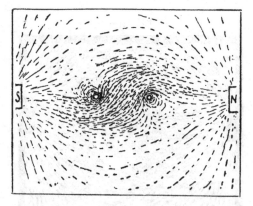

ACTION OF MAGNETIC FIELD ON A WIRE CARRYING A CURRENT.

could shorten themselves, they must, of necessity, twist the
loop of wire round, and cause it to set at right angles to its
present position. Every circuit traversed by a current tends
so to set itself that it shall enclose as many lines of force
as possible. It is obvious that to do this the loop used in
Fig. 330 must turn at right angles to its present position.

On this very principle was constructed the earliest electric
motor of Ritchie, so well known in many forms as a stock
piece of electric apparatus, but little better in reality than
a toy. Joule[1] also devised several forms of electric motor.

A great step in advance was made by Jacobi, who, in 1838,
constructed the multipolar machine of which we give a

---

[1] *Annals of Electricity,* ii. 222, 1838; and iv. 203, 1839.

representation in Fig. 331. This motor, which Jacobi designed for his electric boat, had two strong wooden frames, in each of which a dozen electromagnets were fixed, their poles being set alternately. Between them, upon a wooden disk, was placed another set of electromagnets, which, by the alternate attraction and repulsion of the fixed poles, were kept in rotation; the current which traversed the

FIG. 331.

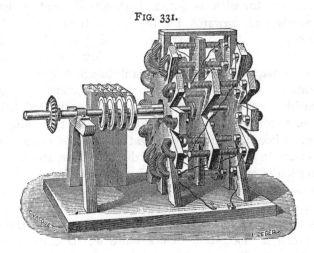

JACOBI'S ELECTRIC MOTOR.

rotating magnets being regularly reversed at the moment of passing the poles of the fixed magnets by means of a commutator, consisting, according to Jacobi's directions, of four brass-toothed wheels having pieces of ivory or wood let in between the teeth for insulation. Jacobi's motor is, in fact, a very advanced type of dynamo, and differs very little in point of design from one of Wilde's[1] most successful forms. (Fig. 216).

A still earlier rotating apparatus, and, like Ritchie's motor, a mere toy, was Sturgeon's wheel (Fig. 200, p. 250), described in 1823. This instrument, interesting as being the forerunner of

---

[1] Wilde's is, however, designed as a generator. Jacobi's, on the contrary, was designed as a motor; though of course it would generate currents if driven round by mechanical power.

Faraday's disk dynamo, is the representative of an important class of machines, namely, those which have a sliding contact merely and need no commutator.

A fourth sort of motors may be named, wherein the moving part, instead of rotating upon an axis, is caused to oscillate backwards and forwards. Professor Henry, to whom we owe so much in the early history of electro-magnetism, constructed, in 1831, a motor with an oscillating beam, alternately drawn backwards and forwards by the intermittent action of an electromagnet. Dal Negro's motor of 1833 was of this class; in it a steel rod was caused to oscillate between the poles of an electromagnet, and caused a crank to which it was geared to rotate in consequence. A distinct improvement in this type of machine was introduced by Page, who employed hollow coils or bobbins as electromagnets, which, by their alternate action, sucked down iron cores into the coils, and caused them to oscillate to and fro. Motors of this kind form an admirable illustration of one of the laws of electromagnetics, first formulated by Gauss, but developed later by Maxwell, to the effect that a circuit acts on a magnetic pole in such a way as to make the number of magnetic lines of force that pass through the circuit a maximum. Once more we have recourse to iron filings to illustrate this abstract proposition of electric geometry.

In Fig. 332 the N. pole of a bar magnet is placed opposite a circuit or loop of wire traversed by a current, and which comes up through the glass at the lower hole and descends at the upper hole. The tendency to draw as many as possible of the magnet's lines of force into the embrace of the circuit is unmistakable. If now we reverse the current, what do we find? Fig. 333 supplies the answer; for now we find that the magnet's lines of force, instead of being drawn in, are pushed out. In fact, in one case the pole is attracted, in the other repelled.

Page's suggestion was further developed by Bourbouze, who constructed the curious motor depicted in Fig. 334,[1]

[1] Taken, by permission of Messrs. Macmillan and Co., from *The Applications of Physical Forces.*

FIG. 332.

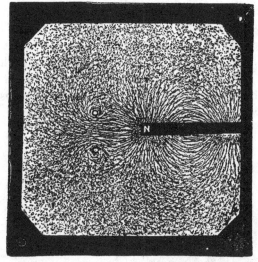

POLE OF MAGNET ATTRACTED INTO A CIRCUIT TRAVERSED BY A CURRENT.

FIG. 333.

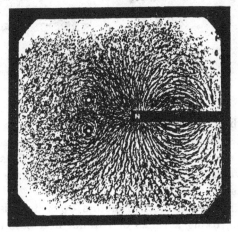

POLE OF MAGNET REPELLED OUT OF THE CIRCUIT WHEN THE CURRENT
IS REVERSED.

which looks uncommonly like an old type of steam-engine.
We have here a beam, crank, fly-wheel, connecting-rod, and
even an eccentric valve-gear and a slide-valve.   But for
cylinders we have four hollow electromagnets; for pistons,
we have iron cores that are alternately sucked in and drawn
out; and for slide-valve we have a commutator, which, by
dragging a pair of platinum-tipped springs over a flat surface
made of three pieces of brass separated by two insulating
strips of ivory, reverses at every stroke the direction of the

FIG. 334.

BOURBROUZE'S ELECTRIC MOTOR.

currents in the coils of the electromagnets.   It is really a
very ingenious machine, but in point of efficiency far behind
many other electric motors.   Unfortunately it does not do
to design dynamo-electric machinery on the same lines as
steam-engines.

Yet a fifth type of electric motors owes its existence to
Froment, who, fixing a series of parallel iron bars upon the
periphery of a drum, caused them to be attracted, one after
the other, by an electromagnet or electromagnets, and thus
procured a continuous rotation.

Lastly, of the various types of motor we may enumerate a class in which the rotating portion is enclosed in an eccentric frame of iron, so that as it rotates it gradually approaches nearer. Little motors, working on this principle of " oblique approach," were invented by Wheatstone, and have long been used for spinning Geissler tubes, and other light experimental work. More recently, Trouvé and others have sought to embody this principle in motors of more ambitious proportions, but without securing any advantage ; for it would be better to bring the armature closer to the pole-pieces of the field-magnet.

It is impossible, within the limits of this work, to deal with a tithe of all the various stages of discovery and invention, or with many interesting and curious machines that have from time to time been tried. It might be told how Page, after inventing his machine in 1834, succeeded in 1852 in constructing a motor of such a size that he was able to drive a circular saw and a lathe by it. Time fails to describe the electric motor of Davidson, which, in 1842, enabled him to propel a carriage, at the speed of four miles an hour, between Edinburgh and Glasgow. An engine which was of 10 horse-power was built in 1849, by Soren Hjörth, at Liverpool.[1]

All these early attempts, however, came to nothing, for two reasons. At that time there was no economical method of generating electric currents known. At that time, moreover, the great physical law of the conservation of energy was not recognised, and its all-important bearings upon the theory of electric machinery could not be foreseen.

While voltaic batteries were the only available sources of electric currents, economical working of electric motors was hopeless. For a voltaic battery wherein electric currents are

---

[1] An excellent account of the early forms of electric motor, both European and American, is to be found in Martin and Wetzler's *The Electric Motor and its Applications*, published in 1887. All readers interested in the subject should also consult the paper on *Electromagnetism as a Motive Power*, by the late R. Hunt in *Proc. Instit. Civil Engineers*, xvi., April 1857, together with the discussion that followed it, in which part was taken by Professor (now Sir William) Thomson, Mr. (now Sir William) Grove, Professor Tyndall, Mr. Cowper, Mr. Smee, and Mr. Robert Stephenson.

generated by dissolving zinc in sulphuric acid is a very expensive source of power. To say nothing of the cost of the acid, the zinc—the very fuel of the battery—costs more than twenty times as much as coal, and is a far worse fuel; for whilst an ounce of zinc will evolve heat to an amount equivalent to 113,000 foot-pounds of work,[1] an ounce of coal will furnish the equivalent of 695,000 foot-pounds.

The fact, however, which seemed most discouraging, and which, if rightly interpreted in accordance with the law of conservation of energy, would have been found to be (on the contrary) a most encouraging fact, was the following :—If a galvanometer was placed in the circuit with the electric motor and the battery, it was found that when the motor was running it was impossible to force so strong a current through the wires as that which flowed when the motor was standing still. Now there are only two causes that can stop such a current flowing in a circuit ; there must be either an obstructive resistance, or else a counter electromotive-force. At first the common idea was, that when the motor was spinning round, it offered a greater resistance to the passage of the electric current than when it stood still. The genius of Jacobi[2] enabled him, however, to discern that the observed diminution of current was really due to the fact that the motor, by the act of spinning round, began to work as a dynamo on its own account, and tended to set up a current in the circuit in the opposite direction to that which was driving it. The faster it rotated the greater was the counter electromotive-force (or " electromotive-force of reaction ") which was

---

[1] A convenient way of regarding the economic question from the point of view of the cost of the voltaic battery is afforded by the following calculation. Supposing the electric motor to convert all the electric energy of the battery without loss into mechanical energy, the amount of zinc used per horse-power in one hour will be almost exactly two pounds divided by the volts of electromotive-force of the cell employed in the battery.

[2] *Mémoire sur l'application de l'électromagnétisme au mouvement des machines*, par M. H. Jacobi (Potsdam, 1835). On p. 45 of this memoir Jacobi points out that the motor, when set into rotation by the current of the battery, becomes by virtue of its motion a magneto-electric apparatus capable of generating a current in a counter-direction in the circuit ; and to this he rightly attributes the limit of uniform speed obtained by the motor.

developed. In fact, the theory of the conservation of energy
requires that such a reaction should exist. Joule,[1] by further
experiment, found that the counter electric action is proportional to the velocity of rotation and to the magnetism of
the magnets.

We know that in the converse case, when we are employing mechanical power to generate currents by rotating a
dynamo, directly we begin to generate currents, that is to say
directly we begin to do electric *work*, it immediately requires
much more power to turn the dynamo than is the case when
no electric work is being done. In other words, there is an
opposing reaction to the mechanical force which we apply
in order to do electric work. An opposing reaction to a
mechanical force may be termed a "counter-force." When,
on the other hand, we apply (by means of a voltaic battery,
for example) an electromotive-force to do mechanical work,
we find that here again there is an opposing reaction ; and
an opposing reaction to an electromotive-force is a " counter
electromotive-force."

The experiment of showing the existence of this counter
electromotive-force is a very easy one. All one requires is a
little motor,[2] a few cells of battery of small internal resistance,
and a galvanometer. They should be connected up in one
circuit, and the deflexion of the galvanometer should be
observed when the motor is held fast, and when it rotates
with small and large loads.

The existence of this counter electromotive-force is of the
utmost importance in considering the action of the dynamo
as a motor, because upon the existence and magnitude of this
counter electromotive-force depends the degree to which any
given motor enables us to utilise electric energy that is
supplied to it in the form of an electric current. In fact,
this counter electromotive-force is an absolute and necessary
factor in the output of the motor, just as much as the
velocity to which (*cæteris paribus*) it is proportional. Lieut.

[1] *Annals of Electricity*, viii., 219, 1842, and *Scientific Papers*, p. 47.
[2] One of any ordinary type—a magneto-machine or a series-wound motor
will answer.

F. J. Sprague has made the suggestion to call it the "motor electromotive-force," thereby emphasising the fact.

In discussing the dynamo as a generator many considerations were pointed out, the observance of which would tend to improve the efficiency of such generators. It is needless to say that many of these considerations, such as the avoidance of useless resistances, unnecessary iron masses in cores, and the like, will also apply to motors. The freer a motor is from such objections, the more efficient will it be. But the efficiency of a motor in utilising the energy of a current depends not only on its efficiency in itself, but on another consideration, namely, the relation between the electromotive-force which it itself generates when rotating, and the electromotive-force or electric pressure at which the current is supplied to it. A motor which itself in running generates only a *low* electromotive-force cannot, however well designed, be an *efficient* or economical motor when supplied with currents at a *high* electromotive-force. A good low-pressure steam-engine does not become more "efficient" by being supplied with high-pressure steam. Nor can a high-pressure steam-engine, however well constructed, attain a high efficiency when worked with steam at low pressures. Analogous considerations apply to dynamos used as motors. They must be supplied with currents at electromotive-forces adapted to them. Even a perfect motor—one without friction or resistance of any kind—cannot give an "efficient" or economical result if the law of efficiency is not observed in the conditions under which the electric current is supplied to it.

Since every dynamo can be also used as a motor, the same classification which has been used in the earlier part of this work for the generators will be adopted also in the chapters which follow.

CHAPTER XXVII.

ELEMENTARY THEORY OF ELECTRIC MOTIVE POWER.

IT will be shown, mathematically, that the efficiency with which a perfect motor utilises the electric energy of the current, depends upon the ratio between the counter electromotive-force developed in the armature of the motor, and the electromotive-force of the current which is supplied by the battery. No motor ever succeeds in turning into useful work the whole of the currents that feed it, for it is impossible to construct machines devoid of resistance, and whenever resistance is offered to a current, part of the energy of the current is wasted in heating the wires that offer the resistance. Let the symbol W stand for the whole electric energy developed per second by a current, and let w stand for that part of the energy which the motor takes up as useful work from the circuit.[1] All the rest of the energy of the current, or $W - w$, will be wasted in useless heating of the resistances.

But if we want to work our motor under the conditions of greatest economy, it is clear that we must have as little heatwaste as possible ; or, in symbols, w must be as nearly as possible equal to W. It will be shown mathematically that the ratio between the useful energy thus appropriated and the total energy spent is equal to the ratio between the counter electromotive-force of the motor and the whole

---

[1] The symbol w must be clearly understood to refer to the value of the work taken up by the motor, *as measured electrically.* The whole of this work will not appear as useful mechanical effect however, for part will be lost by mechanical friction, and part also in the wasteful production of eddy currents in the moving parts of the motor. What proportion of w appears as useful mechanical work depends on the efficiency of the motor *per se,* which we are not here considering. In all that follows immediately we shall suppose such causes of loss not to exist, or the motor will be considered as a perfect motor.

electromotive-force of the battery that feeds the motor. (As it is not wished here to complicate general considerations by introducing into the expression for the efficiency the energy wasted in heat in the field-magnet coils of the motor, we are here supposing that the magnetism of the field-magnets is independently excited.) The proof will be given later. Let us call this whole electromotive-force with which the battery feeds the motor $\mathcal{E}$, and let us call the counter electromotive-force E. Then the rule is

$$w : W = E : \mathcal{E},$$

or, if we express the efficiency as a fraction,

$$\frac{w}{W} = \frac{E}{\mathcal{E}}.$$

But we may go one stage further. If the resistances of the circuit are constant, the current i, observed when the motor is running, will be less than I, the current while the motor was standing still. But, from Ohm's law, we know that

$$i = \frac{\mathcal{E} - E}{R},$$

where R is the total resistance of the circuit. Hence

$$\frac{I - i}{I} = \frac{E}{\mathcal{E}} = \frac{w}{W}.$$

From which it appears that we can calculate the efficiency at which the motor is working, by observing the ratio between the fall in the strength of the current and the original strength. Now as this mathematical law of efficiency has been known for twenty years,[1] it is strange that even in many of the accepted text-books it has been ignored or misunderstood. Another law, discovered by Jacobi, not a law of efficiency at all, but a law of maximum work in a given time, has usually been given instead. It is, indeed, frequent to find Jacobi's law

---

[1] The true law of efficiency was, however, clearly stated by Thomson in 1851, and is recognised in a paper by Joule at about the same date. See also Rankine's *Steam Engine*, p. 546. Professor Anthony, late of Cornell University, informs the author that he has taught the true law for the past fifteen years.

of maximum activity stated as the law of maximum efficiency. Yet as a mathematical expression the true law is implicitly contained in more than one of the memoirs of Joule; it is implied also in more than one passage of the memoirs of Jacobi;[1] it exists in the *Théorie Mécanique de la Chaleur* of Verdet.[2] Yet it remained a mere mathematical abstraction until its significance was pointed out by Sir W. Siemens in his *Address to the Iron and Steel Institute* in 1877.

Jacobi's law concerning the maximum power of an electric motor supplied with currents from a source of given electromotive-force, is the following:—The mechanical work given out by a motor is a maximum when the motor is geared to run at such a speed that the current is reduced to half the strength

---

[1] Jacobi seems very clearly to have understood that his law was a law of maximum working, but not to have understood that it was not a law of true economical efficiency. In one passage (*Annales de Chimie et de Physique*, t. xxxiv. (1852), p. 480), he says :—" Le travail mécanique maximum, *ou plutôt l'effet économique*, n'est nullement compliqué avec ce que M. Müller appelle les circonstances spécifiques des moteurs électromagnétiques." Yet, though here there is apparently a confusion between the two very different laws, in a preceding part of the very same memoir Jacobi says (p. 466) :—" En divisant la quantité de travail par la dépense (de zinc), on obtient une expression très-importante dans la mécanique industrielle : c'est l'effet économique, ou ce que les Anglais appellent *duty.*" Here, again, is a singular confusion. The definition is perfect; but " effet économique" is not the same thing as the maximum power. Jacobi's law is not a law of maximum efficiency, but a law of maximum power; and that is where the error creeps in. It is significant, in suggesting the cause of this remarkable conflict of ideas, that throughout this memoir Jacobi speaks of *work* as being the product of force and velocity, not of force and displacement. The same mistake—common enough amongst continental writers—is to be found in the accounts of Jacobi's law given in Verdet's *Théorie Mécanique de la Chaleur*, in Müller's *Lehrbuch der Physik*, and even in Wiedemann's *Galvanismus*. Now the product of force and velocity is not work, but work divided by time, that is to say rate-of-working, or "activity." This may account for the widely-spread fallacy. Jacobi makes another curious slip in the memoir above alluded to (p. 463), by supposing that the strength of the current can only become = o when the motor runs *at an infinite speed*. We all know now that the current will be reduced to zero when the counter electromotive-force of the motor equals that of the external supply; and if this is finite, the velocity of the motor, if there is independent magnetism in its magnets, need also only be finite. This error—also to be found in Verdet—seems to have thrown the latter off the track of the true law of efficiency, and to have made him fall back on Jacobi's law.

[2] See Verdet, *Œuvres*, t. ix. p. 174, where, however, Verdet makes the very mistake so often made, of supposing that the greatest possible efficiency of a motor, working with a given electromotive-force, is 50 per cent., or is the same as its efficiency when working at the maximum rate.

that it would have if the motor was stopped. This, of course, implies that the counter electromotive-force of the motor is equal to half the electromotive-force furnished by the battery or generator. Now, under these circumstances, only half the energy furnished by the external source is utilised, the other half being wasted in heating the circuit. If Jacobi's law was indeed the law of efficiency, no motor, however perfect in itself, could convert more than 50 per cent. of the electric energy supplied to it into actual work. Now Siemens showed [1] some years ago, that a dynamo can be, in practice, so used as to give out more than 50 per cent. of the energy of the current. It can, in fact, work more efficiently if it be not expected to do its work so quickly. Dr. Siemens, who first made us realise the true physical signification of the mathematical expressions which, until then, had been regarded as mere abstractions, proved in fact, that if the motor be arranged so as to do its work at less than the maximum rate, by being geared so as to do much less work per revolution, but yet so as to run at a higher speed, it will be more efficient; that is to say, though it does less work, there will also be still less electric energy expended, and the ratio of the useful work done to the energy expended will be nearer unity than before.

The algebraic reasoning is as follows:—If $\mathcal{E}$ be the electromotive-force of the mains supplying the current to the motor when the motor is at rest, and i be the current which flows at any time, the electric energy W expended in unit time will be (as expressed in watts) given by the equation—

$$ W = \mathcal{E} i = \mathcal{E}\,\frac{(\mathcal{E} - E)}{R}. \qquad \text{[XLVI.]} $$

Now, when the motor is running, part of this electric energy is being spent in doing work, and the remainder is wasting itself in heating the wires of the circuit. We have already

[1] The matter was also very well and clearly put by Prof. W. E. Ayrton in his lecture on "Electric Transmission of Power," before the British Association in Sheffield, in 1879.

used the symbol w for the useful work (per second) done by the motor. All the energy which is not thus utilised is wasted in heating the resistances. Let the symbol H represent the heat wasted per second. Its mechanical value will be H J, where J stands for Joule's equivalent. Then clearly we shall have

$$W = w + H J.$$

But, by Joule's law, the heat-waste of the current whose strength is i running through resistance R, is expressed by the equation

$$H J = i^2 R.$$

Substituting this value above, we get

$$W = w + i^2 R,$$

whence we get

$$w = W - i^2 R. \qquad \text{[XLVII.]}$$

But by equation [XLVI.] preceding, $W = \mathcal{E} i$, whence

$$w = \mathcal{E} i - i^2 R, \qquad \text{[XLVIII.]}$$

and, writing for i its value $\dfrac{\mathcal{E} - E}{R}$, we get

$$w = \frac{(\mathcal{E} - E) \left\{ \mathcal{E} - (\mathcal{E} - E) \right\}}{R},$$

or

$$w = E \frac{\mathcal{E} - E}{R}. \qquad \text{[XLIX.]}$$

Comparing equation [XLIX.] with equation [XLVI.], we get the following :—

$$\frac{w}{W} = \frac{E (\mathcal{E} - E)}{\mathcal{E} (\mathcal{E} - E)};$$

or, finally,

$$\frac{w}{W} = \frac{E}{\mathcal{E}}. \qquad \text{[L.]}$$

This is, in fact, the mathematical law of efficiency, so long misunderstood until Siemens showed its practical significance.

We may appropriately call it *the law of Siemens.* Here the ratio $\dfrac{w}{W}$ is the measure of the efficiency of the motor, and the equation shows that we may make this efficiency as nearly equal to unity as we please, by so adjusting either the speed of the motor or the magnetism of its field-magnets that E is very nearly equal to $\mathcal{E}$.

Now go back to equation [XLVIII.], which is—

$$w = \mathcal{E}\,i - i^2\,\mathrm{R}.$$

In order to find what value of i will give us the maximum value for w (which is the work done by the motor *in unit time*), we must take the differential coefficient and equate it to zero.[1]

$$\frac{d\,w}{d\,i} = \mathcal{E} - 2\,i\,\mathrm{R} = 0,$$

whence we have

$$i = \tfrac{1}{2}\,\frac{\mathcal{E}}{\mathrm{R}}.$$

But, by Ohm's law, $\dfrac{\mathcal{E}}{\mathrm{R}}$ is the value of the current when the motor stands still. So we see at once that, to get maximum

---

[1] The argument can be proven, though less simply, without the calculus, as follows : write equation [XLVIII.] in the following form :

$$i^2\,\mathrm{R} - \mathcal{E}\,i + w = 0.$$

Solving this as an ordinary quadratic equation, in which i is the unknown quantity, we have

$$i = \frac{\mathcal{E} \pm \sqrt{\mathcal{E}^2 - 4\,\mathrm{R}\,w}}{2\,\mathrm{R}}.$$

To find from this what value of i corresponds to the greatest value of w, it may be remembered that a negative quantity cannot have a square root, and that therefore the greatest value that w can possibly have will occur when

$$4\,\mathrm{R}\,w = \mathcal{E}^2,$$

for then the term under the root sign will vanish. When this condition is observed it will follow that

$$i = \frac{\mathcal{E}}{2\,\mathrm{R}},$$

or the current will be reduced to half its original value.

work per second out of our motor, the motor must run at such a speed as to bring down the current to half the value which it would have if the motor were at rest. In fact, we here prove the law of Jacobi for the maximum rate of doing work. But here, since

$$i = \frac{\mathcal{E} - E}{R} = \tfrac{1}{2} \frac{\mathcal{E}}{R},$$

it follows that ,

$$\mathcal{E} - E = \tfrac{1}{2} \mathcal{E},$$

or

$$\frac{E}{\mathcal{E}} = \tfrac{1}{2} ;$$

whence it follows also that

$$\frac{w}{W} = \tfrac{1}{2}.$$

That is to say, the efficiency is but 50 per cent. when the motor does its work at the maximum rate.[1]

It may be worth while to recall a precisely parallel case that occurs in calculating the currents from a voltaic battery. Every one is familiar with the rule for grouping a battery which consists of a given number of cells, that they will yield a maximum current through a given external resistance when so grouped that the internal resistance of the battery shall, as nearly as possible, equal the external resistance. But this rule, which is true for maximum current (and, therefore, for maximum rate of using up the zinc of one's battery), is not the case of greatest economy. For if external and internal resistance are equal, half the energy of the current will be wasted in the heat of the cells, and half only will be available in the external circuit. If we want to get the greatest economy, we should group our cells so as to have an internal resistance much less than the external. We shall not get so strong a current, it is true; and we shall use up our zincs more slowly ; but a far greater proportion of the energy will be expended usefully, and a far less proportion will be wasted in heating the battery cells. The maximum economy will, of course, be got by making the external resistance infinitely great as compared with the internal resistance. Then all the energy of the current will be utilised in the external circuit, and none wasted in the battery. But it would take an infinitely long time to get through a finite amount of work in this extreme case. The same kind of reasoning is strictly applicable to dynamos used as generators, the resistance of the rotating part of the circuit being the counterpart of the internal resistance of the battery cells. For good economy the resistance of the armature should be very low as compared with that of the external circuit.

GRAPHIC REPRESENTATION OF LAWS OF MOTORS.

Several graphic constructions have been suggested to convey these facts to the eye; one of these enables us, in one diagram, to exhibit graphically both Jacobi's law of maximum rate of working, and Siemens' law of efficiency.[1]

Let the vertical line, A B (Fig. 335), represent the electromotive-force $\mathcal{E}$ of the electric supply. On A B construct a square A B C D, of which let the diagonal B D be drawn. Now measure out from the point B, along the line B A, the counter electromotive-force E of the motor. The length of this quantity will increase as the velocity of the motor increases. Let E attain the value B F. Let us inquire what the actual current will be, and what the energy of it; also what the work done by the motor is.

FIG. 335.

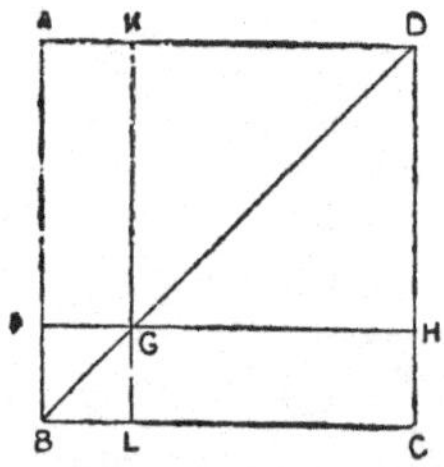

First complete the construction as follows:—Through F draw F G H, parallel to B C, and through G draw K G L, parallel to A B. Then the actual electromotive-force at work in the machine producing a current is $\mathcal{E} - E$, which may be represented by any of the lines A F, K G, G H, or L C. Now the electric energy expended per second is $\mathcal{E} i$; and since $i = \dfrac{\mathcal{E} - E}{R}$,

$$\frac{\mathcal{E}(\mathcal{E} - E)}{R};$$

and the work absorbed by the motor, *measured electrically*, is

$$\frac{E(\mathcal{E} - E)}{R}$$

R being a constant, the values of the two may be written respectively

$$\mathcal{E}(\mathcal{E} - E)$$

---

[1] See paper by the author in the *Philosophical Magazine*, Feb. 1883.

and

$$E (\mathcal{E} - E).$$

Now the area of the rectangle

$$A F H D = \mathcal{E} (\mathcal{E} - E),$$

and that of the rectangle

$$G L C H = E (\mathcal{E} - E).$$

The ratio of these two areas on the diagram is the efficiency of a perfect motor, under the condition of a given constant electro-motive-force in the electric supply.

Turn to Fig. 336, in which these areas are shaded. This figure represents a case where the motor is too heavily loaded, and can turn only very slowly, so that the counter electro-motive-force E is very small compared with $\mathcal{E}$. Here the area which represents the energy expended, is very large; while that which represents useful work realised in the motor is very small. The efficiency is obviously very low. Two-thirds or more of the energy is being wasted in heat.

So far we have assumed that the efficiency of a motor (working with a given constant external electromotive-force) is to be measured electrically. But no motor actually converts into useful mechanical effect the whole of the electric energy which it absorbs, since part of the energy is wasted in friction and part in wasteful electro-magnetic re-actions between the stationary and moving parts of the motor. If, however, we consider the motor to be a *perfect* engine (devoid of friction, not producing wasteful eddy currents, running without sound, giving no sparks at the collecting-brushes, &c.), and capable of turning into mechanical effect 100 per cent. of the electric energy which it absorbs, then, and then only, may we take the electrical measure of the work of the motor as being

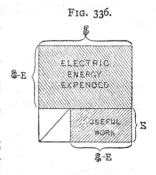

FIG. 336.

a true measure of its performance. Such a "perfect" electric
engine would, like the ideal "perfect" heat engine of Carnot,
be perfectly reversible. In Carnot's heat engine it is supposed that the whole of the heat actually absorbed in the
cycle of operations is converted into useful work ; and in this
case the efficiency is the ratio of the heat absorbed to the total
heat expended. As is well known, this efficiency of the perfect heat engine can be expressed as a function of two absolute temperatures, namely those respectively of the heater
and of the refrigerator of the engine. Carnot's engine is also
ideally reversible ; that is to say, capable of reconverting
mechanical work into heat.

The mathematical law of efficiency of a perfect electric
engine illustrated in the above construction is an equally ideal
case. And the efficiency can also be expressed, when the
constants of the case are given, as a function of two electromotive-forces. We shall return to this comparison a little
later.

The Law of Maximum Activity (Jacobi).

Let us next consider the area G L C H of the diagram
(Fig. 337), which represents the work utilised in the motor.

FIG. 337.

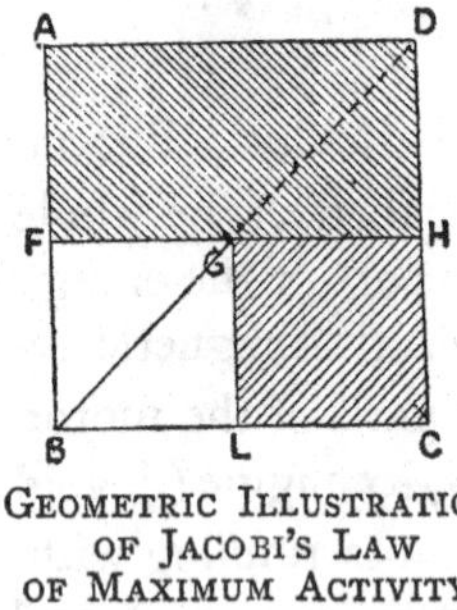

GEOMETRIC ILLUSTRATION
OF JACOBI'S LAW
OF MAXIMUM ACTIVITY.

The value of this area will vary with
the position of the point G, and will
be a maximum when G is midway
between B and D ; for of all rectangles
that can be inscribed in the triangle
B C D, the square will have maximum area (Fig. 337). But if G is
midway between B and D, the rectangle G L C H will be exactly half
the area of the rectangle A F H D;
or, the useful work is equal to half
the energy expended. When this is
the case, the counter electromotive-
force reduces the current to half the strength it would have
if the motor were at rest ; which is Jacobi's law of the

efficiency of a motor doing work at its greatest possible rate. Also F will be half-way between B and A, which signifies that $E = \frac{1}{2} \mathcal{E}$.

Law of Maximum Efficiency (Siemens).

Again, consider these two rectangles when the point G moves indefinitely near to D (Fig. 338). We know from common geometry that the rectangle G L C H is equal to the rectangle A F G K. The area (square) K G H D, which is the excess of A F H D over A F G K, represents therefore the electric energy which is wasted in heating the resistances of the motor. That the efficiency should be a maximum the heat-waste must be a minimum.

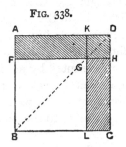

FIG. 338.

GEOMETRIC ILLUSTRATION OF SIEMENS' LAW OF MAXIMUM EFFICIENCY.

In Fig. 336 this corner square which stands for the heat-waste, was enormous. In Fig. 337 it was exactly half the energy. In Fig. 338 it is only about one-eighth. Clearly, we may make the heat-waste as small as we please, if only we will take the point F very near to A. The efficiency will be a maximum when the heat-waste is a minimum. The ratio of the areas G L C H and A F H D, which represents the efficiency, can therefore only become equal to unity when the square K G H D becomes indefinitely small—that is, when the motor runs so fast that its counter electromotive-force E differs from $\mathcal{E}$ by an indefinitely small quantity only.

It is also clear that if our diagram is to be drawn to represent any given efficiency (for example, an efficiency of 90 per cent.), then the point G must be taken so that area $G\,L\,C\,H = \frac{9}{10}$ area A F H D; or, G must be $\frac{9}{10}$ of the whole distance along from B towards D. This involves that E shall be equal to $\frac{9}{10}$ of $\mathcal{E}$, or that the motor shall run so fast as to reduce the current to $\frac{1}{10}$ of what it would be if the motor were standing still. Thus we verify, geometrically, Siemens' law of efficiency. If there is leakage in the line, then, as Professor Oliver Lodge and Mr. G. Kapp [1] have pointed out, this law will require modification, for the higher the counter-electromotive-force of the motor, the higher will be the potential of the line and the greater the loss by leakage.

Further, if the motor be not a "perfect" one, but one whose intrinsic efficiency, or *efficiency per se*, is known, the actual mechanical work performed by the motor can be represented on the diagram by simply retrenching from the rectangle G L C H the fraction of work lost in friction, &c. Similarly, in the case where the electric energy expended has been generated in a dynamo-electric machine whose intrinsic efficiency is known, the total mechanical work expended can be represented by adding on to the area A F H D the proportion spent on useless friction, &c. To make the diagram still more expressive, we may divide the area K G H D into slices proportional to the several resistances of the circuit ; and the areas of these several slices will represent the heat wasted in the respective parts of the circuit. These points are exemplified in Fig. 339, which represents the transmission of power between two dynamos, each supposed to have an intrinsic efficiency of 80 per cent., each having 500 ohms resistance, working through a line of 1000 ohms resistance, the electromotive-force of the machine used as generator being 2400 volts and the counter electromotive-force of the machine used as motor being 1600 volts.

FIG. 339.

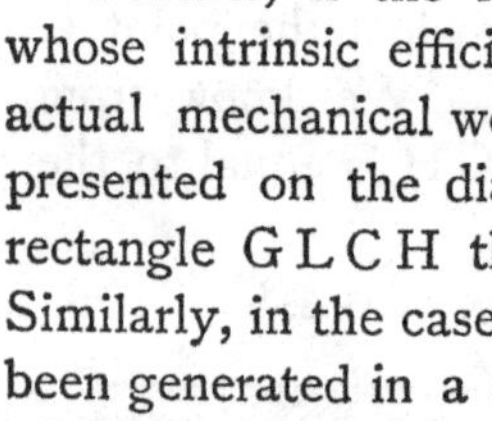

The entire upper area represents the total mechanical work expended. Call this 100. It is expended as follows :—
$a = 20$, lost by friction, &c., in the generator ; $b = 6\frac{2}{3}$, lost in heating generator ; $c = 13\frac{1}{3}$, lost in heating line-wires ; $d = 6\frac{2}{3}$, lost in heating motor ; $e = 10\frac{2}{3}$, lost in friction in the motor ; $w = 42\frac{2}{3}$ is the percentage realised as useful mechanical work.

[1] See Kapp's *Electric Transmission of Energy* (1886), p. 165.

We may now extend this graphic method to two further cases.

Suppose that $\mathcal{E}$ is no longer taken as a constant, but that the work to be done by the motor per second is a constant. For this case we may write equation [XLIX.], p. 521, as

$$E (\mathcal{E} - E) = w R.$$

This equation is graphically represented by the curve P H Q (Fig. 340), in which the values of $\mathcal{E}$ are plotted as abscissæ and those of E as ordinates. From this curve it is at once seen that there will be a certain minimum value of $\mathcal{E}$ which will suffice to give to the motor the prescribed amount of energy per second. The curve is so drawn that it passes through the corner H of all the areas equal to G L C H drawn to fit under the diagonal of the square. Of these areas, which represent equal work done by the motor, the one which has minimum value of $\mathcal{E}$ is the square which fits to the apex of the curve and corresponds to the case where $\mathcal{E} = 2 E$. This result, which was first pointed out by Prof. Carhart,[1] is the converse of Jacobi's law, and like it, involves an efficiency of only 50 per cent. A much higher efficiency is obtained when $\mathcal{E}$ and E are both greater, as indicated by the square drawn through the point h.

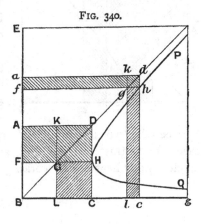

FIG. 340.

Again, suppose that W, the work supplied electrically, is maintained constant; for this case (Fig. 341) we may write equation [XLVI.], p. 520, in the form

$$\mathcal{E} (\mathcal{E} - E) = W R,$$

giving us the curve T H S. In this case $\mathcal{E}$ is a minimum when E is zero, and all the work is wasted in heat; and w,

[1] *American Journal of Science,* xxxi. p. 95, 1886.

the work of the motor, is a maximum only when both $\mathscr{E}$ and E are infinitely great.   This result is also due to Carhart.

It only remains to point out a curious contrast that presents itself between the efficiency of a perfect heat engine and that of a perfect electric engine.   We saw that the one could be expressed as a function of two temperatures, whilst the other could be expressed as a function of two electromotive - forces.   But in the heat engine the efficiency is the greatest when the difference between the two temperatures is a maximum ; whilst in the electric engine the efficiency is the greatest when the difference between the two electromotive-forces is a minimum.   The two cases are contrasted in Figs. 342 and 343, Fig. 342 showing the efficiency of a heat engine working between temperatures T and

FIG. 341.

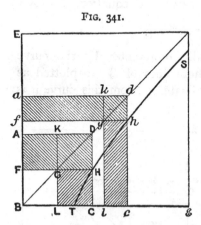

t (reckoned from absolute zero) ; whilst Fig. 343 shows the efficiency of an electric engine receiving current at an electromotive-force $\mathscr{E}$, its counter electromotive-force being E.   Joule's remark, here illustrated, that an electric engine may be readily

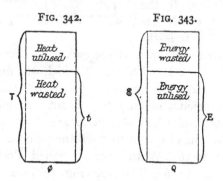

FIG. 342.   FIG. 343.

HEAT ENGINE AND ELECTRIC ENGINE CONTRASTED.

made to be a far more efficient engine than any steam-engine, is amply justified by all experience. The rapid extension in the use of electric motors which has taken place during the past two years, particularly in New England, proves that where power can be produced sufficiently cheap on the large scale, either from natural sources, such as water power, or by large steam-engines burning cheap coal, such power can be electrically distributed to small motors (not exceeding a few horse-power) with an economic result. For the cost of erecting, maintaining, and supplying the electric power to such motors is less than the cost of erecting, maintaining, and supplying fuel and water to small steam-engines of equal power.

It yet remains to be true that the electric engine as a prime mover is more costly to use than the heat engine, for energy in the form of electricity supplied at a high potential is, as yet, dearer than energy in the form of heat supplied at a high temperature.

ELECTRIC TRANSMISSION OF ENERGY.

In all the preceding discussion, it has been assumed that the motor is to be worked with a supply of current furnished *at a constant potential.* It is not only convenient, but useful, to make such a condition the basis of the argument, because this is the condition under which electric energy is now being distributed over large areas in towns and cities for the purposes of lighting and motive power. It would be absurd, in the present stage of electro-technical science, to deal with such a question as the construction and use of motors, without taking into account the practical conditions under which they will be used. But the condition of having a constant electric pressure is not the only condition of supply ; for, as we have seen in preceding chapters, a generator or system of generators may be worked so as to yield a *constant current* at a pressure varying according to the demand.

Now the method of distribution by a constant current is, where power is to be transmitted to long distances, a much more economical method than that of distribution with a constant potential, owing to the fact that for the former method thinner and therefore less expensive conducting wires may be employed. We shall therefore, in further discussing the theory of the different windings of motors, have to take both cases into account. Meantime we may discuss two problems bearing upon the transmission of power by motors, problems which are vital to the understanding of the conditions which motors must fulfil.

It is required to determine the relation between the potential at which the current is supplied to the motor, and the heat-waste in the circuit.

Let ΣR stand for the sum of all the resistances in the part of the circuit between the points at which $\mathcal{E}$ is measured; then, by Joule's law, the heat-waste is (in mechanical measure)

$$H J = i^2 \Sigma R.$$

And, since $i = \dfrac{\mathcal{E} - E}{\Sigma R}$, we may write the heat-waste as

$$H J = \frac{(\mathcal{E} - E)^2}{\Sigma R}.$$

Now suppose that without changing the resistances of the circuit we can increase $\mathcal{E}$, and also increase E, while keeping $\mathcal{E} - E$ the same as before, it is clear that the heat loss will be precisely the same as before. But how about the work done? Let the two new values be respectively $\acute{\mathcal{E}}$ and $\acute{E}$. Then the electric energy expended is

$$\acute{W} = \frac{\acute{\mathcal{E}} \, (\acute{\mathcal{E}} - \acute{E})}{\Sigma R},$$

and the useful work done is

$$\acute{w} = \frac{\acute{E} \, (\acute{\mathcal{E}} - \acute{E})}{\Sigma R}.$$

That is to say, with no greater loss in heating, more energy is transmitted, and more work done. Also the efficiency is greater, for

$$\frac{\acute{w}}{\acute{W}} = \frac{\acute{E}}{\acute{\mathcal{E}}},$$

and this ratio is more nearly equal to unity than $\frac{E}{\mathcal{E}}$, because both $\mathcal{E}$ and E have received an increment arithmetically equal. Clearly, then, it is an economy to work at high electromotive-force. The importance of this matter, first pointed out by Siemens, and later by Marcel Deprez, cannot be overrated. But how shall we obtain this higher electro-motive-force? One very simple expedient is that of driving both generator and motor at higher speeds. The objections to this expedient are the purely mechanical considerations of liability to heating of bearings and centrifugal flying. A second way is to wind the armatures of both machines with many coils of wire having many turns. This expedient has, however, the effect of putting great resistances into the circuit. This circumstance may, nevertheless, be no great drawback, if there is already a great resistance in the circuit—as, for example, the resistance of many miles of wire through which the power is to be transmitted. In this case, doubling the electro-motive-force will not double the resistance. Even in the case where the line resistance is insignificant, an economy is effected by raising the electromotive-force. For, as may be deduced from the equations, when $\mathcal{E} - E$ is kept constant, the effect of doubling the electromotive-force is to increase the efficiency, when the resistance of the line is very small as compared with that of the machines, and to double it when the resistance of the line is very great as compared with that of the machines. It is, in fact, worth while to put up with the extra resistance, which we cannot avoid if we try to secure high electromotive-force by the use of coils of fine wire of many turns. It is true that the useful effect falls off, *cæteris paribus*, as the resistance increases; but this is much more than counter-balanced by the fact that the useful effect increases in proportion to the square of the electromotive-force. A third way, not open to any of these objections, is to increase the magnetic field by employing a more powerful field-magnet.

The advantage derived in the case of the electric transmission of energy from the employment of very high electromotive-forces in the two machines is also deducible from the diagram.

Let Fig. 338, given above, be taken as representing the case where $\mathcal{E}$ is 100 volts and E 80 volts. Now suppose the resistances of the circuit to remain the same while $\mathcal{E}$ is increased to 200 volts and E to 180 volts. (This can be

FIG. 344.

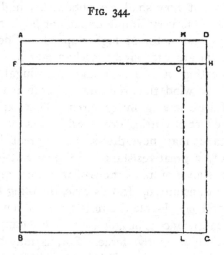

accomplished by increasing the speed of both machines to the requisite degrees.) $\mathcal{E} - $ E is still 20 volts, and the current will be the same as before. Fig. 344 represents this state of things. The square K G H D which represents the heat-waste is the same size as before ; but the energy spent is twice as great, and the useful work done is more than twice as great as previously. High electromotive-force therefore means not only a greater quantity of power transmitted, but a higher efficiency of transmission also. The efficiency of the system in the case of Fig. 338 was 80 per cent. ; in the case of Fig. 344 it is 90 (the dynamos used being supposed:

" perfect ") ; and whilst double energy is expended, the useful return has risen in the ratio of 9 to 4.

In the attempt of M. Marcel Deprez to realise these conditions, in the transmission of power from Miesbach to Munich in 1882, through a double line of telegraph wire, over a distance of thirty-four miles, very high electromotive-forces were actually employed. The machines were two ordinary Gramme dynamos, the magnets being series-wound, similar to one another, but their usual low-resistance coils had been replaced by coils of very many turns of fine wire. The resistance of each machine was consequently 470 ohms, whilst that of the line was 950 ohms.[1] The velocity of the generator was 2100 revolutions per minute ; that of the motor, 1400. The difference of potential at the terminals of the generator was 2400 volts; at that of the motor, 1600 volts. According to Professor von Beetz, the President of the Munich Exhibition, where the trial was made, the mechanical efficiency was found to be 32 per cent. In Deprez's later experiments, in 1886, between Creil and Paris,[2] the potentials were 6004 and 5456 volts respectively; and according to M. Levy the mechanical efficiency as measured by dynamo-meters was 44·81 per cent., the distance being about 36 miles.

M. Deprez has given the rule that the efficiency $\frac{w}{W}$ is obtained, in the case where two identical machines are employed, by comparing the two velocities at the two stations. Or

$$\frac{w}{W} = \frac{n_2}{n_1},$$

where n_1 is the speed of the generator, n_2 that of the motor. There is, however, the objection to this formula, that the electromotive-forces are not proportional to the speeds, unless the magnetic fields of the two machines are also equally intense, and the current running through each machine the same. This is not the case if there is any leakage along the

---

[1] These figures, and those which follow, are given on the authority of the President of the Munich Exhibition, Professor von Beetz.

[2] *Electrician*, xvii. 318, 1886.

line. Moreover, the ratio of the two electromotive-forces of the machines is not the same as the ratio of the two differences of potentials, as measured between the terminals of the machines.

Further, even though the current running through the armatures and field-magnets in the generator which creates the current, and in the motor which utilises the current, be absolutely identical, the intensities of the magnetic fields of the two machines are not equal, even though the machines be absolutely identical in build : the reaction between the armature and field-magnet in the dynamo that is used as a motor is different from that in the dynamo which is being used as the generator. These reactions are considered below.

In an experiment, M. Fontaine,[1] by using several Gramme machines coupled in series at each end of a line, the resistance of which was 100 ohms, succeeded in transmitting 50 horse-power with a mechanical efficiency of 52 per cent.

ECONOMY OF TRANSMISSION.

It can readily be shown that with two series dynamos, the electrical efficiency of transmission, when there is no leakage, is the ratio of the electromotive-forces developed in the armatures of the two machines. To do this we will consider separately the efficiencies of the three parts of the system. Writing E_1 for the electromotive-force developed in the generator, E_2 for that of the motor, r_1 and r_2 for their respective internal resistances, we shall then have

$$\text{Efficiency of generator} \quad .. \quad \eta_1 = \frac{E_1 i - r_1 i^2}{E_1 i} \, ;$$

$$\text{Efficiency of line} \quad .. \quad .. \quad \eta_2 = \frac{E_2 i + r_2 i^2}{E_1 i - r_1 i^2} \, ;$$

$$\text{Efficiency of motor} \quad .. \quad \eta_3 = \frac{E_2 i}{E_2 i + r_2 i^2} \, .$$

[1] *L'Électricien,* x. 707, 1886.

Hence the resulting efficiency of the whole system will be

$$\eta = \eta_1 \times \eta_2 \times \eta_3 = \frac{E_2}{E_1}.$$

If the machines are shunt-wound or compound-wound, or if there is leakage on the line, the currents through the armatures will no longer be alike in the two machines. Writing the respective armature-currents as i_1 and i_2, we shall have in this case, as the electrical efficiency of transmission,

$$\eta = \frac{E_2 \, i_2}{E_1 \, i_1}.$$

As an example we may refer to recent experiments[1] made with Brown's dynamos (Fig. 116), between Kriegstetten and Solothurn, through a conductor of $9 \cdot 23$ ohms resistance. In one experiment, the observed values were—E_1 1231·6, E_2 988·6, i_1 14·204, i_2 14·177. The actual horse-power measured at the two ends showed a mechanical efficiency of 74·7 per cent.

A simple problem in the electric transmission of power is the following :—Suppose that one were desirous of working a motor, so as to do work at the rate of a specified number of horse-power, and that the wire available to bring the current cannot safely stand more than a certain current, without being in danger of becoming heated unduly. It might be desirable to know what electromotive-force such a motor ought to be capable of giving back, and what electromotive-force must be applied at the transmitting end of the wire. Let P stand for the number of horse-power to be transmitted, and i for the maximum strength of current that the wire will stand (expressed in ampères). Then by the known rule for the work of a current, since

$$\frac{E \, i}{746} = P,$$

$$E = \frac{746 \, P}{i}$$

gives the condition as to what electromotive-force (in volts) the machine must be capable of giving when run at the speed it is

eventually to run at as a motor. Moreover, the primary electromotive-force, E_1, must be such that

$$\frac{E_1 - E}{\Sigma R} = i,$$

where ΣR is the sum of all the resistances in the circuit.   Whence

$$E_1 = E + i \Sigma R.$$

Which gives the required primary electromotive-force.

RELATION OF SPEED AND TORQUE OF MOTORS TO THE CURRENT SUPPLIED.

Certain very important relations subsist between the condition of the electric supply and the speed and turning-moment
of a motor.

As mentioned in the paragraphs on p. 130, concerning
the governing of generators, the power transmitted along a
shaft is the product of two factors, the speed and the torque
(or turning-moment). If ω stands for the angular velocity and
T for the torque,[1] then

ω T = mechanical work per second, or " activity,"

and this, measured electrically, is equal (save for the fraction
lost in friction, &c.) to the electric energy absorbed per second,
or, if E is the electromotive-force of the motor, and i the
current through its armature,

E i = electric work per second (in watts [2]).

---

[1] If n be the number of revolutions per *second*, then $2 \pi n = \omega$. Also if F be
the transmitted pull on the belt (or rather the difference between the pull in that
part of the belt which is approaching the driving pulley and the pull in that part
which is receding from the driving pulley) in pounds' weight, and r be the radius
of the pulley, F r = the turning-moment or torque = T, then ω T = $2 \pi n r$ F =
the number of foot-pounds per second transmitted by the belt. This may also be
proved as follows : Horse-power is product of the force into the velocity. The
circumference of the pulley is $2 \pi r$, and it turns n times per second, therefore the
circumferential velocity is $2 \pi r n$, and this, multiplied by F, gives the work per
second. If F is expressed in grammes' weight, and r in centimetres, then
$2 \pi r n$ F will give the activity in gramme-centimetres, and must be divided by
$7 \cdot 6 \times 10^6$ to bring it to horse-power, and must be multiplied by 981×10^{-7} to
bring it to watts.

[2] Since 1 volt = 10^8 C.G.S. units, and 1 ampère = 10^{-1} C.G.S. units,
1 watt (or volt-ampère) will be = 10^7 C.G.S. units of work per second = 10^7 ergs
per second = $10^7 \div 981$ gramme-centimetres per second.

Now we know that if the current running through a series dynamo be constant, the electromotive-force it develops is almost exactly proportional to its speed. It therefore follows that if E is proportional to ω, T will be proportional to i. This is abundantly verified in the case of a series motor by experiments. When a Siemens series dynamo was arranged to lift a load of 56 lbs. on a hoist, it lifted this load at the rate of 212 feet per minute, developing a counter electromotive-force of 108·81 volts. The applied electromotive-force was 111 volts and the resistance of the circuit was 0·3 ohm. The effective electromotive-force was therefore 2·19 volts and the current 7·3 ampères. When the resistance of the circuit was increased to 2·2 ohms the speed fell to 169 feet per minute, the counter electromotive-force to 94·94; the effective electro-motive-force, $\mathcal{E} - E$, was therefore 16·06 volts and the current 7·3 ampères as before. When 4·8 ohms were inserted the speed fell to 141 feet per minute, and E to 76 volts. $\mathcal{E} - E$ was 35 volts, and the current 7·3 ampères as before. *With the same load, the same current,* whatever the speed. The figures given on p. 562, relative to the Immisch motor, also illustrate this point. The speed of a given series-wound motor depends solely on the electromotive-force of the generator and on the resistance of the circuit.

The fact that the torque of a series motor depends only on the current, is of advantage in the application of motors to propulsion of vehicles (such as tram-cars) which at starting require for a few seconds a power greatly in excess of that needed when running. To start, a large current must be turned on. One convenient way of arranging this is to use two motors, coupled habitually in series. When starting, they are, by moving a commutator, coupled in parallel. This doubles the electromotive-force for each, and at the same time halves the resistance. For a few seconds a very strong current flows—much stronger than that which the motors would stand for any prolonged work—and so provides the needful additional torque.

In the series motor, when supplied at constant potential E is not proportional to the speed, because the field-magnetism is not constant, but falls off as E increases, being (if unsaturated) nearly proportional to $\mathcal{E}$ − E. It therefore will not run at a constant speed. Neither will it run at a constant speed if supplied with a constant current.

In the case of the shunt motor, if supplied at constant potential, its field-magnetism is constant, and E will be very nearly proportional to the speed. But $\mathcal{E} - r_a i_a = \mathrm{E}$; and if r_a is (as it should be) very small, E will be very nearly equal to $\mathcal{E}$, and therefore will be nearly constant. Hence the speed also will be nearly constant. If supplied with a constant current, the speed of the shunt motor increases with the load. All this will be considered in detail later.

GRAPHIC METHODS APPLIED TO MOTOR PROBLEMS.

Of the following problems, the first two relate to series motors, and are due to Dr. E. Hopkinson and to Mr. Alexander Siemens.[1]

Given a system of distributing mains supplying electricity at a constant potential $\mathcal{E}$, it is required to construct a motor which when working with a given load shall make n revolutions per minute.

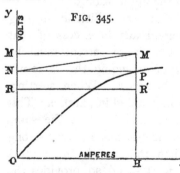

FIG. 345.

Taking, in Fig. 345, as usual, the currents as abscissæ, and the electromotive-forces as ordinates, draw O M to represent the potential $\mathcal{E}$ of the main in volts. Now, makers of dynamos know from experience what percentage of the electrical energy supplied to a machine of the type they make is absorbed in maintaining the magnetic field. Take a point N in O M such that N M ÷ O M represents this

[1] *Journal of the Society of Arts*, April 1883.

percentage. Also it is known what percentage of the energy thus taken up by the armature, and converted into mechanical work, is wasted in friction at the bearings and brushes. Take the point R such that N R ÷ O N represents the percentage so wasted. From O *x* cut off O H such that the area O H·R′ R represents the actual mechanical output of the motor in watts. For example, if the motor is to realise 1 horse-power, then the area O H R′ R must equal 746 watts. Then O H represents the current (in ampères), and H P is the counter electromotive-force which the motor must exercise. The motor, then, must be such that, running at *n* revolutions per minute, its characteristic will pass through P. The economic coefficient will obviously be equal to H P ÷ H M, and the nett efficiency to H R ÷ H M. The energy spent in magnetisation is measured by the area N P M′ M. The tangent of the angle P N M′ represents the resistance of the armature, and of the magnets in the case of a series motor.

It is possible to work out similar problems for a shunt-wound motor, and also for the case of a distribution with a constant current.

Given a motor needing a certain current and a certain electromotive-force to enable it to do its work, it is required to construct a suitable generator, the distance between the machines being represented by an electrical resistance of R *ohms.*

Let O P P′ (Fig. 346) be the characteristic of the motor running at its required speed; P H being the volts, and O H the current needed for it. Draw P N horizontally; and draw N M′ from N at an angle such that its tangent represents the sum of the resistances of the motor and line. Then M′ H represents the difference of potential between the terminals of the generator. Then produce

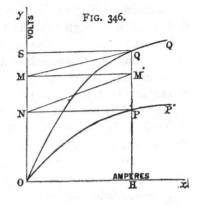

FIG. 346.

H M' to Q so that H M' ÷ H Q shall represent the economic coefficient of the machine of the type that is to be used as generator. Then the proper generator will be that which when running at the proper speed will have a characteristic running through O and Q; and the tangent of the angle M' M Q represents the resistance of the armature and field-magnet of the generator.

The following problem is due to Dr. J. Hopkinson:—

Given the characteristic of the generator, it is required to determine the maximum work which can be transmitted when the electromotive-force of the generator depends on the current passing through its armature.

This problem arises from the failure of Jacobi's law of maximum activity to take into account the reactions in the magnetism of the field arising from the armature. Were it not for these reactions the activity of the motor would be greatest when the counter electromotive-force of the motor was equal to half that of the generator. Let O P B (Fig. 347) be the characteristic of the generator. From this curve another must be derived in the following manner. Take any point P, and draw a tangent P T. Draw T N parallel to O *i* cutting P M in N. Produce M P to L such that L P = P N. This operation repeated for successive points along the characteristic will give the derived curve wanted. Next draw O A at such an angle that the tangent of its slope is equal to twice the resistance of the whole circuit. Draw the ordinate A C, cutting the characteristic at B, and bisect it at D. The work of the generator is represented by the area of the rectangle O C B R: that part which is wasted in heat is represented by the area O C D S; that utilised in the motor S D B R. It will be seen that the efficiency will be less than 50 per cent. in this case.

Another graphic method of comparing the power and efficiency of a motor has been proposed by Mr. Gisbert Kapp. The speeds being taken as abscissæ, the electric horse-power absorbed is plotted out vertically, the number of watts

divided by 746 being taken for the ordinates. Fig. 348 shows
the result in the curve A A, the shape of which will vary with
the type, a series-wound motor in this case. A second curve
B B is then plotted, the ordinates in this case being the

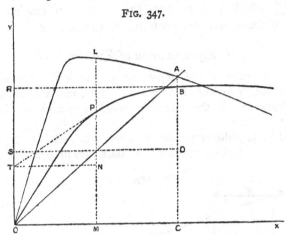

FIG. 347.

mechanical horse-power observed at different speeds. In
both cases the variation of speed is obtained by loading a
brake dynamometer with various loads. With a great load
the speed is small, the applied electromotive-force very great.

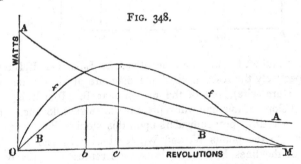

FIG. 348.

With no load. a certain maximum speed O M is obtained
at which (owing to the counter electromotive-force developed)
very little current passes. Between these two extremes there
will be a point *b*, corresponding to a certain speed O *b*, for
which the activity is a maximum. Next divide the values

of the mechanical work B, by those of the electrical energy spent E at the corresponding speed. These quotients will of course be the commercial efficiency at different speeds. These values are plotted out to an arbitrary scale in the curve ff, which shows that the maximum efficiency is attained at the speed corresponding to the point c.

Graphic Representation of Transmission.

A convenient mode of representing graphically the relative amounts of energy expended at the transmitting end and utilised at the receiving end is the following, which is due to von Hefner Alteneck :—

Let (Fig. 349) the perpendicular lines $A E_1$ and $B E_2$ represent respectively the electromotive-forces at the transmitting and receiving

FIG. 349.

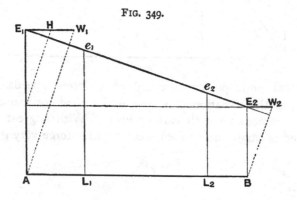

machines; and let the horizontal lengths $A L_1$, $L_1 L_2$, and $L_2 B$ represent respectively the resistances of the machine at A, the line (including return wire), and of the machine at B. Join $E_1 E_2$: the tangent of slope of this line will represent the current flowing. From A and from B drop perpendiculars upon this sloping line, and produce them to the points W_1 and W_2, level with E_1 and E_2. The lengths of the lines $E_1 W_1$ and $E_2 W_2$ will represent relatively the energy transmitted and received. For, by the construction each is proportional to the respective electromotive-force and to the slope of $E_1 E_2$. The energy lost in heat may, on the same scale, be represented by the length of the line $E_1 H$.

For a further geometrical discussion of the problem of electric transmission of power, see a paper by Reignier, in *La Lumière Electrique*, xxiii. 352, 1887.

CHAPTER XXVIII.

REACTION BETWEEN ARMATURE AND FIELD-MAGNETS IN A MOTOR.

IN Chapter V., pp. 85 to 109, the reactions between the armature and field-magnets of a dynamo were considered in detail, but attention was confined solely to that which occurs when the dynamo is used as a generator. In that case the current induced in the armature coils tended to magnetise the armature core in a direction nearly at right angles to the direction in which the field-magnets magnetised it, and in consequence there was a resultant magnetisation at an oblique angle. This obliquity compelled a certain angular lead to be given to the brushes in the sense of the rotation ; and the necessary result of the forward lead of the brushes was to cause the polarity of the armature current to tend partially to demagnetise the field-magnets. Reference to Fig. 60, p. 93, will show that wherever the brushes are placed there is a tendency to form corresponding poles, and these armature poles tend to produce in the iron pole-pieces of the field-magnets an opposite polarity to their own, and therefore to weaken the field. Now in a motor this is not so. A current supplied from an external source magnetises the armature and makes it into a powerful magnet, whose poles would lie, as in the dynamo, nearly at right angles to the line joining the pole-pieces, were it not for the fact that in this case also a lead has to be given to the brushes. Suppose, as in most of the drawings in this book, that the S. pole of the field-magnets is on the left, and the N. pole on the right. Also that the current so traversed the armature that it caused the highest point to be a S. pole and the lowest point a N. pole. Clearly, in this case, the armature will rotate right-handedly, because the S. pole at the top will be repelled

2 N

from the S. pole on the left and attracted toward the N. pole
on the right. A still more important effect will be that these
two polarities will attract each other. The N. pole on the
right tends to induce a S. pole in the part of the armature
nearest to it; and there will be a strong resultant S. pole at
an oblique position on the right of the highest point. Fig. 350
shows the course of the lines of force in the mutual field; and
shows also how the armature's magnetism reacts on the field-
magnets, adding to its lines of force (those which are dotted

FIG. 350.

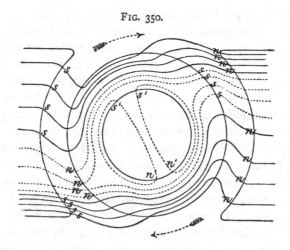

MAGNETIC REACTIONS BETWEEN FIELD-MAGNETS AND ARMATURE
IN MOTOR.

are supposed to be due to the armature), and perturbing its
field. Two consequences are at once apparent. The lead to
be given to the brushes must be a forward lead [1] if the proper
advantage is to be taken of the mutual strengthening of
the two magnetic forces. Also, since the armature polarity
strengthens that of the field-magnet, it is possible for a motor
to be worked without any other means being taken to mag-

[1] That is to say, in the motor, as in the dynamo, the brushes are displaced a
little in the direction of the rotation, if it is desired to get the most powerful rotation, but are displaced a little in the contrary sense if it is desired to work with
the least sparking.

netise the field-magnets, the armature will induce a pole in the field-magnet and then attract itself round toward this induced pole. This principle has been used for many years in small motors, having been apparently first applied by Wheatstone.

But if a forward lead is given to the brushes in order thus to obtain the most powerful torque, then it will be obvious that at the instant that each coil passes the brush it will be actively cutting across the magnetic lines of the mutual field, and there will be much sparking in consequence. If the motor is not to spark, it is essential that there should not be a forward lead, and indeed if the armature's reaction is powerful enough to perturb the magnetism of the field-magnets (as in the figure), then there must be given a backward lead if sparking is to be avoided. For, if the armature reaction causes obliquity in the direction of the resultant field, the neutral point will be displaced backwards correspondingly. From this it also follows that if a forward lead is given to the brushes of a motor in order to get a more powerful rotation, the motor will inevitably spark at the brushes. In the first edition of this work, Fig. 350 was given with a forward lead, or in the position where the mutual attraction of armature and field-magnet was as great as possible : the figure has now been corrected to the position of minimum sparking. In such a position (which is the one that is chosen in practice) the motive power relatively to weight would be small. With a forward lead the motive power relatively to weight would be much larger, but accompanied by destructive sparking. Minimum of sparking may be reconciled with high efficiency, but only by one way. That way is to design and construct motors so that the armature shall not perturb the magnetic field due to the field-magnets. This can be accomplished by following out the very same principles of design and construction which were found to be correct guides in the case of dynamos used as generators. *The field-magnets must be made very powerful in proportion to the armature.* If they are, then there will be no perturbations, no obliquity in the resultant magnetic field, no lead to the brushes, and no sparking. Only in the few cases when minimum weight is of more importance than high efficiency, can this rule be departed from, and then it becomes necessary to take special means, such as providing auxiliary poles, or specially shaped pole-pieces, to prevent the sparking which would follow from having small or weak field-magnets.

In several respects it is even more important that the rules laid down for the good design of generators should be observed for motors. Eddy-currents must be even more carefully eliminated. In a generator the self-induction in the sections of the armature coil and the eddy-currents in the core are antagonistic; in the motor they tend to increase one another. This remark is due to Mr. Mordey. Also the greatest attention must be paid to proper mechanical arrangements for transmitting to the shaft the forces that are thrown by the magnetic field upon the conducting wires around. Substantial driving-projections, well keyed to the shaft, should project between the wires at intervals.

A careful comparison should be made between Figs. 58 and 350, which exhibit the magnetic fields of the generator and the motor respectively. In one the armature is mechanically driven round while the magnetic forces in the field tend to pull it back. In the other, the magnetic forces of the field tend to drag it round, and it is thereby enabled to do mechanical work. In one case there is an opposing mechanical reaction tending to stop the steam-engine. In the other there is set up an opposing electrical reaction (the induced counter electromotive-force) tending to stop the current.[1] In both cases the rotation is supposed to be taking place in the same sense—right-handedly. In both the effect is to displace the lines of force of the field, but in the generator the mechanical rotation acts as if it dragged the magnetism round, whilst in the motor, the reciprocal magnetic reactions act as if they tried to drag

---

[1] The law of the electrical reaction resulting in a generator from the mechanical motion is summed up in the well-known law of Lenz, that *the induced current is always such that by virtue of its electro-magnetic effect it tends to stop the motion that generated it.* In the converse case of the mechanical reaction resulting, in a motor, from the flow of electric energy, it is easy to formulate a converse law, viz. that *the motion produced is always such that by virtue of the magneto-electric inductions which it sets up it tends to stop the current.*

round the magnetism of the armature and succeeded in producing mechanical rotation. In both drawings there is a lead
given to the brushes. In the dynamo as generator we found
that the effect of self-induction in the armature was to
increase the lead. In the motor, on the contrary, the effect
of self-induction is to decrease the lead. If a motor is set
with no lead, and if the armature be very powerful relatively
to the field-magnets, it will run in either direction according
as it may be started. If the current be reversed in the
armature part of the circuit only, the motor will usually
reverse its rotation, but. will also require the lead to be
reversed to run as strongly as before. If instead of reversing
the current in the armature the magnetism of the field-magnet
be reversed, a similar result will follow. If both are reversed
at the same time, the motor will go on rotating as if nothing
had happened.

Dynamos wound and connected for working as generators
of continuous currents may be used in all cases as motors,
but with some difference. A series dynamo set to generate
currents when run right-handedly (and therefore having a
forward right-handed lead), will, when supplied with a current
from an external source, run as a motor, but runs left-handedly
against its brushes. To set it right for motor purposes requires
either that the connexions of the armature should be reversed,
or that those of the field-magnet should be reversed (in either
of which cases it will run right-handedly), *or* else the brushes
must be reversed and given a lead in the other direction (in
which case it will run left-handedly). A shunt dynamo set
ready to work as a generator will, when supplied with current,
run as a motor in the same direction as it ran as a generator ;
for if the current in the armature part is in the same direction
as before, that in the shunt is reversed, and *vice versâ.* A
compound-wound dynamo, set right to run as a generator,
will run as a motor in the reverse sense, against its brushes, if
the series part be more powerful than the shunt, and with
its brushes if the shunt part be the more powerful. If the
connexions are such (as in compound dynamos) that the field-
magnet receives the sum of the effects of the shunt and series

windings when used as a generator, then it will receive the difference between them when used as a motor. There are certain advantages in using a differentially-wound motor, as will appear hereafter.

Some remarks on the action of alternate-current machines as motors are to be found on pages 480 and 564.

REVERSING GEAR FOR MOTORS.

A motor, as will be seen from the preceding discussion, can be reversed by the operation of reversing the current through the armature, and at the same moment reversing the lead. But reversing the current can also be accomplished by rotating the brushes through 180°. Consequently, both these actions may be accomplished by the single operation of advancing the brushes through $180° - 2\phi$, where ϕ is the original angle of lead. But as the brush would then slant in the wrong direction, it is better to provide a second set of brushes. This is, indeed, Hopkinson's method of reversing. He employs two pairs of brushes, each pair being capable of moving about a common pivot, so that either the pair having a lead in one direction, or the pair having a lead in the other direction can be let down upon the collector. A reversing gear designed by Mr. A. Reckenzaun for the motors of the launch *Electricity* is shown in Fig. 351. In it there are two pairs of brushes; the two upper are fixed to a common brush-holder, which turns on a pivot, and can be tilted by pressing a lever handle to right or to left. The two lower brushes are also fixed to a holder. Against each brush-holder presses a little ebonite roller, at the end of a bent steel spring, fixed at its middle to the handle. The result of this arrangement is that, by moving the lever, the brushes can be made to give a lead in either direction, and so starting the motor rotating in either direction. Such a reversing gear is obviously a most useful adjunct for industrial applications of motors, and if the difficulties of sparking at the brushes caused by the sudden removals of them from the collector be obviated, must prove much better than any mechanical device to reverse the motion by transferring it from the axle of the motor through a train of gearing to some other axle. One great advantage of electric motors is, that they can be easily fixed directly on the spindle of the machine which they are to drive—an advantage not lightly to be thrown away.

FIG. 351.

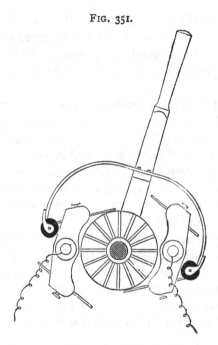

REVERSING GEAR FOR ELECTRIC MOTOR.

Another form of reversing gear has been designed by Professors Ayrton and Perry. It consists of a double collar upon the spindle of the motor; in one, the inner collar, having a pin fitting into a spiral groove in the spindle, and being free to move relatively to the spindle. Any displacement along the spindle given to the inner collar through the outer one causes the pin in the former to move along the groove, and the collar rotates through a certain angle. This collar in Ayrton and Perry's motor carries the brush-holders, and therefore by rotating alters the lead. The motor shown in Fig. 355

is fitted with this gear, though useless for driving a fan. Other forms of reversing gear for small motors have been designed by the author, who cuts the segments of the collector or commutator spirally, and therefore obtains a change of lead by simply sliding the brushes forward or backward parallel to the axle of the motor.

In Reckenzaun's motor (Fig. 356) the reversing gear consists of two pairs of brushes which are mounted so as to slide on guides or ways, reversal being accomplished by shifting a lever which slides forward one pair whilst it draws the other pair back.

It is also theoretically possible to construct a motor which shall reverse by simply reversing the current in the armature part ; for this end the pole-pieces must be so shaped that when no angular lead is given to the brushes, the angle between the diameter of commutation and the effective pole in the pole-piece shall be that required for steady running. If this can be found then merely reversing the polarity of either part will reverse the motor.

If the field-magnets of a motor are so powerful relatively to the armature that no lead has to be given to the brushes, the rotation can be reversed by reversing the polarity of either part. In Immisch's larger motors, the reversing-gear, which is very substantial, removes one pair of brushes and puts down at the same diametral points a second pair, reversed in position and polarity.

CHAPTER XXIX.

SPECIAL FORMS OF MOTOR.

SOME of the earlier typical forms of motor have been described at the beginning of this section. Many of the dynamos described in Chapters VII. to XIII. preceding, are used as motors, and all can be so used : indeed all good modern dynamos, though designed primarily as generators, make far more efficient motors than any of the earlier electromagnetic engines of Jacobi, Froment, and Page. Gramme in 1875 devised a special form of dynamo, suitable for the electric transmission of power, drawings of which appeared in former editions of this work. It had four pole-pieces surrounding the ring, and four brushes. For special purposes, however, small motors of various types have been introduced in recent years.

Deprez's Motor.—In 1879 Marcel Deprez introduced a very convenient form of small motor, consisting of a simple shuttle-wound Siemens armature placed longitudinally between the parallel limbs of a steel magnet. It had a two-part commutator, and consequently had the defect of possessing a dead-point. He obviated this defect by employing two armatures, one 90° in advance of the other, so that while one was at the dead-point the other should be in full action.

Griscom's Motor.—A very convenient small motor adapted for sewing-machines, has been devised by Griscom (Fig. 352), and is well known both in the United States and in Europe. Like the motor of Deprez, this machine has a simple shuttle armature ; but its field-magnets are of malleable iron surrounded with coils united in series with the armature. According to Professors Ayrton and Perry a useful power of about ·015 horse, with an efficiency of about 13 per cent., is the greatest that the Griscom motor of normal size can maintain.

Beside the little motors of Griscom, Howe, Deprez, Cuttriss, and others, adapted to work sewing-machines, and instruments requiring very small power, there are in the market larger motors, for driving lathes and heavier machinery, though not yet so well known to the public. There is, indeed, such an immense field for useful industrial application of

FIG. 352.

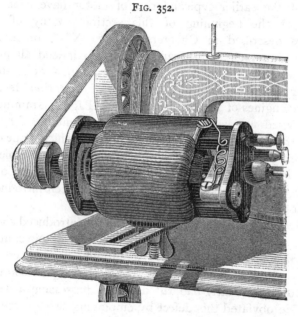

GRISCOM'S ELECTRIC MOTOR.

electric motors, now that electric currents are regularly supplied from central stations, that many inventors have turned their attention to this branch.

De Meritens' Motor.—Another pattern of motor (Fig. 353) has been invented by De Meritens, who employs a ring armature very like that of Gramme, but places it between very compact and light field-magnets, which form a framework to the machine.

A long series of experimental tests of a De Meritens' motor has been published by Professors Ayrton and Perry,[1] from

FIG. 353.

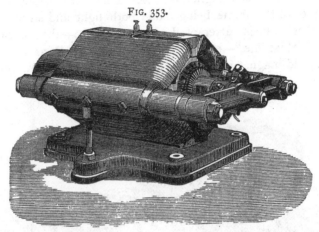

DE MERITENS' ELECTRIC MOTOR.

which it appears that one weighing 72 lbs. and giving out $\frac{3}{4}$ of a horse-power had an efficiency of 50 per cent.

FIG. 354.

ARMATURE AND FIELD-MAGNET OF AYRTON AND PERRY'S MOTOR.

Ayrton and Perry's Motor.—Professors Ayrton and Perry in 1882 devised an ingenious motor which is extremely com-

[1] *Journal of the Society of Telegraph-Engineers and Electricians*, vol. xii., No. 49, 1883.

pact and of considerable power in proportion to its weight. In this motor (Figs. 354 and 355) the armature is fixed, and within it the field-magnet rotates. This construction, which permits of the frame being made both light and strong, had previously been attempted in a dynamo—the so-called "Topf-Maschine"—exhibited by Siemens at Paris in 1881. The field-magnet of Ayrton and Perry's motor shown sepa-

Fig. 355.

ELECTRIC MOTOR DRIVING VENTILATING FAN.

rately in Fig. 354, is of the simple shuttle-wound type. The armature is an enlarged ring of the Pacinotti kind; having protruding teeth, between which the coils are wound, and is built up of flat toothed rings of sheet charcoal iron. The brushes rotate with the field-magnet, and the commutator or collector is fixed. The reversing gear described on p. 551 is adapted for use with this machine. Fig. 355 shows one of these motors driving a fan. These motors are usually series-wound; but we owe to these indefatigable workers the theory[1] of adapting the compound-winding for the purpose of making motors self-governing. They have introduced the differential system of winding into a few of their machines. Their motor weighing 37 lbs. and giving out $\frac{1}{5}$ of a horse-power at 1570 revolutions per minute showed a nett efficiency of 40 per cent. The armature is made large and powerful relatively to the field-magnets, in order to utilise their mutual reactions to obtain the most powerful rotation relatively to the weight. The following data are taken from results of tests made on various forms of Ayrton and Perry's motors :—

Weight of Motor in lbs.	Horse-power actually given out at the Rotating Shaft.
37	0·35
55	0·50
75	0·75
96	1·50
125	2·20

Reckenzaun's Motor.—Mr. A. Reckenzaun has devised a motor which is both strongly-built and powerful, and at the same time extremely light (Fig. 356). It is well fitted for use in boats and vehicles. The field-magnets are constituted by the iron framework. The armature is made of a number of small iron links, from 300 to 3000 according to the size of the machine, wound with wire and united by bolts into a strong, light, polygonal frame, thus securing good ventilation. They are usually fitted with two pairs of brushes and a reversing gear. One of these motors, weighing 124 lbs., series-wound, having a resistance of 0·564 ohm between its terminals, and

---

[1] The theory of Professors Ayrton and Perry includes much more than this, and extends to all the cases mentioned below of shunt or series windings combined with an independent constant magnetism, or with an independent magneto machine on the same axle. Their memoir in the *Journal of the Society of Telegraph Engineers* (May 1883) is a mine of suggestive information. The specification of their British patent includes nearly all the possible combinations for self-regulation, as well as many other methods for governing motors by periodic and centrifugal governors like those of steam-engines.

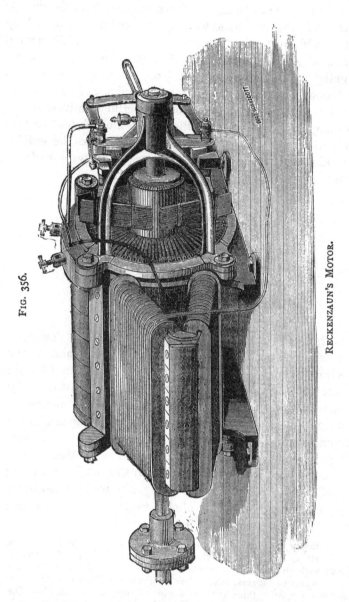

FIG. 356.

RECKENZAUN'S MOTOR.

designed to be run normally with an electromotive-force of
70 volts, gave the following tests :—

Revolutions per Minute.	Current in Ampères.	E.M.F. in Volts.	Horse-power on Dynamometer.
1158	24·8	43·3	0·86
1436	25·6	53·3	1·07
2184	26·2	71·6	1·54
1724	39·4	70	2·13
1554	47·6	76	2·4
1931	50	95	2·87

This is at the normal rate of 558 foot-pounds of work per minute for every pound of dead weight in the motor itself. A similar motor, but wound with finer wire to run at 120 volts, gave 550 foot-pounds per minute per pound of dead weight.

This motor may be compared with a Siemens dynamo weighing 519 lbs., which at 906 revolutions per minute gave, according to Professors Ayrton and Perry, 4·96 horse-power, with an efficiency of 74·6 per cent. Another Siemens "D₂" machine, employed as a motor in the electric launch shown at the Vienna Exhibition in 1883, gave 7 horse-power on the

FIG. 357.

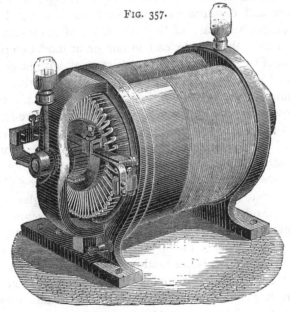

GRAMME'S MOTOR.

shaft while absorbing 9 electric horse-power ; an efficiency of
about 78 per cent.   It weighed 658 lbs. ; there being 1 lb. of
dead weight for each 351 foot-pounds per minute exerted at
the shaft.

Gramme's Motor.—A compact form of motor, due to M.
Gramme, is shown in Fig. 357.   The ring armature revolves
between the polar ends of an electromagnet which is constituted by a back plate of cast iron, and two parallel limbs
having a D-shaped section.   The end of the shaft and the
poles of the magnet are received by a front .plate of gun-
metal.   One of these machines, wound with wire of 1 millimetre diameter, running at 83·5 volts with 8·1 ampères gave
39 kilogrammetres per second as its mechanical output.
Another motor, built more lightly, and weighing 10·4 kilogrammes, gave 40 kilogrammetres per second output, or
nearly 800 foot-pounds per minute per pound weight of the
machine.

S. P. Thompson's Motor.—For light work the author has
designed a small motor which has some advantages over those
of Deprez and of Griscom.   The field-magnets, which also
constitute the bed-plate of the motor, are of malleable cast
iron, of a form that can be cast in one or at most two pieces.
The form of them is that of a Joule's magnet, with large pole-
pieces, and wound with coils, arranged partly in series, partly
as a shunt, in certain proportions, so as to give a constant
velocity when worked with an external electromotive-force of
a certain number of volts.   For the armature a simple form is
employed which without any unnecessary complication obviates dead-points.   This result is attained by modifying the
old Siemens armature by embedding, as it were, one of these
shuttle-shaped coils within another, at right angles to one
another.   And having duplicated the coils, the segments of
the commutator must also be duplicated, so that it becomes
a four-part collector.   There are no solid iron parts in the
armature, but the cores are made of thin pieces of sheet iron,
stamped out and strung together.   For larger motors an
armature having a larger number of sections and a greater
moment of inertia is preferable.   For small motors a four-part

armature suffices. Even if the impulses be intermittent, the mechanical inertia of the moving parts will steady the motion. In the case of generators, we found that to produce steady currents we had to multiply coils on the armature in many separate paths, grouped round a ring or a drum, involving a complicated winding, and a collecting apparatus consisting of many segments. In motors no such necessity exists, provided only we arrange the coils so that there shall be no dead-points. For large motors it may be advisable to multiply the paths and segments for other reasons—as, for example, to obviate sparking at the collectors—but the inertia of the moving parts spares us in small machines the complication of parts which is expedient in the generator for securing steady running.

Fig. 358.

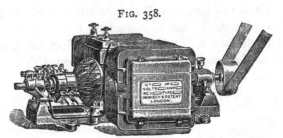

IMMISCH'S MOTOR.

Immisch's Motor.—Mr. Moritz Immisch has constructed motors in many sizes ; they are remarkable for their excellent mechanical construction, high efficiency, and great power in proportion to weight. Only those of smallest size have ring armatures, all others have drum armatures, and the field-magnets are usually of the double-horseshoe type, as shown in Fig. 358. A section of the machine, ½ of the full size, is shown in Fig. 359. The armature is 15 centimetres in diameter, and at 1400 revolutions per minute gives eight horse-power on the shaft, with an efficiency of 85 per cent. It weighs only 350 lb.

The armature core consists of thin iron disks insulated with thin asbestos, having at the ends, and at intervals, thicker disks provided with projected driving-teeth, all the disks being securely keyed to the shaft. The windings are carefully insulated with Willesden paper protected with indiarubber varnish.

2 O

The arrangement of the armature presents certain peculiarities, amongst which are the commutator connexions. The bars of the commutator or collector are grouped alternately in two sets, side by side, so that a single broad brush, or two brushes

FIG. 359.

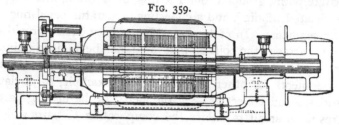

IMMISCH'S MOTOR (Section).

side by side in one holder, will always cut out of circuit the coil of the armature that is connected from the bar of the front set to the bar of the back set; the effect being exactly the same as if an ordinary collector were used with two pairs of connected brushes, those of each pair being set apart to a distance equal to the breadth of a single bar. The effect of cutting out the coils as they reach the neutral point, is, according to the inventor, to diminish cross-magnetising influences and obviate changes of lead; the brushes being set, once for all, in the required position. Some brake tests made in the presence of the author on a small ½-H.P. motor, weighing about 20 kilogrammes (of which only about 3·5 kilogrammes were of copper), gave following results:—

FIG. 360.

IMMISCH'S ½ HORSE-POWER MOTOR.

Revolutions per Minute.	Volts.	Ampères.	E.H.P. absorbed.	H.P. given out.	Efficiency.
1420	40	10·5	·565	·354	·625
1840	49	10·5	·69	·46	·666
1920	50	11	·74	·48	·645
2280	61	12	·97	·57	·587

This motor is shown in Fig. 360.

A large 30-H.P. motor, weighing 850 kilogrammes, gave following results :—

Revolutions per Minute.	Volts.	Ampères.	E.H.P. absorbed.	H.P. given out.	Efficiency.
660	500	49	33	29·8	·90
680	500	48	32·2	29·5	·91
675	500	49	33	29·5	·89

Mr. Immisch[1] has designed forms of reversing gear and of brush-holder that are noticeable for excellent mechanical qualities.

Sprague's Motor.—Lieut. F. J. Sprague, of New York, has produced several forms of motor of excellent design and con-

FIG. 361.

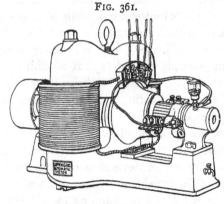

SPRAGUE'S MOTOR.

struction, many hundreds of them being in use in the States. One form of these machines is shown in Fig. 361. For further details the reader is referred to the accounts published in the technical press.[2] Sprague's method of winding the field-magnets with a differential compound winding is identical with that invented in 1883 by Ayrton and Perry, depending upon

[1] For further details about these motors, see *Electrical Review*, Nov. 12, 1886; April 1 and Nov. 4, 1887; *Electrician*, Oct. 22 and 29, 1886; March 11 and June 17, 1887; *Industries*, March 11, 1887; *La Lumière Électrique*, xxiv. 261, and xxv. 458, 1887.

[2] *Electrical World*, October, 1886; also Martin and Wetzler's treatise on *The Electric Motor*, 157–75.

the use of a coil in series with the armature to demagnetise
and weaken the field.   Many other ingenious methods of
governing, and practical applications, have been worked out
by Sprague.

Other Motors.—Almost every kind of continuous current
dynamo can be used, with proper adjustment of the brushes,
as a motor, though as a general rule motors require to be
more solidly built than generators.   Makers have their own
special ways of adapting their machines to the varied purposes
for which motive power is required.   Many forms have lately
been introduced and patented, particularly in the United
States.   Few of these possess originality, save in details and
methods of governing.   Daft's motor, which is much used for
locomotive power, is a modified Gramme machine.   Edgerton's
motor and Pendleton's motor are revivals of a discarded pattern of Immisch's motor.   A small motor designed by Messrs.
Curtis, Crocker and Wheeler,[1] for use in arc-light circuits, has
for its armature winding a single spiral of ribbon wire wound
by special machinery, so that when slipped over the core it lies
edgeways and fills almost the entire gap-space between the
core and the pole-piece.

Dynamos of the open-coil class can also be used as motors.
Both the Brush and the Thomson-Houston dynamo have
been so used.   Brush has devised a centrifugal governor to
adjust the lead of the brushes for his machines.

Alternate-current Motors.—Running motors with an alternate-current supply presents certain practical difficulties which
have not yet met with any very satisfactory solution.   The motors
described above, having solid iron in their field-magnets,
are not at all suitable for alternate-current work.   Even the
device of laminating the field-magnet core does not obviate
unequal retardation in the armature and field-magnet parts.
It has been independently proposed by Whitney, and by
Rechniewski, to construct laminated motors and dynamos, by
stamping out from thin sheet-iron the cores both of field-
magnet and armature.   Alternate-current dynamos would

---

[1] *Electrical World*, February, 1887 ; and *Engineering*, xliv. 83, 1887.

make excellent alternate-current motors, were it not for the difficulty of starting them. When once they have got up speed so that their own alternations correspond precisely to the alternations of the supply-current, they are beautifully self-regulating (see Chap. XXIV.). But the difficulty is to get them to run up to this point. Professor Elihu Thomson,[1] who has made a special study of the phenomena of alternating currents, has devised several special forms of motor for alternate-current work. The latest suggestion for alternate-current motors comes from Professor Ferraris,[2] who proposes to introduce into two sets of coils wound at right angles, which constitute a symmetrical external field-magnet, two alternate currents differing in phase by one quarter wave, thus giving a polarity that is successively displaced in a rotatory manner.

The following references deal with the most recent developments of the subject of electric motors :—

Wietlisbach, on Motors : *Elektrotecnnische Rundschau,* 1887, p. 57.

Jullien's Motor : *La Lumière Électrique,* xxiv. 495, 1887.

Wheeler, on Small Motors : *Electrical World,* ix. 4, 93, and 203; and *Electrical Review,* xx. 324.

Baxter's Motor : *Electrical World,* ix. 210 and 279.

Diehl's Motor : *Electrical World,* ix. 86.

Hochhausen's Motor : *Electrical World,* ix. 193.

Cleveland Motor : *Electrical World,* ix. 67.

Thomson-Houston Motors: *Electrical World,* x. 51.

Siemens' Small Motor : *Elektrotechnische Zeitschrift,* viii. 436, 1887.

For the various applications of electric motors to tram-cars, electric railways, electric launches and other boats, balloons, and other kindred purposes, the reader is referred to the technical journals, and to the following works:— Du Moncel and Geraldy's *Electricity as a Motive Power;* Raineri's *La Navigazione Elettrica ;* Hospitalier's *Modern Applications of Electricity ;* but especially to Martin and Wetzler's *The Electric Motor.* Kapp's *Electric Transmission of Energy* is particularly recommended for useful study.

---

[1] *Electrical World,* ix. 274, 1887.

[2] *Industries,* May 18th, 1888. A somewhat similar device by Mr. N. Tesla of New York is announced (see *Electrical World,* x. 281, June 2nd, 1888) while this matter is going through the press.

CHAPTER XXX.

THEORY OF ELECTRIC MOTORS (CONTINUOUS CURRENT).

IN treating of the dynamo as a generator, it was assumed that the mechanical power could be supplied under one of the two standard conditions, on the one hand of *constant speed* (and torque varying with the electrical output), or else on the other (p. 130) of *constant torque* (and speed varying with the output). One of these two conditions being prescribed, algebraic expressions had then to be found for the two corresponding factors of the electric output, namely the *electromotive-force* and the *current*, under varying conditions of resistance in the circuit. Also we investigated these conditions which would result in making one or other factor of the electric output constant. It was found convenient to study the relation between the two factors of output by the aid of the curves known as *characteristics*.

Similarly, in treating the dynamo as a motor it will be assumed that such arrangements of electric supply can be made, that the electric power can be furnished under one of the two standard conditions, on the one hand of *constant potential* (and current varying with the mechanical output of the motor), or on the other of *constant current* (and potential varying with the mechanical output). One of these two conditions being prescribed, we shall then have to find algebraic expressions for the two corresponding factors of the mechanical output, namely the *speed* and the *torque*, under varying conditions of load on the shaft. Also we shall investigate what are the conditions which will result in making one or other factor of the mechanical output constant : in other words, we shall ascertain what are the conditions of self-regulation to make the motor run at constant speed or with constant torque.

Lastly, it will be found convenient to study the relation between speed and torque by the aid of curves, which, by analogy, we may call *mechanical characteristics*.

GENERAL EXPRESSIONS FOR TORQUE AND SPEED.

The work imparted per second to the shaft of the motor may be expressed either in electrical or mechanical measure. In the former case it is the product of the motor's electromotive-force (i. e. the counter electromotive-force opposing the electromotive-force of supply) into the current flowing the armature; in the latter case it is the product of angular speed into torque. So we may write

$$w = E\,i_a = \omega\,T = 2\,\pi\,n\,T\,;$$

and (average) $E = n\,C\,N$ exactly as in a dynamo that is being used as a generator (see p. 337). Hence

$$2\,\pi\,n\,T = n\,C\,N\,i_a,$$
$$2\,\pi\,T = C\,N\,i_a\,;$$

and finally the average value of the torque will be

$$T = i_a \frac{C\,N}{2\,\pi} \quad . \quad . \quad . \quad . \quad . [a].$$

From this it appears that if N is constant, the torque is simply proportional to the current in the armature.

To develop this expression further, we must remember that i can be calculated in terms of the electromotive-force of supply $\mathcal{E}$, as measured at the terminals of the machine, and the internal resistance of the circuit through the armature part, which we may call r; and then

$$i_a = \frac{\mathcal{E} - E}{r}\,;$$

whence it follows that

$$T = \frac{C\,N}{2\,\pi} \cdot \frac{\mathcal{E} - n\,C\,N}{r} \quad . \quad . \quad . [\beta].$$

From this it follows that when the speed becomes so great that $n\,C\,N = \mathcal{E}$, there will be no torque. In fact, when there is no resisting force on the shaft the motor runs empty at its highest speed, namely, such as will make the counter electro-motive-force as nearly as possible equal to the electromotive-force of supply. The maximum value of T, supposing N constant, is obviously when $n = 0$.

An expression for the speed can be obtained from the preceding :

$$2 \pi \, \text{T} \, r = \text{C N} \, \mathcal{E} - n \, \text{C}^2 \, \text{N}^2 \, ;$$

$$n = \frac{\mathcal{E}}{\text{C N}} - \frac{2 \pi \text{T} \, r}{\text{C}^2 \, \text{N}^2} \quad \cdot \quad \cdot \quad \cdot \quad [\, \gamma \,].$$

In equation [a] T will be expressed in dyne-centimetres if i_a is in absolute units of current (see p. 18) ; if i_a is given in ampères, then the value must be divided by 10 if T is to be obtained in dyne-centimetres, or by 9810 if it is to be obtained in gramme-centimetres, or by $13\cdot56 \times 10^7$ if the torque is to be expressed in pound-feet (i. e. so many pounds' weight acting at a radius of one foot).

In equation [γ], in order that n may be expressed in revolutions per second, the value of $\mathcal{E}$, if given in volts, must be multiplied by 10^8 ; that of r, if in ohms, by 10^9 ; whilst T must be reduced to dyne-centimetres. If T is given in pound-feet, its value must be multiplied by $1\cdot356 \times 10^7$.

[The three equations [a], [β], and [γ] are true, not only for motors, but for generators, the $\mathcal{E}$ of the formulæ being in the latter case replaced by e. This will give negative values for T, the significance of the sign being that the torque due to the action of the magnetic field on the conductors carrying the armature current is such as to oppose the driving.]

It will be noticed that if only r is very small, and N relatively very large, the second term may be neglected, and the speed will then depend on the first term only. It will be the smaller as N is greater : this being the simple converse of the corresponding fact that the more powerful the magnetic field the less need be the speed of the dynamo to give the desired output. We may also notice that if N is constant, the speed is proportional to $\mathcal{E}$: it will be constant if the condition of supply is that of constant potential, but will be variable if $\mathcal{E}$ varies. Another important point is that, if a motor is to do its work at a slow speed, C should be great as well as N.

We must next inquire how n and T are affected by the fact that the value of N depends upon the construction and winding of the field-magnet of the motor, and by the conditions of supply. We shall consider the following kinds of machine :—

A. *Magneto Motor* and *Separately-excited Motor.*
B. *Series-wound Motor.*
C. *Shunt-wound Motor.*
D. *Compound-wound Motor.*

In each instance we shall have to take into account the conditions of supply according as $\mathcal{E}$ or i is constant.

MAGNETO MOTOR AND SEPARATELY-EXCITED MOTOR.

It is here assumed that N is constant, in other words, that the perturbing reactions of the armature may be neglected. This will only be true when the lead at brushes is nearly zero, and the field-magnet powerful. Under these circumstances the general formulæ already found require small modification. The only internal resistance is that of the armature r_a.

Case (i.): $\mathcal{E}$ *constant.*

In this case formula [γ] gives the desired relation, from which the *mechanical characteristic* may be plotted out, as in Fig. 362. It is a straight line cutting the axis of n at a point representing to scale that speed at which n C N $= \mathcal{E}$; and it slopes downwards at an angle such that the tangent of the slope is equal to $2 \pi r_a \div C^2 N^2$, or is proportional to the internal resistance. In the case of the separately excited motor, increase in the exciting current, strengthening the field,

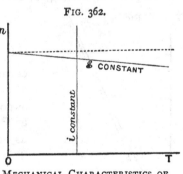

FIG. 362.

MECHANICAL CHARACTERISTICS OF
MAGNETO MOTOR.

will obviously make the sloping line more nearly horizontal, as well as lowering the speed as a whole.

Case (ii.) : *i constant.*

In this case, as reference to formula [*a*] shows, the torque is constant, being independent of speed and of internal resistance. The mechanical characteristic of the machine under these conditions is a vertical straight line.

If we attempt to take into account the reactions of the armature, we must remember that the effect of the armature current is to demagnetise, if there is a backward lead, and to magnetise, if there is a forward lead. A forward lead, then, would tend to make the sloping line, at constant $\mathcal{E}$, rise as the torque increased, whilst a backward lead would tend to make it slope still more.

SERIES MOTOR.

The fundamental equations are as before, with the addition of the following :—

$$r = r_a + r_m,$$

and, as on p. 402, where i' is the diacritical current,

$$N = \overline{N}\, \frac{i}{i + i'}.$$

Putting this value of N into the expression [*a*], on p. 567, for the torque, and writing for brevity $\dfrac{C\overline{N}}{2\pi} = Y$, we have

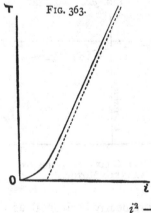

T   FIG. 363.

$$T = Y\, \frac{i^2}{i + i'}.$$

This relation between torque and current is given graphically in Fig. 363. For values of i that are small as compared with i', T varies nearly as i^2; whilst for large values of i, as magnetic saturation advances, T is nearly proportional to i. The equation may also be written in the quadratic form

$$i^2 - \frac{T}{Y}\, i - \frac{T}{Y}\, i' = 0,$$

the solution of which is

$$i = \frac{T}{2Y}\left\{1 \pm \sqrt{1 + \frac{4Y}{T}i'}\right\}.$$

Now from the expressions given above we have

$$n = \frac{\mathcal{E}i}{2\pi T} - \frac{ri^2}{2\pi T};$$

$$n = \frac{\mathcal{E}i}{2\pi T} - \frac{r(i+i')}{2\pi Y};$$

and inserting the value of i as above

$$n = \frac{\mathcal{E}}{4\pi Y}\left\{1 + \sqrt{1 + \frac{4Y}{T}i'}\right\} - \frac{rT}{4\pi Y^2}\left\{1 + \sqrt{1 + \frac{4Y}{T}i'}\right\} - \frac{ri'}{2\pi Y}.$$

To simplify this hyperbolic expression, it is permissible for large values of T to neglect the second term under the root sign, giving approximately,

$$n = \frac{\mathcal{E} - ri'}{2\pi Y} - \frac{rT}{4\pi Y^2}.$$

Case (i.) : $\mathcal{E}$ *constant.*

If $\mathcal{E}$ is constant, then, as the last equation shows, for large values of T the values of n are equal to a certain constant less a quantity proportional to T; or the mechanical characteristic at this point (when the magnets are well saturated) is, for all large values of T, approximately a straight line, as shown in Fig. 364.

Case (ii.) : i *constant.*

Here, clearly, saving for secondary effects the magnetisation will be constant, hence the torque will also be constant, as in Fig. 364. With a load exceeding a certain amount, the motor will not start; with a lesser load it will race until friction and eddy-currents make up the difference.

Use of two Series Motors in Transmission.

It is known that if two similarly constructed series-wound machines are used—one as generator, the other as motor—the arrangement is almost perfectly self-regulating, the speed o the motor at the receiving end being almost constant if that of the dynamo at the transmitting end is constant. Every addition to the load put upon the motor, tending to check the speed, causes an increase of current to flow, and so throws a proportionate additional work upon the generator, which in turn takes more power from the steam engine to keep up its speed. As we have shown above, the torque of the motor T_2 will depend, in the given machine, on the current alone, and on the current will depend the torque at the dynamo T_1. Mr. Kapp has further shown [1] how, if there is a resistance in the line, the arrangement may still be made self-regulating by

Fig. 364.

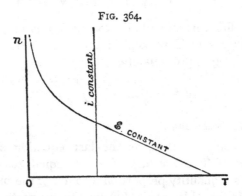

MECHANICAL CHARACTERISTICS OF SERIES MOTOR.

choosing as generator and motor two machines so wound that, comparing their characteristics for the prescribed speeds, the difference in their electromotive-forces corresponding to a given value of current shall be equal to the electromotive-force requisite to drive that particular current through the resistance in the whole circuit.

The late Sir C. W. Siemens [2] drew attention in 1880 to the singular properties of the combination of a generating dynamo and a magneto motor, instancing a locomotive motor which, when descending an incline, quickens its speed and actually

---

[1] See Kapp's *Electric Transmission of Energy*, p. 176.
[2] *Journal Soc. Telegr. Engineers,* ix. 301, 1880.

becomes a generator of currents, paying back the spare power into store. He also remarked how two trains driven by motors, running on the same pair of electric rails, tend to regulate one another, the one on a descending portion of the road transmitting power to the other as though "connected by means of an invisible rope."

SHUNT MOTOR.

The fundamental conditions are as follows :—

$$T = i_a \frac{C\,N}{2\,\pi};$$

$$i_a = i - i_s;$$

$$N = \bar{N}\,\frac{\mathscr{E}}{\mathscr{E} + \mathscr{E}'};$$

$$E = \mathscr{E}\left(1 + \frac{r_a}{r_s}\right) - r_a i.$$

From the first three of these we get

$$T = \frac{C}{2\,\pi}\left(i - \frac{\mathscr{E}}{r_s}\right)\bar{N}\,\frac{\mathscr{E}}{\mathscr{E} + \mathscr{E}'};$$

and, transposing and writing Y for $C\,\bar{N} \div 2\,\pi$,

$$i = \frac{T}{Y}\cdot\frac{\mathscr{E} + \mathscr{E}'}{\mathscr{E}} + \frac{\mathscr{E}}{r_s};$$

and from the last of the four

$$n = \frac{1}{C\,N}\left\{\mathscr{E}\left(1 + \frac{r_a}{r_s}\right) - r_a i\right\}.$$

Inserting the value of i, we have

$$n = \frac{\mathscr{E} + \mathscr{E}'}{2\,\pi\,Y}\left\{1 + 2\,\frac{r_a}{r_s} - \frac{r_a\,T}{Y}\cdot\frac{\mathscr{E} + \mathscr{E}'}{\mathscr{E}^2}\right\}. \qquad [\delta].$$

Case (i.) : *$\mathscr{E}$ constant.*

The last equation shows that a shunt motor supplied at constant potential, will have a speed that would be constant and independent of the torque if it were not for internal resistance ; and further, that the consequent falling off as the torque increases, will be the less as the field-magnetism is the more powerful.

As an example, a Victoria shunt motor tested by Mr. Mordey in which the load was varied from $91\cdot8 \times 10^7$ to $1357\cdot2 \times 10^7$ dyne-centimetres, only decreased its speed from $16\cdot25$ to $15\cdot75$ revolutions per second.

It is instructive to contrast the self-regulating power of a shunt dynamo with the self-governing power of a shunt motor. The former, when driven at a constant speed, generates electric power at a nearly constant potential; the latter, when supplied

FIG. 365.

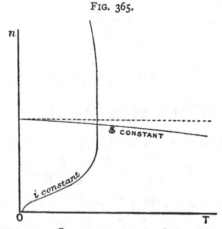

MECHANICAL CHARACTERISTICS OF SHUNT MOTOR.

from the mains at a constant potential, would furnish mechanical power at a nearly constant speed; and in both cases the departure from absolute constancy is proportional to the internal resistance of the armature coils, and to the output, electrical or mechanical, of the machine for the time being.

Case (ii.): *i constant.*

The determination of this case is more complicated, though the general considerations are simple enough. If the motor is standing still when the current is turned on, nearly all the current will go through the armature, next to none through the shunt; hence there will be little magnetism, and therefore almost no torque; such a machine will not start itself with any load on. But if it be once started, its counter electromotive-force will cause the current in the armature to decrease, whilst that round the shunt increases. The torque will therefore then increase with the speed, but not indefinitely, for as the magnetism advances in its degree of saturation, the increase of N will no longer compensate for the decrease of i_a, and from that point onwards the torque will decrease if the speed is allowed to increase. And, hypothetically, the speed should increase until the motor's own electromotive-force exactly equals the difference of potentials due to the whole of the constant current flowing through the resistance of the shunt, under which circumstances there will be no current through the armature and zero torque. Fig. 365, which, like the preceding, is taken from Dr. Frölich's work, gives the mechanical characteristics for the two cases.

The consideration of compound-wound motors is discussed in the next Chapter.

As an example of the calculations of torque on p. 567, we may take the case of the Edison-Hopkinson machine described on p. 203. At full load, with 320 ampères, the value of N was about 1,085,000, and C was 40; hence T was about 2210×10^6 dyne-centimetres, or about 163 pound-feet.

It may be convenient also to remember that the mechanical drag on any single conductor of the armature may be calculated by the rule that the force per unit of length is proportional to the product of the strength of the current into the intensity of the magnetic field. If i is the current (in absolute units), H the intensity of the field (lines per square centimetre), and l length of conductor in the field (in centimetres), then the force in dynes will be

$$f = i \, H \, l.$$

In the Edison-Hopkinson machine the area of polar surface was taken as 1600 sq. cm., making average H = 6800. At full load, i (in any one conductor) = 16; $l = 48 \cdot 3$ cm. This makes the average force on each conductor while in the field $5 \cdot 6$ million dynes, or roughly, about 12 pounds' weight. Of course this force is removed and again put on twice in each revolution as commutation occurs.

CHAPTER XXXI.

Government of Motors.

It is extremely important that electric motors should be so arranged as to run at a uniform speed, no matter what their load may be. For example, in driving lathes, and many other kinds of machinery, it is essential that the speed should be regular, and that the motor should not "race" as soon as the stress of the cutting tool is removed.

One of the earliest attempts to secure an automatic regulation of the speed, was that of M. Marcel Deprez, who in 1878 applied an ingenious method of interrupting the current at a perfectly regular rate by introducing a vibrating brake into the circuit. The motor employed had a simple 2-part commutator whose rotation timed itself to the makes-and-breaks of the current. One of Deprez's motors thus governed was shown in Paris some four or five years ago. It ran at a perfectly uniform speed, quite irrespective of the work it was doing. Whether it was lifting a load of 5 kilogrammes from the ground, or was letting this load run down to the ground, or ran without any load at all, the speed was the same. This method is however inapplicable to large motors.

Centrifugal Governing.

Another suggestion, equally impracticable on the large scale, was to adopt a centrifugal governor to open the circuit whenever the motor exceeded a certain speed. A motor so governed runs spasmodically fast and slow. It is also possible for a centrifugal governor to be employed to vary the resistance of a part of the circuit; for example, to work an automatic adjustment to shunt part of the current of a series machine

from its field-magnets (resembling the automatic regulator of
the Brush dynamo, p. 127), or to introduce additional resistance
into the field-magnet coils of a shunt-wound machine, in proportion to the speed.

Professors Ayrton and Perry have also proposed several
forms of "periodic" centrifugal governor, a device by which
in every revolution power is supplied during a portion of the
revolution only, the proportion of the time in every revolution during which the power is supplied being made to vary
according to the speed. The main difficulty with such
governors is to prevent sparking. But there is a still more
radical defect in all centrifugal governors ; they all work too
late. They do not perform their functions until the speed has
changed. The perfect governor will not wait for the speed to
change.

Dynamometric Governing.

The author has devised another kind of governor which is
not open to this objection. He proposes to employ a dynamometer on the shaft of the motor to actuate a regulating
apparatus which may consist either of a periodic regulator
to shunt or interrupt the current during a portion of each
revolution, or of an adjustable resistance connected in part of
the circuit. The dynamometric part may take the form of a
belt dynamometer (such as Alteneck's) or of a pulley dynamometer (such as Morin's or Smith's). In the latter case, which
is the more convenient, a loose pulley runs on the motor shaft
and is connected by a spring arrangement with a fixed pulley.
The rotation of the motor will drag round the fixed pulley in
advance of the loose pulley, and the angular advance will be
proportional to the turning moment or torque. The amount
of such angular advance determines the action of the regulating part. The regulator in this case is therefore worked,
not according to the speed of the motor, but according to the
load it is carrying. Any change in the load will instantly
act on the dynamometric governor before the speed has time
to change.

2 P

Electric Governing.

Another method of governing, not requiring any rotating parts, has been proposed by the author. He uses as field-magnets a double set of poles, set at different angles with respect to the brushes of the motor. One pair of magnetic poles, having a certain lead, is actuated by series coils, the other pair, having a different lead, by shunt coils. When both shunt and series are working, there will, of course, be a resultant pole having some intermediate lead. If the load of the motor is diminished, it will tend to run faster, increasing the current in the shunt part, decreasing it in the series part, and therefore altering the effective lead and preventing the increase of speed.

In 1880 a motor was patented by Mr. G. G. André in which the field-magnets were wound in two separate circuits, one of thick and the other with thin wire, the current dividing between them, and the armature was connected as a bridge across these circuits, exactly as the galvanometer is connected across the circuits of a Wheatstone's bridge. Motors governed on this principle have more lately been constructed by Mr. F. J. Sprague ; they show remarkably good regulation.

Another method of governing, due to Mr. C. F. Brush, consists in building up the field-magnet coils in sections, and by varying the number of sections in circuit, or the mode of their connexion, he proposes to obtain an automatic regulation.

The method of constructing a motor with coils in sections so that a movable internal core may be successively attracted as successive sections are switched in, has been made use of by Deprez in constructing an electric hammer. This principle of construction had been employed by Page in his motors many years previously.

The method of automatic regulation that is most perfect in theory, is undoubtedly that imagined by Professors Ayrton and Perry, and expounded in their paper in the *Journal of the Society of Telegraph-Engineers.*[1] The theory of self-regulation propounded by them demands the most careful

[1] Vol. xii., May 1883.

attention. It is expounded in the following pages ; but it is only fair to readers to mention that in order not to have any confusion in the use of symbols, the notation of Professors Ayrton and Perry has not been followed : the symbols employed have the same meaning as in the chapters which precede. Neither has the author adopted that part of Professors Ayrton and Perry's argument which regards one part of the motor as acting as a brake to another part. This way of regarding the matter, though doubtless propounded originally with good reasons, does not commend itself to the minds of students generally. The author prefers to regard the use of a shunt winding opposing a series winding—which is the final result of this method of regulation—as simply a differential winding designed to produce a certain result.

Theory of Self-governing Motors.

In the two Chapters on Self-regulating Dynamos, on pp. 417 to 446, were set forth the methods of solving the problem how to arrange a dynamo so that it shall feed the circuit with electric energy under the condition either of a constant potential or of a constant current, when driven at a constant speed. The solution to that problem consisted in the employment of certain combinations for the field-magnets, which gave an initial magnetic field independent of the current that might be flowing in the main circuit.

Now it is not hard to see that this problem may be applied conversely, and that motors may be built with a combination of arrangements for their field-magnets, such that, when supplied with currents under one of the two standard conditions of distribution, their speed shall be constant whatever the load. It will be evident without any numerical calculations that the windings must oppose one another—one must tend to demagnetise the field-magnet, the other to magnetise. Take the case of a shunt motor supplied at a constant potential $\mathcal{E}$, and running at a certain speed with a certain load. If the load is suddenly removed the motor will begin to race, its racing will increase the counter electromotive-force developed and will partly cut down the armature current. But the decrease of current will not be quite adequate to bring back the speed, because of the internal resistance of the armature which has prevented the whole energy of the armature current from being utilised as work. A demagnetising series coil of appropriate power wound on the field-magnet will, however, effect what is wanted, for then, with any reduction of load, the corresponding reduction of current can take place, the resulting increase in the field-magnetism being sufficient to get the required larger counter electromotive-force without any increase in speed. The combination may, therefore, rightly be regarded as a differential winding. For constant-current distribution a differential magnetisation is also needed ; and in this case also the demagnetising coil must be connected in series with the armature.

The following synoptical table contrasts the arrangements for self-regulating generators with those of self-governed motors :—

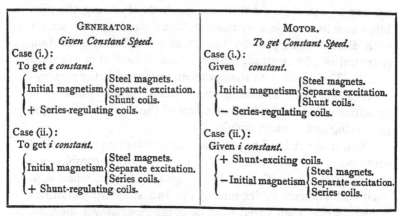

GENERATOR.	MOTOR.
Given Constant Speed.	*To get Constant Speed.*
Case (i.):	Case (i.):
To get *e constant.*	Given *constant.*
Initial magnetism { Steel magnets. Separate excitation. Shunt coils. + Series-regulating coils.	Initial magnetism { Steel magnets. Separate excitation. Shunt coils. − Series-regulating coils.
Case (ii.):	Case (ii.):
To get *i constant.*	Given *i constant.*
Initial magnetism { Steel magnets. Separate excitation. Series coils. + Shunt-regulating coils.	+ Shunt-exciting coils. −Initial magnetism { Steel magnets. Separate excitation. Series coils.

The method of compounding a dynamo to give constant current has not, as remarked on p. 438, become as yet a practical one; neither has the corresponding case of the differentially wound motor.[1]

---

[1] In the former editions the case of the constant-current motor was erroneously stated ; a negative sign in the equations, which would have indicated a negative speed, having been overlooked. Professor Ayrton, who pointed this out to the author, has himself indicated the correct solution of the case now given.

In discussing the theory of the self-governed motor, we shall follow the same general lines as in discussing the theory of the self-regulating generator, namely, find an equation expressing the desired condition of constancy.

Case (i.)—*Given : & constant.*

(*a*) *Magneto Motor with Series-regulating Coil.*—Using the same notation as previously, we have for the counter electromotive-force developed in the armature—

$$E = n\,C\,N\,;$$

also

$$E = \mathscr{E} - (r_a + r_m)\,i.$$

Now N is made up of two parts, viz. :—N_1 the permanent part, and another part depending on the series coil which we may write $q\,S\,i$; where q has the same signification as on p. 422, and is equal to $4\,\pi$ divided by ten times the sum of the magnetic resistances. Its value therefore depends upon the permeability, and therefore upon the degree of saturation of the iron of the magnetic circuit. Reserving this point for further consideration, we may write

$$N = N_1 - q\,S\,i.$$

(If we had written $+$ instead of $-$, we should find the solution coming out with the negative sign, indicating that the windings must be so arranged that the current in the series coil circulates in the negative or demagnetising sense. We write the negative sign, however, as we already know that this must be so.) Substituting the value of N in the fundamental equation, we have

$$E = n\,(C\,N_1 - C\,q\,S\,i)\,;$$

and equating this to the other value of E in the second equation above, we find

$$n = \frac{\mathscr{E} - (r_a + r_m)\,i}{C\,N_1 - C\,q\,S\,i}. \qquad [\text{LI.}]$$

Having thus obtained an expression for the speed, we must examine the various parts of the expression to see which are variable and which constant, and so deduce a relation which shall make n constant. Now in both numerator and denominator there are two terms, the first of which is a constant, whilst the second of each contains the variable i. A little consideration will show that the fraction cannot have a constant value unless the two coefficients of the variable in the second terms bear the same ratio to one another as do the two constants which stand as the first terms; or n cannot be constant unless

$$\frac{\mathcal{E}}{C\,N_1} = \frac{r_a + r_m}{C\,q\,S},$$

or

$$\frac{\mathcal{E}}{N_1} = \frac{r_a + r_m}{q\,S}, \qquad [\text{LII.}]$$

which is the desired equation of condition.

If this condition be observed (and it will be noted that the quantity of series winding required is proportional, as in the self-regulating dynamo, to the internal resistance of the machine), then the speed will be constant and of the value

$$n = \frac{\mathcal{E}}{C\,N_1} = \frac{r_a + r_m}{C\,q\,S}. \qquad [\text{LIII.}]$$

From the first of these relations we see that the speed at which the machine is thus governed to run, is the same speed as that at which if driven as a generator on open circuit, it will yield an electromotive-force equal to that of the supply at the mains. When running as an unloaded motor, it ought of course to turn so fast as to reduce the current through its armature to a minimum, which it can do by running at this speed. It is evident that by making the permanent part of the magnetism strong enough, the critical speed, that is to say, the speed for which the motor is self-governing, may be made as low as desired. As the load on the motor is increased, the flow of current through the armature must be increased, and this increased current cannot flow unless in some way the counter electromotive-force of the armature be diminished. As the speed is to be kept up, this is accomplished by the lowering of the magnetism which occurs in consequence of the increased current flowing through the demagnetising coils. The quantity denoted by q, which depends on the permeability of the iron, ought of course to be taken at an average value between the two extremes which it has at maximum load and at zero load. The magnetism is a maximum when the motor is running empty. When the load is greatest, if the motor is running at say 80 per cent. efficiency, E will be 80 per cent. of $\mathcal{E}$; that is to say, N will be 80 per cent. of N_1. It is between these limits in the magnetisation that the value of q must be averaged. It is evident from equation [LIII.], that if the motor is already provided with a given series winding, there can be found a value of $\mathcal{E}$, for which the condition of self-governing can be still fulfilled. All the preceding argument applies also to separately-excited motors.

Practical Determination of the Proper Potential and Winding for Given Motor.[1]

Suppose a motor having a certain steel magnet and a certain armature be given, and let it be required to determine the potential at which it will give a certain constant speed, and the winding that must be adopted for the demagnetising coil. Two experiments are needful. First, run the motor on open circuit as a generator at the given speed, and observe the potential at its terminals. That is the number of volts $\mathcal{E}$ with which it must be supplied. Secondly, connect in series with the armature a resistance a little less than r_a (say five-sixths as great) to represent r_m. Prepare some accumulators to give the proper potential $\mathcal{E}$ of supply. Wind a temporary coil of S turns on the field-magnet by which to excite these temporarily with a demagnetising power from some separate source of current that can be varied. Provide an ampère-meter to measure this exciting current and a second ampère-meter to observe the armature current. Then, by means of a

[1] It should be pointed out that this process differs from that suggested by Professors Ayrton and Perry in their paper on electromotors, in *Journ. Soc. Telegr.-Eng.*, May 1883. Their method depends on the volume left on the bobbins of the field-magnets, which is assumed to be constant.

suitable dynamometric brake, put on the axle of the motor its normal maximum load. Supply its armature with current at potential $\mathcal{E}$, and vary the magnetisation in the temporary coil until the proper speed is once more attained : note the actual armature current i_a and the temporary exciting current i'_m. Then it is clear that the number of turns required will be

$$ S = \frac{S' \, i'_m}{i}, $$

and it only remains to choose a wire such that S turns will have a resistance equal to r_m, or less if there is room in the available space.

(*b*) *Shunt Motor with Series-regulating Coil.*—This really includes two cases : namely, shunt and long-shunt. We take the latter case, which is· the usual one. The fundamental equations will now be

$$
\begin{aligned}
\text{E} &= n\,\text{C}\,\text{N}, \\
\text{E} &= \mathcal{E} - (r_a + r_m)\,i_a, \\
\mathcal{E} &= r_s\,i_s, \\
\text{N} &= q\,\text{Z}\,i_s - q\,\text{S}\,i_a.
\end{aligned}
$$

From these we get successively

$$ n = \frac{\text{E}}{\text{C}\,q\,\text{Z}\,i_s - \text{C}\,q\,\text{S}\,i_a}, $$

$$ n = \frac{\mathcal{E} - (r_a + r_m)\,i_a}{\text{C}\,q\,\text{Z}\,i_s - \text{C}\,q\,\text{S}\,i_a}, $$

$$ n = \frac{r_s\,i_s - (r_a + r_m)\,i_a}{\text{C}\,q\,\text{Z}\,i_s - \text{C}\,q\,\text{S}\,i_a}, $$

which cannot be constant unless the respective coefficients of i_s and i_a in the numerator bear the same ratio to one another as do the corresponding ones in the denominator, giving as the equation of condition

$$ \frac{r_s}{\text{C}\,q_1\,\text{Z}} = \frac{r_a + r}{\text{C}\,q_2\,\text{S}}. $$

The values of q are here distinguished by suffixes for the following reason :—If the above relation is fulfilled, then when the motor runs with zero load there will be practically no current through the armature, and the counter electromotive-force $n \, C \, q_1 \, Z \, i_s$ will be equal to $\mathcal{E}$. At this stage of things, when the shunt coil is in full work, the magnetism is at its maximum, and q_1 (which is proportional inversely to the total magnetic resistance) will be at its minimum value. On the other hand, when the motor is driving its maximum load the armature current, and therefore the demagnetising effect of the regulating coils is at its maximum, the magnetisation will be at its lowest running value, and q_2 will be at its maximum. This being promised, the equation of condition may be written

$$\frac{Z}{S} = \frac{r_s}{r_a + r_m} \cdot \frac{q_2}{q_1}. \qquad \text{[LIV. (a).]}$$

In the former editions of this work saturation was left out of account, and the formulæ were only approximate, the equation having the form

$$\frac{Z}{S} = \frac{r_s}{r_a + r_m}, \qquad \text{[LIV.]}$$

which is Ayrton and Perry's rule for the winding of the self-governing motor. It may be remarked, however, since in a well-designed motor the resistances in the armature-circuit are very small, and the efficiency as a whole high, the demagnetising effect of the series coils, even at full load, need only reduce the magnetisation by a small percentage, so that q_2 may be nearly equal to q_1. Moreover, if a backward lead be given to the brushes, the armature itself may act partially as a demagnetising series coil, and so compensate for alteration in the permeability. It is quite certain that motors wound differentially in the proportion indicated in equation [LIV.] are very nearly self-governed. Some excellent motors by Lieut. F. J. Sprague are wound according to this rule. One very curious property of this method of winding is as follows:—Suppose the motor to be standing still and the current turned on. The ampère-turns due to the shunt will be equal to $\mathcal{E} \, Z \div r_s$, whilst those due to the series coil will be $\mathcal{E} \, S \div (r_a + r_m)$; and these, according to equation [LIV.] will be equal, and they are of opposite sign. There should then be no magnetism excited at all. But if there is any lead at the brushes, the magnetising tendency of the armature will come into play ; and if the brushes have a considerable negative lead, the effect will be to magnetise the field-magnet in the wrong sense, and then the motor starts the wrong way. The defect might be remedied by cutting out the series coil or reversing it, until the motor has got up its speed. The latter course is preferable, as the additional torque of the series motor is of great advantage in overcoming the statical resistance to motion experienced at starting.

It is obvious that the number of shunt-turns should theoretically be such that the motor, driven on open circuit at the given speed, shall generate an electromotive-force equal to $\mathcal{E}$.

Determination of the Windings.

As in the preceding case, a temporary coil must be wound and separately excited, a resistance equal to the future r_m being added to the armature resistance. Two experiments are required. Run the motor first with no load at the brake, using the proper potential $\mathcal{E}$, and excite the temporary coil, observing the number of ampère-turns that are needful to bring the speed down to the required n. The number of ampère-turns in this case is equal to $Z i_s$, where i_s is the current which economy dictates should be used in the shunt. Secondly, run the motor with its fullest load at the brake, and again excite the field-magnet with such a number of ampère-turns, that the speed is constant at n. From this and the previous experiment S can be calculated.

Case (ii.)—*Given : i constant.*

As this case is not yet practical, it may be treated very briefly.

(a) *Shunt-wound Motor with Initial Negative Permanent Magnetism.*

The fundamental expressions are

$$E = n\,C\,N = \mathcal{E} - r_a i_a = r_s i_s - r_a i_a = (r_s + r_a)\,i_s - r_a i,$$
$$N = q\,Z\,i_s - N_1,$$

where N_1 is the permanent negative magnetism; whence

$$n = \frac{(r_s + r_a)\,i_s - r_a\,i}{C\,q\,Z\,i_s - C\,N_1}.$$

As i_s is a variable, this cannot be true unless Z is such that

$$\frac{r_s + r_a}{Z\,q} = \frac{r_a\,i}{N_1},$$

which is therefore the equation of condition; and

$$n = \frac{r_a\,i}{C\,N_1},$$

or, the speed will be such that were the shunt not acting, the induced electromotive-force due to the permanent magnetism would just suffice to send a current equal to i through the short-circuited armature.

(*b*) *Shunt-wound Motor with Series-demagnetising Coil.*

$$E = n\,C\,N = r_s\,i_s - (r_a)\,i_a = (r_s + r_a)\,i_s - r_a\,i,$$
$$N = q_0\,Z\,i_s - q_1\,S\,i;$$

whence

$$n = \frac{(r_s + r_a)\,i_s - r_a\,i}{C\,q_0\,Z\,i_s - C\,q_1\,S\,i}.$$

This cannot be constant unless

$$\frac{r_s + r_a}{q_0\,Z} = \frac{r_a}{q_1\,S};$$

whence

$$\frac{Z}{S} = \frac{r_s + r_a}{r_a} \cdot \frac{q_1}{q_0}. \qquad\text{[LVI.]}$$

Here q_0 refers to the state of magnetisation (see top of p. 422, and p. 429) when the load is zero, and q_1 to the value when the load is an average between zero and maximum.

The efficiency of a differentially-wound motor cannot be expected to be as high as that of one which is non-differentially wound, since the energy expended in the former case in magnetising the field-magnets is greater, relatively to the amount of magnetisation produced. Thus if the demagnetising coil has a value in ampère-turns of one-fourth of the magnetising coils, the energy required to give the necessary magnetisation will be 50 per cent., or half as much again as in the case of simple direct magnetisation.[1] Professors Ayrton and Perry have therefore suggested another solution not open to this objection.

(c) *Shunt-wound Motor with Initial Positive Permanent Magnetism and an Auxiliary Battery.*

A few accumulator cells introduced in the armature part, having their poles so directed that their electromotive-force assists that of the current supplied by the main circuit, will answer the purpose. Let the electromotive-force and resistance of this battery be called respectively E_b and r_b. The equations then become

$$\mathscr{E} + E_b - E = (r_a + r_b) i_a,$$
$$E = n\, C\, q\, Z\, i_s + n\, C\, N_1,$$
$$\mathscr{E} = r_s i_s,$$
$$i_a = i - i_s;$$

whence

$$n = \frac{(r_s + r_a + r_b)\, i_s + E_b - (r_a + r_s)\, i}{C\, q\, Z\, i_s + C\, N_1};$$

where Z must be such that

$$n = \frac{r_s + r_a + r_b}{C\, q\, Z},$$

and E_b must be such that

$$n\, C\, N_1 = E_b - (r_a + r_b)\, i;$$

or that

$$\frac{E_b - n\, C\, N_1}{r_a + r_b} = i,$$

the electromotive-force of the auxiliary battery be just so great that, after subtracting the back electromotive-force generated (at that speed) in the armature by the permanent part of the magnetism alone, the remaining volts are exactly equal to the volts lost in sending the given current through the resistance

---

[1] Some elaborate observations upon compound-wound motors, when worked with constant potentials and constant currents, have been published by Dr. Frölich in the *Elektrotechnische Zeitschrift* for June 1885.

in the armature part of the circuit. There are two obvious corollaries to this problem : namely (1) that a series coil wound non-differentially might be used instead of permanent magnetism; (2) that fewer accumulator cells might be used if the permanent magnetism is zero.

OTHER METHODS OF GOVERNING MOTORS.

A further suggestion for governing motors is due to Mr. Mordey and Mr. C. Watson. They wind the armature with two windings, having separate commutators. One winding—the main one—is the ordinary armature circuit of the motor, and is supplied with current from the external source, causing the armature to revolve. The other winding, which may be called the regulating armature-winding, is small in amount, and is disposed over, or side by side with, the main motor-winding. This additional winding is not connected to the mains or source of current, but to the field-winding by means of a special commutator or collector and brushes. It will be observed that this additional armature-winding, revolving in the field, constitutes a generator of current. The regulating action is as follows :—When a tendency to increase in speed results from a diminution of the load, the additional armature-winding tends to increase the strength of the field by supplying more current to the field-coils, and thus raises the opposing electromotive-force of the motor, diminishes the amount of current received from the mains, and so reduces the speed to its normal rate. Again, an increase of the load, tending to reduce the speed, is counteracted by a lessening of the magnetising current produced by the additional winding, a consequent lowering of the opposing electromotive-force of the motor, and an increase of the current received from the mains. This method has given very good results, but has been rendered unnecessary, except in special cases, by the simpler plan to be next described. It will be seen that as it is summative it does not require so great an expenditure of energy in the fields as a differential winding; nor is it open to the objection that the motor may start in the wrong direction.

On the other hand, it has the drawback of requiring an additional commutator.

Still more important is the discovery by Mr. Mordey[1] that if a pure shunt motor is constructed upon perfect designs— that is to say, having very small resistance of armature and very large resistance of shunt, and having also field-magnets which are very powerful relatively to the armature, and an armature properly laminated and sectioned so as to reduce eddy-currents and self-induction to a minimum—such a shunt dynamo if supplied from mains at a constant potential will run at a constant speed, whatever the load.[2] The following tests showed a constancy to within $1\frac{1}{2}$ per cent. for all loads within working limits.

Potential at Terminals.	Current (Ampères).	Horse-power at Brake.	Revolutions per Minute.	Torque (pound-feet).
68·4	44	1·1	1125	5·15
68·4	126	7·4	1120	33·4
68·4	165·5	10·36	1115	48·8
68·4	180	11·14	1110	53

With a lower electromotive-force the same motor regulated almost equally well, but at a lower speed.   It was observed that, especially when the motor was giving out small horse-power, that the speed was increased by weakening the field.

Another possible mode of governing constant-current motors is by providing a variable magnetic shunt, in the converse of the manner suggested by Trotter for constant-current generators, or by the modification of that method suggested by the author (p. 446). Messrs. Siemens and Halske have patented, as an approximately governed motor for constant-current work, a shunt motor having a few series coils wound so as to add to the magnetisation. Mr. C. W. Hill proposes to provide armatures with special connexions for changing the number of active convolutions whilst the machine is running, thereby varying the counter electromotive-force.

---

[1] See *Philosophical Magazine*, January 1886.

[2] This might have been foreseen from the equations of p. 585, in which if $r_a + r_m = 0$, the condition of regulation will give $S = 0$.

Mechanical Characteristics of Compound Differential Motors.

It may be convenient here to consider the graphic representation of the regulations between speed and torque in motors provided with mixed windings.[1]

The curves for constant-potential supply are shown in Fig. 366, wherein the letters S and Z refer to series windings and

FIG. 366. FIG. 367.

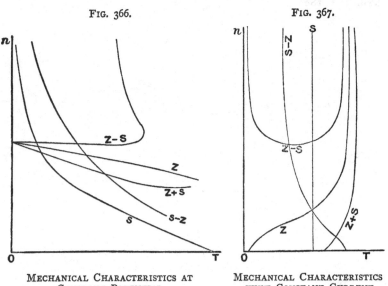

MECHANICAL CHARACTERISTICS AT CONSTANT POTENTIAL.

MECHANICAL CHARACTERISTICS WITH CONSTANT CURRENT.

shunt windings respectively. The forms of the curves for mixed windings differ somewhat according to the proportions of the two sets of coils. The important case is that of the differential winding marked Z – S, having a few series turns to correct the droop of the pure shunt-winding; and it will be noted that up to a certain limit the speed is nearly constant, but that there is a maximum value to the torque. In the case of constant-

---

[1] The author is indebted to Frölich's *Die Dynamoelektrische Maschine* for the curves of motors with mixed windings. Similar curves have been deduced by Rechniewski, see *Séances de la Société de Physique*, 1885, p. 197.

current supply, as the curves of Fig. 367 show, the only winding which gives any approximation to a constant speed is the differential winding with large shunt and small series coil. For, as in the case of the constant-current generator, the variation of the magnetism has to be carried through an enormous range, defying any averaging of the magnetic permeability.

An elegant graphic method of treating the problem of self-government of motors is given by M. Picou in *La Lumière Électrique*, xxiii. 114, 1887.

CHAPTER XXXII.

TESTING DYNAMOS AND MOTORS.

TESTS to be applied to dynamos are of two kinds, viz. those which relate to the design and construction of the machines, and those which relate to their performance. Under the former are included tests of the resistance of the various coils and connexions, and of the insulation of the working parts. Under the latter are comprised the tests of efficiency under various loads, and of electrical output or activity at different speeds.

Testing Construction.

The resistance of the various parts of the armature coils, of the field-magnet coils, and of the various connexions, may be tested in the ordinary manner, by means of a Wheatstone's bridge, or by one of the recognised galvanometer methods. The only point of difficulty lies in measuring such small resistances as those of armatures and of series coils, which are often very small fractions of an ohm. In this case probably the best method of proceeding is the following. By means of a few cells of accumulators send a strong current through the coil or armature whose resistance is to be measured, interposing in the circuit an ampère-meter. While this current is passing, measure, by means of a sensitive voltmeter, the fall of potential between the two ends of the coil. By Ohm's law, the number of volts of fall of potential divided by the number of ampères of current will give the resistance in ohms. Additional accuracy may be secured by connecting in the circuit a strip of stout German silver, as recommended by Lord Rayleigh, of known resistance, and comparing the fall of potential between the two ends of the strip with the fall of potential in the coil. The ratio of the two falls of potential will equal the ratio of the resistances.

It ought on no account to be forgotten that the internal resistance of a dynamo when warm after working for a few hours is considerably higher than when it is cold. Tests of resistance ought therefore to be made both before and after the dynamo has been running.

The insulation-resistance of the various parts should be tested with care, and should be repeated at intervals. In particular, measurements should be made of the insulation-resistance between the terminals of the machine and its metal bed-plate, and between the segments of the collector and the axle. Nowhere is the insulation more likely to deteriorate than at the collector. Occasionally a dynamo, which otherwise may appear to be in good condition, will mysteriously lose its power of furnishing the usual current, in consequence of invisible short-circuiting having occurred in the layers of insulating matter between the segments of the collector. A film of charred oil, possibly with the assistance of small particles of metal worn off the brushes, will afford a ready passage to the current, which will thus leak away instead of being delivered in full strength to the circuit. The perfection of the magnetic circuit may be tested in two ways. If the design or construction of the magnetic circuit is such as to give rise to magnetic leakage, it is desirable to ascertain the proportion of leakage. The actual amount of magnetic leakage may be measured by placing exploring coils over different parts of the magnetic circuit in the manner used by J. and E. Hopkinson. They passed a single convolution of wire around the magnet-frame at different points, and joined it to a suitable galvanometer. The throw imparted to the galvanometer when the current in the magnetising coil was suddenly turned off or on, is a measure of the number of magnetic lines enclosed by the exploring coil. The other way of examining the perfection of the magnetic circuit is to join up a known suitable resistance to the terminals of the machine, and then to run it at a slow speed, gradually increasing the number of revolutions until it excites itself. (The method is of course inapplicable to many alternate-current machines.) The least-speed of self-excitation is, *cæteris paribus*, a measure of the goodness of the magnetic circuit. See Appendix XI.

Testing Performance and Efficiency.

The testing of the efficiency and working capacity of a dynamo, whether working as generator or as motor, is a more serious matter, and involves both electrical and dynamometrical measurements.

In the case of the dynamo generating currents, measurements must be made (*a*) of the horse-power expended, and (*b*) of the energy of the electric currents realised.

In the case of the motor doing work, measurements must be made (*a*) of the electric energy consumed, and (*b*) of the mechanical horse-power realised.

Measurement of Horse-power.

There are four general methods of measuring mechanical power :—

(*a.*) *Indicator Method.*—By taking an indicator diagram from the steam-engine which supplies the power.

(*b.*) *Brake Method.*—By absorbing the power delivered by the machine, at a friction brake such as that of Prony, Poncelet, Appold, Raffard, or Froude.

(*c.*) *Dynamometer Method.*—By measuring in a transmission dynamometer or ergometer, such as that of Morin, von Hefner-Alteneck, Ayrton and Perry, or of F. J. Smith, the actual mechanical power of the shaft or belt.

(*d.*) *Balance Method.*—By balancing the dynamo or motor on its own pivots and making it into its own ergometer.

To these must now be added a fifth method, namely :—

(*e.*) *Electrical Method.*—By making the motor drive the dynamo which supplies it, measuring electrically the work given out in the one, or absorbed by the other, and then measuring, either mechanically or electrically, the difference.

(*a.*) *Indicator Method.*—The operation of taking an indicator diagram of the work of a steam-engine is too well known to engineers to need more than a passing reference. This method is, however, not always applicable, for in many cases the steam-engine has to drive other machinery, and heavy shafting for other machinery. In such cases the only remedy is to take two sets of indicator diagrams, one when the dynamo is at work, the other when the dynamo is thrown out of gear, the difference being assumed to represent the horse-power absorbed by the dynamo.

(*b.*) *Brake Method.*—The friction-brake of Prony is well known to engineers, but the same can hardly be said of the more recent forms of friction dynamometers. Various improvements have been introduced in detail from time to time by Poncelet, Appold, and Deprez. In Prony's method the work is measured by clamping a pair of wooden jaws round a pulley on the shaft; the torque on the jaws being measured directly by hanging weights on a projecting arm with a sufficient moment to prevent rotation. If p is the weight which at a distance l from the centre balances the tendency to turn, then the friction-force f multiplied by the radius r of the pulley will equal p multiplied by l.

This may be written,

$$\text{Torque} = f r = p l.$$

From which it follows that

$$f = \frac{p l}{r}.$$

If n be the number of revolutions *per second*, then $2 \pi n$ is the number of radians per second, or in other words the angular velocity for which we use the symbol ω, and $2 \pi n r$ is the linear velocity v at the circumference. Now the work per second, or "activity," is the product of the force at the circumference into the velocity at the circumference, or

$$w = f v = \frac{p l}{r} \cdot 2 \pi n r = 2 \pi n p l.$$

If p is measured in pounds' weight, and l in feet, then, remembering that 550 foot-pounds per second go to one horse-power, we have

$$\text{Horse-power absorbed} = \frac{2 \pi n p l}{550};$$

or, if p is expressed in grammes' weight, and l in centimetres, it must be divided by $7 \cdot 6 \times 10^6$ to bring it to horse-power.

The later improvements imported into the Prony brake are of great importance. Poncelet added a rigid rod at right angles to the lever, and attached the weights at the lower end. Appold substituted for the wooden jaws a steel strap giving a more equable friction, and therefore having less tendency to vibration. Raffard [1] substituted a belt differing in breadth, and therefore offering a variable coefficient of friction, according to the amount wrapped round the pulley. Further modifications of this kind of brake dynamometer have been made by Professor James Thomson, Professor Unwin, M. Carpentier, and by Professors Ayrton and Perry. The friction of a turbine wheel was also applied as a dynamometer brake by the late W. Froude.

As all these brake dynamometers measure the work by destroying it, it will be seen that though they are admirably adapted to measure the work furnished by a motor, they cannot, except indirectly, be applied to measure the work supplied to a dynamo. Some experience in working with these machines is essential if reliable results are to be obtained ; but with the more modern forms of instrument, such as those of Poncelet and Raffard, the results are very good. The great secret of success is to keep the friction surfaces well lubricated with an abundant supply of soap and water.

(*c.*) *Dynamometer Method.*—The Prony brake was styled above a brake dynamometer ; but the true dynamometer for measuring transmitted power does not destroy the power which it measures. Transmission dynamometers may be divided into two closely allied categories : those which measure the power transmitted along a belt, and those which measure power transmitted by a shaft.

In the case of transmitting power by a belt, the actual

---

[1] For further accounts of these instruments the reader is referred to Weisbach's *Mechanics of Engineering* ; Spons' *Dictionary of Engineering*, Article "Dynamometer" ; Smith's *Work-measuring Machines* ; a series of articles in the *Electrician*, 1883–4, by Mr. Gisbert Kapp ; *Proc. Inst. Mech. Eng.*, 1877, p. 237 (Mr. Froude) ; *Rep. Brit. Assoc.*, 1883 (Prof. Unwin) ; *Journ. Soc. Telegr.-Eng. and Electr.*, xlix., vol. xii. p. 346 (Profs. Ayrton and Perry).

force which drives is the difference between the tension [1] or pull in the two parts of the belt. If F′ is the pull in the slack part of the belt before reaching the driven pulley, and F the pull in the tight part of the belt after leaving the driven pulley, then F − F′ represents the net pull at the circumference, and (F − F′) × *r* is the torque T, or turning moment. Then if *n* is the number of revolutions *per second*, the angular velocity ω will be equal to 2 π *n*. This gives us as the work per second, or activity,

$$w = \omega \, T = f v = 2 \, \pi \, n \, r \, (F - F').$$

As before, if F is expressed in pounds' weight and *r* in feet, the expression must be divided by 550 to bring to horsepower : or must be divided by 7·6 × 10⁶ if the quantities are expressed in grammes' weight and centimetres.

A dynamometer which can be applied to a driving belt, and actually measures the difference F − F′ in the tight and slack parts of the belt, has been designed by von Hefner-Alteneck, and is commonly known as Siemens's dynamometer. [2] Other forms have been devised by Sir F. J. Bramwell, W. P. Tatham, [3] W. Froude, T. A. Edison, and others. Nearly all of these instruments introduce additional pulleys into the transmitting system, causing additional friction.

Much more satisfactory are those transmission dynamometers which measure the power transmitted by a shaft. In nearly all instruments of this class there is a fixed pulley keyed to the shaft, and beside it a loose pulley connected with it by some kind of spring arrangement so set that the elongation or bending of the spring measures the angular advance of the one pulley relatively to the other ; this angular advance is proportional to the transmitted angular force or torque. To this

---

[1] The word tension, though used by engineers as synonymous with "pulling force," ought not to be so used ; it ought to be confined to the exact meaning of force per unit of area of cross-section.

[2] One form of the Siemens dynamometer is described by Hopkinson, *Proc. Inst. Mech. Eng.*, 1879. A more modern form is described by Schröter, *Bayerisches Industrie- und Gewerbeblatt,* 1883.

[3] *Journ. Franklin Institute,* Nov. 1886.

class of instrument belongs the well-known dynamometer of Morin, in which the displacement of the loose pulley is resisted by a straight bar spring, the centre of which is attached to the driving shaft. Modifications of the Morin instrument have been devised by Easton and Anderson, Heinrichs,[1] Ayrton and Perry,[2] and the Rev. F. J. Smith. The dynamometer of Mr. Smith claims special notice on account of its completeness and accuracy. In this instrument, a sketch of which is given in Fig. 368, there are as before two pulleys, the front one fixed to the shaft, the hinder one running loose upon it. But the displacement of the loose pulley is measured in a different manner from the methods of the earlier constructors. The loose pulley carries a bevel-wheel (not shown) which gears into two other bevel-wheels,[3] which of necessity move in proportion to the angular displacement. Each of these two bevel-wheels is furnished with a shallow cylindrical drum, over which is coiled a gut or steel tape attached to a cross-head. The shaft is of forged Whitworth steel, tubular at its ends and link-shaped between. The cross-head passing through the link, and being attached to a spring, any displacement of the loose pulley will cause the steel tapes to be wound up on the drums, and will extend the spring. A light rod of steel passing through the hinder end of the link, through the shaft, actuates the pointer of a dial, thus enabling the transmitted torque to be read off directly. A speed-counter is also attached. Mr. Smith has also added to this machine an integrating apparatus which traces on a drum a continuous record of the work done ; the speed of rotation of the recording drum, having a known ratio to the speed of driving, and the displacement of the recording pen, which writes the ordinates, being proportional to the torque. The position of the spring, coincident with the axis of the dynamometer, obviates all fear of error arising from

[1] See *Engineering*, May 2, 1884, and *Electrical Review*, April 26, 1884, for an excellent account of a series of tests carried out with great care and ability for Mr. Heinrichs, by Messrs. Alabaster, Gatehouse and Co.

[2] *Journ. Soc. Telegr.-Eng. and Electr.*, vol. xii. p. 163, 1883.

[3] Similar devices of geared cog-wheels had been applied, though in an entirely different way, in the dynamometers of Hachette and of White.

Fig. 368.

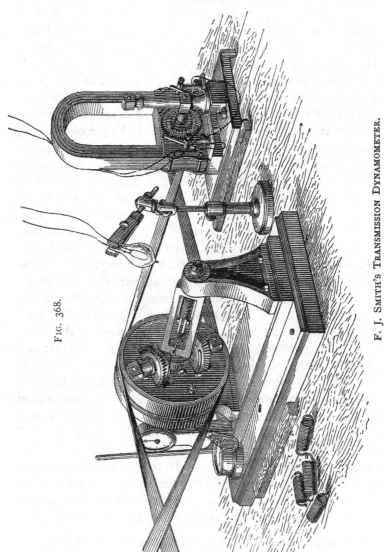

F. J. Smith's Transmission Dynamometer.

centrifugal force. Mr. Smith has suggested a novel method of calibrating the readings of the dynamometer, which appears to have some advantages over the common method of hanging weights to the loose pulley. It is as follows :—Let the transmission dynamometer be driven by some steady prime mover—a water-wheel is probably the steadiest of all—and let the dynamometer itself drive another shaft on which a Prony brake or Appold brake is applied. Then if the work done against friction at the brake is measured, and the speed is known also, it is known what the transmitted torque is. Every transmission dynamometer ought to be similarly calibrated. Recently Mr. Smith has greatly simplified his dynamometer, obviating the costly slotting of the shaft and avoiding the use of the bevel-wheels. The loose pulley is connected to the shaft by a strong steel spring in the form of a long enwrapping spiral.

(*d.*) *Balance Method.*—The following method was devised by Mr. Smith when testing some small Trouvé motors at the Paris Exposition of 1881. With small motors there arises the difficulty that the ordinary means of measuring the work they perform introduce relatively large amounts of extraneous friction. The motor to be tested is placed with its armature spindle between centres, or on friction wheels, and the weight

FIG. 369.

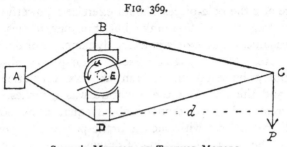

SMITH'S METHOD OF TESTING MOTORS.

of the field-magnets and frame is very carefully balanced with counterpoise weights. In Fig. 369, B D represents the field-magnets and frame of the motor duly counterpoised, and E is the armature. When the current is turned on, the armature tends to rotate in one direction, and the field-magnets in the other; the angular reaction being of course equal to the angular action. If the reaction which tends to drive the field-magnets round, be balanced by applying a force P (for example that of a spring balance) at the point C of the frame A B C D, then the moment of this force P d measures the torque, exactly as in the Prony brake. Hence it will be seen that the motor has become its own dynamometer, the magnetic friction between the armature and the field-magnet being substituted for the mechanical friction between the pulley and the jaws. A modification of the balance method, due to Hermann Müller, consists in swinging the dynamo in a cradle, pendulum fashion, from the driving shaft, and estimating the power absorbed by the displacement from the vertical line.

M. Marcel Deprez and Professor C. F. Brackett have proposed to apply the balance method to dynamos in action. Professor Brackett places the dynamo in a sort of cradle, balanced on centres that lie in the axis of rotation, and measures the angular reaction-force or torque between the armature and field-magnets, and multiplying this by the angular velocity $2 \pi n$, obtains the value of the power transmitted to the armature. It may seem incredible that the invisible magnetic field in the narrow space between the armature and the pole-pieces should exercise a powerful drag like the drag of a friction-brake, but it is nevertheless true. This mechanical drag exists in every dynamo. For example, Messrs. Crompton's dynamo described on p. 155 absorbed about thirty horse-power. It ran at 1000 revolutions per minute, and the radius of the armature was 4 inches. The circumferential drag was therefore about equal to 300 pounds' weight! No wonder with such a brake-power acting on the armature, the "coils," consisting of drawn copper rods of 0·3 square centimetre section, were found, after the first run, to be raked out of their places by the drag of the magnetic field upon them. This undesirable result was only prevented from recurring by the insertion of boxwood wedges kept in their places by an external strap of thin brass wires.

All these several dynamometric methods necessitate the use of a speed indicator to count the number of revolutions *n*, which enters as a factor into the calculation of horse-power. Too great care cannot be taken, especially in testing *small* machines, that no unnecessary friction be thereby introduced. A flexible connexion, such as a piece of dentist's spring, between the axle of the machine and the axle of the counter appears to be desirable. The number of revolutions *per second n* being known, the angular velocity $\omega = 2\pi n$ can be calculated. This only requires to be multiplied by the torque $T = Fr$ to give the activity or work-per-second *w*. And if T is expressed in pound-feet, then,

$$\text{Horse-power} = \frac{2\pi n F r}{550} = \frac{\omega T}{550}.$$

(*e.*) *Electrical Methods.*—There are several varieties of this modern method of testing, and they involve the use of two, or in some cases three machines. J. and E. Hopkinson[1] propose to use two similar machines, one as generator, the other as motor, connected together both electrically and mechanically. The power given out by the former machine, and that absorbed by the latter are measured electrically. The motor is made to use its mechanical power to aid in driving the generator, and the small additional power required to drive the generator (supplied by a steam-engine) is measured mechanically by a dynamometer. Modifications of this method for the purpose of obviating all mechanical measurements have been suggested by Lord Rayleigh,[2] Captain Cardew,[3] M. Menges,[4] Mr. Ravenshaw,[5] and Mr. Swinburne.[6]

All these methods are far more accurate than the rough mechanical methods of earlier date, and each has its ad-

[1] *Phil. Trans.*, 1886, ii. p. 347. See also *Electrician*, vol. xvi. p. 347, 1886; and *Electrical Review*, vol. xviii. pp. 207 and 230, 1886.
[2] *Electrical Review*, vol. xviii. p. 242, 1886.
[3] *Ibid.*, vol. xix. p. 464, 1886 ; and *Electrician*, vol. xvii. p. 410, 1886 : and vol. xxi. p. 275, 1887.
[4] *Electrician*, vol. xvi. p. 371, 1886.
[5] *Electrical Review*, vol. xix. pp. 424 and 437, 1886.
[6] *Ibid.*, vol. xxi. pp. 181 and 215, 1887.

vantages, but Hopkinson's method requires two similar machines, and Cardew's requires three machines, one of which must be powerful enough to run the other two. In Swinburne's method the loss of power due to resistance of conductors is calculated, and this deducted from the whole loss of power in the machine gives the "stray power" made up of losses due to eddy-currents, friction, and magnetic hysteresis, which cannot conveniently be separately determined. This stray power is determined by using the machine as a motor, the field-magnets being separately excited so that the armature has the same magnetic induction as at full load, the electro-motive-force applied to it being such as to drive it at its normal speed. Only a small generating dynamo is required to furnish the current for this. When matters are so arranged that the machine to be tested runs at its normal speed, the power used in driving the machine (which is measured electrically by taking readings of the volts on the armature and the ampères flowing through it, and multiplying up, thus giving the power in watts), is equal to the stray power at full load.

For detailed accounts of tests on dynamos the student is referred to the following sources:—*Report of Committee of Franklin Institution*, 1878 ; *Crystal Palace Exhibition* 1882, *Report*, by Lieut. F. J. Sprague, United States Commissioner ; also see the Inaugural Address of Prof. W. G. Adams in *Journal Soc. Telegr -Engin. and Electr.*, 1884 ; Thesis of J. W. Howell on tests made at Stevens Institute, reprinted in a volume on *Incandescent Electric Lighting*, published New York, 1883, by Messrs. Van Nostrand ; *Official Report of Munich Electric Exhibition*, 1882 ; also the tests made by Messrs. Alabaster, Gatehouse and Co., reported in *Engineering*, May 2, 1884, and *Electrical Review*, April 26, 1884 ; also *Report of Jurors, Cincinnati Industrial Exposition* 1883 ; also Prof. W. G. Adams' Inaugural Address, *Journal of Society of Telegraph-Engineers and Electricians*, vol. xiv. p. 4, 1885 ; also Reports of ₁Electrical Exhibition at Philadelphia 1884, published in *Journal of the Franklin Institution*, 1885.

APPENDICES.

APPENDIX I.

MEASUREMENT OF COEFFICIENTS OF SELF-INDUCTION AND OF MUTUAL INDUCTION.

SINCE the appearance of the last edition of this work much has been done in the way of devising practical methods for measuring the coefficients of self-induction and of mutual induction.

Coefficient of Mutual Induction.

When there are two circuits in proximity so that the magnetic lines due to currents in either circuit thread through the other circuit, there is said to be mutual induction between the two. Any variation of the current in the one induces electromotive-force in the other ; and the ratio between the induced electromotive-force and the rate of change in the inducing current is called the coefficient of mutual induction. If i_1 be the current in the first circuit, and its rate of variation be $\dfrac{d\,i_1}{d\,t}$, and if E_2 be the electromotive-force induced thereby in the second circuit, then the relation between them may be stated in the form

$$- \mathrm{M}\,\frac{d\,i_1}{d\,t} = E_2\,;$$

where M stands for *the coefficient of mutual induction.* It will obviously vary with the configurations of the two circuits, and will be increased by bringing them nearer together, or by increasing the number of spirals in either or both of them, and also by providing a medium of great magnetic permeability, such as an iron core, to more effectually convey the magnetic lines (due to the current in the first circuit) through the convolutions of the second circuit. If there is iron thus acting to convey the magnetic induction, M will not be constant, but will vary with the permeability of the medium. M may only be considered a constant when iron and other magnetic media are absent.   Assuming the constancy of the permeability, then M is numerically equal to the number of magnetic lines of the first circuit, multiplied by the number of times they are enclosed by the second circuit, under the condition that the first circuit carries unit current.

The most convenient method of measuring the coefficient of mutual induction is that due to Professor G. Carey-Foster,[1] depending on the use of a condenser of known capacity.   Another method, involving the use of earth-inductor coils of known area, better adapted for the measurements of large coefficients of induction between circuits of low resistance, has been proposed by Mr. Bosanquet.[2] The reader is referred to the original papers for the details of the methods.

Coefficient of Self-induction.

Since the magnetic lines due to a current in a circuit thread through the convolutions of the circuit itself, any variation in the current induces an electromotive-force in the circuit itself.   The ratio between this self-induced electromotive-force and the rate of change in the current which causes it, is called the coefficient of self-induction.   If the rate of change in the current be called $\frac{di}{dt}$, and if E be the resulting self-induced electromotive-force, then the relation between them may be stated in the form

$$- L \frac{di}{dt} = E ;$$

where L stands for *the coefficient of self-induction*.   For a given form of coil it is proportional to the square of the number of convolutions. It is increased by introducing an iron core; but if iron, or other magnetic substance whose permeability is variable, is present, L will not be constant.   If, in any case, the permeability be assumed to be constant, then L is numerically equal to the number of magnetic lines due to the current, multiplied by the number of times they are enclosed by the second circuit, under the condition that the circuit carries unit current.

Several practical methods of measuring coefficients of self-induction have lately been devised, of which two may be mentioned. Professors Ayrton and Perry[3] have constructed a special instrument,

---

[1] *Phil. Mag.* [5] xxiii. 121, 1887 ; see also *Electrician*, xviii. 89, 104, 381, 1887.

[2] *Phil. Mag.* [5] xxiii. 412, 1887 ; see also *Electrician*, xviii. 381, 1887.

[3] *Journal Soc. Telegr.-Engineers and Electricians*, xvi. 292, 1887 ; see also *Electrician*, xix. 17, 39, 58, 83, 1887.   The newest form of secohmmeter is described in *Electrician*, xxi. 75, 1888.

known as a *secohmmeter*, the principle of which consists in successively performing a cycle of operations of making and breaking the current in the circuit in question, and of making and breaking a short circuit of a galvanometer, by means of a special commutator which can be worked at definite speeds. M. Ledeboer[1] has devised a method depending upon the introduction into the circuit, of a suspended-coil galvanometer and of resistances such as will just render it aperiodic. M. Ledeboer has also[2] applied a rotating commutator in a cumulative method, closely resembling that of Ayrton and Perry. A most important series of generalisations of the bridge methods of measuring coefficients of self- and mutual induction has been elaborated by Mr. W. E. Sumpner,[3] who has also still more recently investigated[4] the variations of the coefficients of mutual and self-induction.

APPENDIX II.

ON THE ALLEGED MAGNETIC LAG.

IT is often stated on high authority that the cause of the lead which must be given to the brushes of a dynamo is due to a sluggishness in the demagnetisation of the iron core of the armature. The pole induced in the iron core when the armature is standing still is exactly opposite the inducing pole of the field-magnet. The pole induced in the core as it rotates when the dynamo is at work is not immediately opposite the field-magnet pole, but is observed to be a little in advance, apparently dragged round by the rotation. Those who take the view—the erroneous view as I shall show—propounded above, maintain that the induced pole is displaced by reason of a lag in the magnetism of the iron, which they say is due to the slowness with which the iron demagnetises.

Now, as a matter of fact, the sluggishness which is well known to take place in the magnetising and demagnetising of an ordinary electromagnet (with a solid iron core) is not due to any slowness in the magnetism itself, but to the fact that in the mass of the iron eddy-currents are set up whenever the magnetising current is turned off or

---

[1] *La Lumière Électrique*, xxi. 6, 1886.

[2] *Ibid.*, xxiv. 152, 1887.

[3] *Journal Soc. Telegr.-Engineers and Electricians*, xvi. 344, 1887; and *Electrician*, xix. 127, &c., 1887.

[4] *Electrician*, xx. 713, 1888.

on; and these eddy-currents, while they last, oppose the change in the magnetisation. If the iron core be built up of a bundle of thin varnished iron wires, so that no such eddy-currents can be set up, then there is no sluggishness or lag in the magnetising or demagnetising. In fact, this is how the cores are constructed for induction coils in which rapid changes in the magnetism are essential.

Now, for the sake of obviating wasteful eddy-currents, the armature cores of all dynamos are built up either of wires or thin laminæ of iron. If there are no eddy-currents we should expect no lag. The displacement of the induced pole cannot, then, be due to this cause.

Moreover, it is well known that iron demagnetises more rapidly than it magnetises. If there is no greater delay in demagnetising than in magnetising, the explanation given of an alleged lag ceases to have any meaning in its application to the displacement in the induced pole.

With the view of arriving at a more accurate knowledge of this matter, a number of experiments, some of which will be briefly narrated, have been made from time to time. The author is also indebted to Mr. W. M. Mordey for a set of most valuable notes on the movement of the neutral point, which are, on account of their importance, printed, with Mr. Mordey's kind permission, *entire* in Appendix III.

The author took a short iron cylinder and placed it between the poles of a magnet. The intervening magnetic field was then examined by means of iron filings. It resembled Fig. 50, p. 79. The cylinder was then rotated rapidly by a multiplying gear: but no change in the lines of force, nor any displacement in the induced polarity could be detected.

It is of significance to remark that in dynamos built, as some have been, without iron in the armature, it is still found necessary, when they are running and feeding a circuit, to give a forward lead to the brushes. It is certain that in this case the displacement is not due to magnetic lag.

Mr. F. M. Newton has made some hitherto unpublished experiments on one of his dynamos. The field-magnets were separately excited from an independent source, and the armature was driven round at a moderate speed. A piece of cardboard having a hole to admit the passage of the axle, was placed against the pole-pieces, covering one face of the armature. Iron filings were then sprinkled against the card, and arranged themselves along the lines of force. So long as the circuit was open, the pattern delineated by the filings resembled Fig. 50, p. 79, the field being quite straight. But when the circuit was closed so that a current traversed the armature circuit, immediately the lines of filings took up positions resembling Fig. 63, p. 95, the induced poles being displaced in the direction of the rotation. There could be no clearer proof wanted than this that the lead is not due to sluggishness in the demagnetisation of the iron, but that it is due to the diagonal direction of the resultant magnetisation as explained on p. 92.

Lastly, in some recent experiments of Mr. Willoughby Smith[1] upon the induction of electromotive-force in disks of iron and copper set to rotate between the poles of a magnet, it was found that whilst with disks of copper and silver there was a considerable lag, the point of maximum electromotive-force being considerably in advance of the position opposite the inducing poles, there was with an iron disk a far higher electromotive-force and no lag at all.

Experiments made by MM. Bichat and Blondlot (*Comptes Rendus*, 1882) showed that the magnetic lag if it existed at all must be less than 0·000033 sec. in duration.

It would be useless to discuss the question further.

APPENDIX III.

MOVEMENT OF NEUTRAL POINT.

Notes by Mr. W. M. MORDEY.

Experiments with a Victoria Dynamo.

1. Ran at a steady speed of 1100 revolutions per minute—this being the normal speed for this particular machine. Separately excited the fields. Circuit of armature open. Placed the brushes on collector in usual way, moving them till total absence of sparking indicated that they were in the neutral position. Found that this position indicated *no lead*—as far as it was possible to ascertain with the necessarily rough conditions. Width of brush and of "neutral point" made it impossible to be sure of perfect accuracy ; but as far as it went there was found no evidence of *lead*. No current was collected.

The exciting current was now varied—slowly, quickly, gradually, every way, between 4 and 13·7 ampères—the normal exciting current being 6·4 under ordinary working conditions, but no change what-

---

[1] Willoughby Smith, *Volta- and Magneto-Electric Induction.* A lecture delivered at the Royal Institution, June 6th, 1884, see *Electrical Review*, xv. p. 83.

ever could be detected in the position of neutral point, speed kept steady at 1100 revolutions throughout.

The experiments were repeated with small brushes and a voltmeter, always with same results.

2. Brushes as in (1). Current in fields (separately excited) steady at 6·4 ampères. No current collected. Speed varied, up and down, from 250 to 1300 revolutions per minute. Could find nothing to indicate any movement of neutral point.

3. The brushes were then moved a little, first forward, then back, so that they were just off the neutral points, and a very slight sparking was perceptible.

Still keeping exciting current steady, the speed was varied as above, the object being to see whether the alteration of speed would bring the neutral point *up to,* or take it further *away from* the brushes. No alteration occurred, except a very slight increase of sparking at higher speeds, caused probably *not* by an alteration in the position of the neutral point, but by the increase of electromotive-force and larger current as the brush left the segments.

Experiment 3 repeated with constant speed and varying field. Result the same.

The previous notes refer to the case of an armature revolving in a magnetic field, but without the generation of current.

When Current is being Generated.

5. *Experiment.* (*a*). A "Victoria" machine made for 150 volts and 48 ampères was separately excited with a steady current in the fields, and run at full (normal) speed on three equal groups of 50 volt lamps, the groups being in *series.*

(*b*). One group was short-circuited, and the speed was reduced to get the same current and 100 volts on two groups of lamps.

(*c*). Another group was short-circuited, and speed again reduced till the machine gave 50 volts and 48 ampères.

The position of the brushes was the same in all three cases, and the neutral point was evidently the same throughout.

Character of Spark.

The character of the spark at the commutator of a Brush machine affords, to a practised eye, a very accurate index to the correctness of the position of the brushes.

In this machine the coils which are traversing the neutral

point are out of circuit, and if the brushes are even slightly out of the proper position the coils which traverse the neutral position are cut out of circuit a little too soon, or too late, as the case may be; and this causes an alteration in the length and character of the spark.

6. *Experiment:* An ordinary No. 7 (16-light) Brush machine was run on a circuit consisting of resistance coils of suitable size; the current was kept constant at 10 ampères by switching in resistance as the speed rose; the speed was gradually increased from 160 revolutions to 1400 revolutions (the normal speed being 800). The difference of potential at the terminals rose from 90 to 1000. It was found that no change whatever was required in the position of the brushes, and that consequently the alteration of speed did not affect the neutral point.

7. When a Brush machine is run with less than its full complement of lamps, and is made to give the ordinary current, by having its speed reduced, the position of the brushes is always the same. This is the case even with a 40-light machine, throughout its whole working range.

The foregoing notes are of simple experiments made without any special apparatus or precautions, the object being to find whether there is any *practical* movement of the neutral line produced by speed. The conclusions to which they appear to point are :—

1. When no current is being generated, neither speed nor strength of field have any effect on the position of the neutral point, which is the same (sensibly) whether the armature is at rest or revolving.

2. When the strength of the field, and the current in the armature, are constant, the position of the neutral line is likewise constant, speed having no effect on it. This is even the case with a cast-iron armature, like that of the Brush machine.

3. The eddy-currents do not appear to have the effect that might have been expected.

APPENDIX IV.

ON THE INFLUENCE OF POLE-PIECES.

THERE exists a singular dynamo which is best described as a kind of perverted Gramme machine, being, in fact, a double machine with two Gramme rings, each of which, however, has only one pole-piece to furnish it with the requisite magnetic field. Its inventor, Mr. Ball, calls it a uni-polar dynamo, and considers it a great improvement to omit one of the pole-pieces from each armature. Quite apart from the question whether this particular machine may or may not be well constructed in itself, is the more important general question whether or not the omission of one pole-piece or polar surface adds to the

FIG. 370.

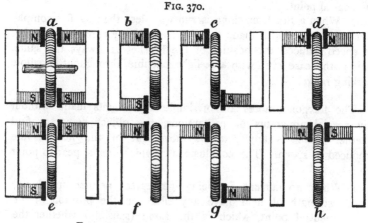

EXPERIMENTS ON INFLUENCE OF POLE-PIECES.

power or efficiency of the machine. All the evidence goes the other way. The omission of the pole-piece—leaving not even a polar face to concentrate the lines of magnetic force—always diminishes the strength of the magnetic field, and increases the tendency to useless scattering of the magnetism. An ordinary 2-pole Gramme dynamo tested by Mr. C. Lever, gave at a speed of 1250 revolutions per minute, 20 ampères at 150 volts, i. e. 3000 watts. When the lower pole-piece and magnet core were removed, it gave at the same speed and with the same resistances only 14 ampères at 90 volts, i. e. 1260 watts.

A set of experiments made in 1884 by Mr. Mordey, with a Schuckert dynamo, are still more conclusive. Eight different arrangements of poles were tried, as indicated in Fig. 370.

The speed was maintained constant at 1180 revolutions per minute. The field-magnets in all cases were separately excited by a current of 4·73 ampères, being the ordinary working field-current of that machine. The external circuit was throughout of 2·2 ohms resistance. The eight results were as follow :—

Experiment.	Potential at Terminals (volts).	Current (ampères).	Watts in External Circuit.
a	69·8	31·22	2179
b	54·5	24·8	1155
c	46·6	21·3	994
d	46·3	21·2	982
e	22·	10·	220
f	21·7	9·9	215
g	1·4	0·6	0·9
h	0	0	0

Experiment *d* corresponds to the case which Ball has patented as an improvement, the ring being here magnetised at one side only, the south pole-pieces being removed. It will be observed that the machine thus *improved* gives only 45 per cent. of its former output !

The author is informed by Mr. D. E. Laire of Cornell University that he has reversed the top magnet of a Ball dynamo so as to magnetise one armature between the two pole-pieces instead of half-magnetising two armatures with one pole-piece each : the result being that the dynamo now yielded from the one armature precisely as high an electromotive-force as previously had been got from the two armatures coupled in series. Ball's method of using two half-magnetised armatures therefore needlessly doubles the internal resistance of the machine besides entailing cost of construction and trouble of attending to two commutators and two sets of brushes.

APPENDIX V.

ON THE INFLUENCE OF PROJECTING TEETH IN RING-ARMATURES.

IN Chapter IV., the author described some experiments showing that the Pacinotti form of ring-armature with teeth possesses decided advantages over the Gramme form of ring without teeth, remarking that in those experiments he had assumed the cost of construction, the liability to heat, and other circumstances of a practical kind to be equal in the two cases.

In Chapter V., the author shows how the employment of armatures with projecting teeth may give rise to wasteful eddy-currents in the field-magnet. It might have been concluded from this that toothed armatures are on this account to be condemned for practical purposes.

But the advantage possessed by the toothed armatures in affording excellent paths for the magnetic lines of force, and thus reducing the magnetic resistance of the gap between armature-core and field-magnet, is one not lightly to be thrown away ; and additional information on the subject confirms the original opinion of the general superiority of the Pacinotti construction.

Mr. Crompton found, in experimenting with armatures of Burgin's kind with protruding corners of iron (p. 152), and with armatures of his own later pattern (p. 153), that the Burgin form was preferable when the field-magnets were relatively weak ; for with a weak field the projecting corners of iron appeared to gather up and concentrate the magnetic lines of force. With strong field-magnets he found little difference. Mr. Esson, in working upon the "Phœnix" dynamo (p. 161), found a similar result—namely, that with an unsaturated field the toothed armature gives a higher electromotive-force than a smooth armature. He has also found that the tendency of the projecting teeth to generate eddy-currents may be to some extent overcome by making the teeth numerous, and is fully overcome by increasing slightly the clearance between the armature and the polar surface. The gain in using teeth more than compensates for any loss entailed in the use of a wider air-gap ; while, at the same time, ventilation is improved, and there is less risk of damage to the coils. The presence of teeth is also a gain from the point of view of ventilation, as they help to radiate from the interior of the coil the heat that would otherwise be confined.

Lastly, in view of the great peripheral drag (p. 602) exerted by the magnetic field on the coils of the armature, tending to drag them round on the core, the existence of iron teeth projecting between the coils is a pure gain ; though, of course, proper precautions must be taken to prevent any short-circuiting between the coils and core by the interposition of suitable insulation of a durable kind.

The effect of the teeth in concentrating the magnetism of the field-magnet is greater when the latter is unsaturated than when it is well saturated. This causes the characteristic curve to be more bent than it would otherwise be, and renders the machine a little less easy to "compound." But here the effect only differs in degree from the effect produced in general by increasing the magnetic capacity of the armature in any dynamo. It has been objected that, if the space occupied by protruding teeth was filled with copper wires, an equal electromotive-force would be generated. Iron is however cheaper than copper. Moreover, as fewer turns of copper wire are required in the armature, the internal resistance is less, and the machine is of itself more nearly self-regulating.

APPENDIX VI.

ELECTRIC GOVERNORS.

THE subject of electric governors for steam engines was briefly alluded to in Chapter VI. No centrifugal governor attached to the steam-engine can keep the speed of the dynamo truly constant; for it does not act until the speed has become either a little greater or a little less than the normal value. Few mechanical governors will keep the speed within 5 per cent. of its proper value, under sudden changes of load. Hence the suggestion which underlies all electrical governors, that the admission of steam from the boiler to the engine should be controlled by the electric current itself, the speed of driving being varied according to the demands of the circuit. Numerous suggestions of a more or less practical nature have been made by Lane-Fox, Andrews, Richardson, and others. The three forms which will here be described are those which are in practical use.

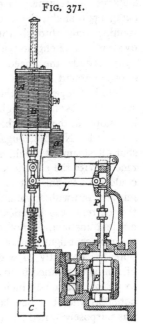

FIG. 371.

RICHARDSON'S ELECTRIC GOVERNOR.

Richardson's Governor.—This governor is used to maintain either a constant current or a constant potential. In the former case its coils are included in the main circuit and are of thick wire: in the latter they are arranged as a shunt to the mains and are of fine wire. The arrangements are shown in Fig. 371.

The valve which admits steam to the engine is a double-beat equilibrium valve E; its stalk passes upwards and is acted upon by a plunger P, which is pressed down by the shorter end of a lever L, which is in turn connected with a long vertical spindle having a weight

C at its lower end, and at its upper end carrying the iron core B, surrounded by the solenoid A. A spring S counterpoises the slight upward pressure of the steam on the valve. When the current passes through the solenoid A, it lifts the core B to a certain height and admits to the engine a sufficient quantity of steam to drive the engine at the speed requisite to maintain the current. Should the resistance of the circuit be increased by the introduction of additional lamps, the core B will fall a little, thereby turning on more steam, until the speed has risen to that now necessary. For additional safety a separate electromagnet *a* is added, which when in action holds up the heavy iron block *b*. Should the circuit from any cause be broken, the block *b* instantly descends and cuts off the steam. In some experiments made at Lincoln in 1883 in the author's presence on a Brush 16-light machine fitted with a Richardson's governor, the following results were attained:—Seventeen arc lamps being alight, six were suddenly switched off: in four seconds the speed of the engine came down from 138 to 107, and the current which was 10·2 ampères had returned to exactly the same value. Seventeen lamps being again alight, the whole were short-circuited, leaving the current running only through the governor and the field-magnets of the dynamo. The engine pulled up in less than one stroke, and in fourteen seconds the speed had come down to 24, the engine just crawling round at a speed sufficient to keep the magnets charged. In another experiment the circuit of the whole seventeen lamps was suddenly broken, the engine running at 140. In fifty-five seconds it had stopped, the steam having been cut off in less than a quarter of a second. No centrifugal governor could have so instantaneously shut off steam: it would not have acted until the engine began to race. With the electric governor the steam was cut off before racing could even begin. At all speeds from 25 up to 146 revolutions per minute, and with any number of lamps from none to seventeen alight, the current was practically kept at a constant value in a most efficient manner. Another of these governors connected with an incandescent-light system working at 92 volts was found to keep the potential correctly to within 1 per cent., even though the number of lamps was varied from 91 to 31, and the boiler-pressure from 32 to 55 lbs. per square inch. It also maintained an absolutely constant potential when but one lamp was alight, though the boiler-pressure was purposely varied from 31 to 55 lbs. per sq. inch.

Willans' Governor.—This instrument has been applied with great success at Victoria Station and elsewhere. In common with Richardson's governor, it employs the attraction exerted by a solenoid on an iron core to actuate an equilibrium valve; but the action is indirect, the solenoid core operating on the small valve which controls a hydraulic piston, the latter in turn controlling the large steam valve. The arrangement is shown in Fig. 372, where T is the large

FIG. 372.

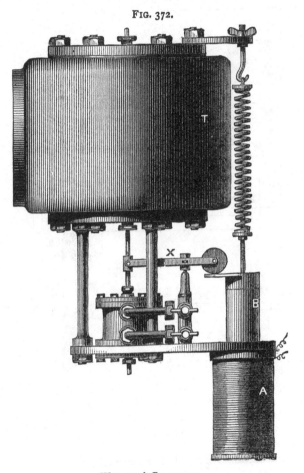

WILLANS' GOVERNOR.

piston throttle-valve. The throttle-valve spindle passes downwards and is connected direct to the piston of the hydraulic relay. The solenoid A attracts its core B suspended on a spring. The position of B determines that of the lever X, which is connected at one point to the spindle of the throttle-valve and at another to that of the small controlling valve.   If the potential at the mains falls, less current flows round A, in consequence B rises, and its projecting ear-piece raises the lever X, admitting more water above the controlling piston,

Fig. 373.

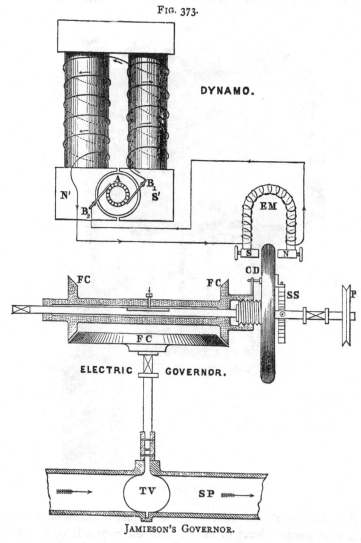

JAMIESON'S GOVERNOR.

which consequently sinks, drawing down the throttle-valve with great power and admitting more steam to the driving engine.   A comparatively small solenoid, actuated by but o·3 ampère of current and absorbing only about 32 watts of power, may thus bring a force of many pounds to bear upon the steam valve, and will control with ease a 60 horse-power engine.

Jamieson's Governor. — This consists of a copper disk C D (Fig. 373) revolving between the poles of a small electromagnet E M, and actuating a throttle-valve, T V, by means of a spring and a cone gearing. The pulley P is driven by a band from the dynamo shaft. Its spindle passes loose through the copper disk, to which its motion is communicated through the spiral spring S S. The disk has projecting at one side a screw which engages in a sliding sleeve having two friction-cones F C. Between these is a third friction-cone connected with the throttle-valve : the latter being turned whenever either of the two rotating cones is pressed against the third cone. The action of this governor is as follows. Suppose the normal number of lamps to be alight : there will be a certain current flowing through the electromagnet E M, and the copper disk rotating between its poles will experience a certain drag owing to the reaction of the eddy-currents generated in it. This drag is balanced by the spiral spring, which is therefore under a certain normal strain. Should the number of lamps on the circuit be diminished, more current will be thrown round the electromagnet, which will consequently exercise a greater drag upon the copper disk. The effect of this will be to coil up the spiral spring and to bring the left-hand driving cone into contact with the large horizontal cone, which in consequence will turn and close the valve sufficiently to reduce the flow of steam to such a point that the engine will drive the dynamo more slowly, the dynamo will send a smaller current through the electromagnet, the electromagnet will exercise a less drag on the disk, the spiral spring will uncoil, and the friction-cones will fall out of gear, leaving the throttle-valve in its latest position until some further change is needed.

Elihu Thomson's Dynamometric Governor. — This is a dynamometric governor, which alters the lead of the brushes in correspondence with any change which may occur in the driving force. The belt which passes from the driving pulley to the dynamo is under unequal tension in its two parts, the difference of tension between the upper and lower portions being greater according to the torque on the shaft of the dynamo. With greater tension, there will be a greater upward force exerted upon a small tenting pulley, the rise of which will shift the brushes forward. No information as to the practical working of this interesting form of governor is yet to be obtained.

Goolden and Trotter's Governor. — This governor, used to maintain a constant current, is a mechanical device, driven from the shaft of the dynamo, for automatically introducing resistance into part of the exciting circuit. The mechanical gear is controlled by the core of a solenoid introduced into the circuit. It is figured in the *Electrical Review*, xix. 289, 1886.

Other governors have been devised by Statter (see p. 444), Menges, and others.

Further information respecting electric governors and their actual applications in various installations of electric lights may be found in the following papers :—A. Jamieson, *Electric Lighting for Steamships*, Proc. Inst. Civil Engineers, vol. lxxix., Session 1884–5, part i. ; P. W. Willans, *The Electric Regulation of the Speed of Steam Engines*, Proc. Inst. Civil Engineers, vol. lxxxi., Session 1884–5, part iii.

APPENDIX VII.

Frölich's Theory.

SINCE the publication of the original edition of this work, several important additions have been made to the theory of the dynamo. Dr. Frölich, upon whose work of 1881 the theory propounded in 1884 by the present author, and now revised in Chapter XIX., was avowedly based, published in 1886 a volume under the title *Die dynamoelektrische Maschine*, giving more full developments to his theoretical work. It may therefore be convenient to give here an abstract of Frölich's theory, as well as a *résumé* of his more recent studies.

Frölich's theory is based upon (1) Faraday's law of induction, (2) Ohm's law, (3) a curve expressing certain results of experiments made with a series-wound dynamo.

The electromotive-force E is proportional to the number of revolutions per second n and to the "effective magnetism" M of the machine ; so that

$$E = M\,n. \tag{1}$$

By the "effective magnetism" Frölich understands what we should term "the number of magnetic lines of force cut in one revolution," a quantity which depends on the construction and number of coils of the armature, as well as on the construction and degree of excitation of the field-magnet, and is equal to our CN (p. 337).

By Ohm's law the current i is equal to the electromotive-force E divided by the whole resistance in the circuit R.

$$i = \frac{E}{R};$$

whence

$$i = \frac{M\,n}{R}.$$ [2]

But since, in a given dynamo, M is a function of i and not directly of n or R, if we divide both sides of this equation by M, we obtain on the left a function of i only, and on the right n/R. The current then depends only on the *ratio* between the speed and the resistance. It then remains for experiment to determine whether the current is or is not proportional to the values of this ratio. Accordingly experiments are made at different speeds and with different resistances, and the results are examined by plotting out as ordinates the values of i and as abscissæ the corresponding values of n/R. The curve so obtained—and which Frölich names as the *current-curve*—is depicted in Fig. 374. It will be seen that with

FIG. 374.

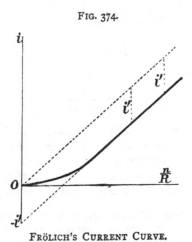

FRÖLICH'S CURRENT CURVE.

small values of n/R the current is inappreciable; but that when n/R has reached a certain value the corresponding values of i increase regularly from that point, the curve consisting of an oblique line, which, however, does not pass through the origin, but, if prolonged backwards, crosses the vertical axis at a point marked $-i'$. In other words, i is a simple linear function of n/R,[1] and its values may be represented by an equation of the form

$$ i = c\,\frac{n}{R} - d, \qquad\qquad [3] $$

where c and d are constants, the nature of which will be discussed later. We have now to compare equation [2] which expresses well-established theory, with equation [3] which merely expresses certain observed facts. They both contain the quantity $\dfrac{n}{R}$ and may be re-written as follows :

$$ \frac{i}{M} = \frac{n}{R}\,; $$

$$ \frac{d+i}{c} = \frac{n}{R}\,; $$

whence

$$ \frac{d+i}{c} = \frac{i}{M} $$

and

$$ M = \frac{c\,i}{d+i} $$

now form two new constants, $a = \dfrac{d}{c}$, and $b = \dfrac{1}{c}$, and we may write the last expression as

$$ \cdot M = \frac{i}{a+b\,i}\cdot $$

This equation, giving the "effective magnetism" in terms of the current and the two constants a and b, is nothing more or less than the equation of an electromagnet such as is discussed in Chapter XIV. It comes out here, however, as the result of applying the laws of Faraday and Ohm to the observed facts of a series dynamo.[2]

---

[1] This is true for the series dynamo only, not for shunt dynamos or compound dynamos. Further, it is found that the curves taken while n is constant and R is varied, do not quite agree with curves taken while R is constant and n is varied.

[2] It is extremely interesting and suggestive to see how Dr. Frölich arrived thus at the equation that has formed the basis of so much work of later date. It was on finding that so simple an equation for the magnetism of an electromagnet was adequate to express the facts in the case of the series-wound dynamo, that induced the author of this book to apply an equation of the same form to build up the equations of the shunt dynamo in Chapter XIX. It should also be remarked

Frölich next remarks that if the values of a and b are deduced from experiments made on different machines, they are different in different cases because into each of them the number and size of coils in the armature comes in as an element. M, in fact, as used up to this point by Frölich, corresponds to the product C N as those symbols are used in Chapters XV. to XXI. of this book ; and Frölich's a is the reciprocal of S the number of turns of wire on the field-magnet, and his b is analogous to the saturation-constant σ. It may be remarked incidentally that the reciprocal of b is numerically the maximum value to which M can attain, when i is indefinitely increased, and his d is the half-saturated or *diacritical* current.

Frölich now makes the remark—with which the author of this work entirely disagrees—that it is better to express the effective magnetism of a dynamo as a relative, and not as an absolute quantity : and he therefore proceeds to so modify the formula that the maximum magnetism for each individual machine shall be called $= 1$ when saturated. Accordingly we write $b = 1$, and the expression may then be still further modified by writing m for $\frac{1}{a}$, giving us :

$$M = \frac{i}{a+i} = \frac{\frac{1}{a}i}{1 + \frac{1}{a}i} = \frac{m\,i}{1 + m\,i}. \qquad [4]$$

Here m is the number of turns of wire on the field-magnet (or some power of that number). The product $m\,i$ represents the number of ampère-turns. If values are assigned to $m\,i$, and the corresponding values of M are calculated and plotted out, we get the " characteristic " curve of that electromagnet, or, as Frölich calls it, the " curve of magnetism."

Now substitute in equation [2] the value of M as expressed in equation [4], and we get :

$$i = f\frac{n}{R} - \frac{1}{m}, \qquad [5]$$

---

that the equations for the series dynamo in Chapter XIX. are similarly built up from the law of the electromagnet, and are not deduced from the "current-curve" which forms the starting-point in Frölich's way of stating his theory. He deduces the law of the electromagnet from observations on the dynamo. I have built up the law of the dynamo from the law of the electromagnet. Though my method is more logical, Frölich's is the original one, and has the merit not only of being anterior, but of being anterior in a field where others, including such names as Mascart, Schwendler, Herwig, Meyer and Auerbach, had sought in vain for a feasible solution. The utmost credit is therefore due to Dr. Frölich.

which is the equation for the current of a series dynamo. The symbol f stands for a constant " the introduction of which," says Dr. Frölich, " is necessitated by the removal of constant b." [1]

The term $\dfrac{1}{m}$ which relates to the power of the field-magnet coils to magnetise, acts on the machine in such a way as to cause the current to be less than it would otherwise be at the given speed. We see also that, if R is given there will be a certain minimum speed below which the machine will give no current. For this reason Dr. Frölich gave the name of " the dead turns " to the quantity $\dfrac{1}{m}$. (A better name would have been " dead current," for it is equal to i' in Fig. 374.)

So far we have been following the theory as propounded by Dr. Frölich in 1880–1. That which follows has been only published so recently as 1885.

Each of the two terms in the right-hand half of the equation [5] possesses a physical meaning : both of them are particular values of current. The first, $f\dfrac{n}{R}$, is the maximum current which the machine could give with given speed n and resistance R, when the field-magnet was saturated to magnetism $= 1$. This may be seen from the " current-curve," Fig. 374. If in this diagram we draw a line through the origin parallel to the " current-curve," the equation of that line will be

$$I = f\frac{n}{R},$$

or it is the line of maximum current for the given value of n/R. The value of the actual current for any given value of n/R is obtained by subtracting from the corresponding value of I the quantity i', which has the value

$$i' = \frac{1}{m}.$$

Now, as will be seen by reference to equation [4], i' or $\dfrac{1}{m}$ is that

---

[1] That it should be needful to reintroduce the constant b under this form shows that b ought not to have been removed—in other words, that Dr. Frölich's plan of taking relative instead of actual values for the magnetism is a mistake. Further, if b has to be introduced under this form it ought to be multiplied into both terms of the expression.

value of current which will excite the field-magnet to the degree $M = \frac{1}{2}$.[1]

Dr. Frölich then asserts that a similar division into two terms will occur in the expressions for electromotive-force and for potential at terminals, not of series-wound dynamos only, but also of shunt and compound dynamos : and he formulates this rule :—In order to find the value of any electrical quantity (except electric energy) for a machine of given winding, consider the machine under two different circumstances (without altering the given speed or given resistance) : 1st, when the field-magnets are independently excited to saturation $(M = 1)$; 2nd, when the currents in the field-magnet coils have such values as will evoke half-magnetism $(M = \frac{1}{2})$; calculate the desired electric quantities for both these circumstances, *the difference between the two values is then equal to the real value which was to be found.*

Following out this principle, Dr. Frölich gives (March 1885), but without proofs, the following resulting expressions. [The notation has been here adapted to agree with that used in Chapters XV. to XXII.]

Series Dynamo.

$$i_a = i = \frac{fn}{r_a + r_m + R} - \frac{1}{S}.$$

Shunt Dynamo.

$$i_a = \frac{fn}{r_a + \dfrac{r_s R}{R + r_s}} - \frac{1}{Z}\frac{R + r_s}{R}.$$

$$i_s = \frac{fn}{r_a \dfrac{R + r_s}{R} + r_s} - \frac{1}{Z}.$$

$$i = \frac{fn}{r_a \dfrac{R + r_s}{r_s} + R} - \frac{1}{Z}\frac{r_s}{R}.$$

$$E = fn - \frac{1}{S}(r_a + r_m + R).$$

$$E = fn - \frac{1}{Z}\left(r_a \frac{R + r_s}{R} + r_s\right).$$

$$e = \frac{fn}{1 + \dfrac{r_a + r_m}{R}} - \frac{R}{S}.$$

$$e - \frac{fn}{1 + r\left(\dfrac{1}{r_s} + \dfrac{1}{R}\right)} - \frac{r_s}{Z}.$$

---

[1] This corresponds partially to the anterior proposition of the author that the saturation-coefficient was the reciprocal of that number of ampère-turns that would saturate the field-magnet to the "diacritical point" of half-permeability; a proposition of extreme importance, which the author showed to be applicable to all cases of dynamos, series, shunt, and compound wound.

Compound Dynamo (*Short Shunt*).

$$i = \frac{fn}{r_a + (R+r_s)\dfrac{r_a+r_s}{r_e}} - \frac{\dfrac{1}{R+r_m}}{\dfrac{Z}{r_e}+\dfrac{S}{R+r_m}}.$$

$$e = \frac{fn}{\left(1+\dfrac{r_m}{R}\right)\left(1+\dfrac{r_a}{r_e}\right)+\dfrac{r_a}{R}} - \frac{\dfrac{R}{R+r_m}}{\dfrac{Z}{r_e}+\dfrac{S}{R+r_m}}.$$

Compound Dynamo (*Long Shunt*).

$$i = \frac{fn}{r_a+r_m+R\left(1+\dfrac{r_a+r_m}{r_e}\right)} - \frac{\dfrac{1}{R}}{\dfrac{Z}{r_e}+S\left(\dfrac{1}{R}+\dfrac{1}{r_e}\right)}.$$

$$e = \frac{fn}{1+(r_a+r_m)\left(\dfrac{1}{R}+\dfrac{1}{r_e}\right)} - \frac{1}{\dfrac{Z}{r_e}+S\left(\dfrac{1}{R}+\dfrac{1}{r_e}\right)}$$

Dr. Frölich regards the latter formulæ for compound winding (the sufficiency of which he has proved by finding them to agree exactly with the results of observation of Siemens' dynamos) as marking an essential step forward. They are undoubtedly an advance upon the approximate equations used in 1884 by the author, which, however, led the author to entirely correct practical deductions.

Dr. Frölich was led by his theory to discover a method of compound winding giving "a fairly constant current" in the external circuit, the machine being wound as a shunt machine with a few turns of series coil connected so as to be traversed by the current in the opposite sense—a differential compound winding.

The next portion of Dr. Frölich's work is devoted to the application of the theory to winding machines so as to produce a constant potential at terminals. He points out that in both kinds of compound winding, the potential at terminals may be expressed in the form

$$e = A\,\frac{R}{R+\alpha} - B\,\frac{R}{R+\beta}.$$

He then shows that if $\beta > \alpha\dfrac{A}{B}$, e will be a maximum with small values of R, and that if $\beta < \alpha\dfrac{A}{B}$, e will be a maximum with large values of R; giving the conclusion that the condition of self-regulation is

$$\beta = \alpha\frac{A}{B}.$$

For the further deductions the reader is referred to the *Elektro-technische Zeitschrift*, April 1885, and to Dr. Frölich's book.

Dr. Frölich next considers the case of compound-wound motors, and shows that because the magnetism is not proportional to the current, machines constructed according to the theory of Professors

Ayrton and Perry, with differential windings, will not realise the condition of running at a constant speed. He considers that for working from mains supplied at a constant potential, electric motors should be shunt - wound rather than differentially - wound ; and that it is even preferable in practice to add a few turns of series winding to add to the magnetism given by the shunt coils, though, according to him, neither of these will give a really constant speed.

Finally Frölich returns to the "current-curve" from which he started, and shows that if the residual magnetism of the core be taken into account and equation [4] is written as

$$M = \frac{\mu + mi}{1 + mi},$$

the expression for i which results when this new value is introduced into equation [2], gives, when plotted out, the precise form of the current curve. Dr. Frölich's treatment of the subject has been criticised by Professors Meyer and Auerbach,[1] and Professor von Waltenhofen,[2] to whom Dr. Frölich has recently[3] replied.

APPENDIX VIII.

CLAUSIUS'S THEORY.

THIS theory, published in *Wiedemann's Annalen*, xx. 353–390, 1883 (translated in *Phil. Mag.* [V.], xvii. 50 and 119, 1884), is based upon an electrodynamic method of treatment. Clausius regards the dynamo as consisting essentially of two circuits, one surrounding the iron of the field-magnet, the other surrounding the iron core of the armature part. He treats the magnetised iron as though, on Ampère's theory, it consisted of innumerable closed electric circuits. Suppose, then, that currents i, i_1, i_2, &c., are flowing in a system of closed circuits s, s_1, s_2, &c., and that these are to act on a closed circuit σ in which unit current is flowing. Then taking ds as an element of any one of these many circuits, $d\sigma$ as an element of the circuit σ, and $(s\sigma)$ as the angle between these elements, and r their distance apart, we may write the following expression for the

[1] *Elektrotechnische Zeitschrift*, vii. 240, 1886.
[2] *Zeitschrift für Elektrotechnik*, 1885, 549: and 1886, 450; see also *Elektrotechnische Zeitschrift*, viii. 166 and 389, 1887.
[3] *Elektrotechnische Zeitschrift*, viii. 161, 217, and 394, 1887.

electrodynamic potential, on the circuit σ, of the given current system

$$W = \int\!\!\int \frac{i\cos(s\,\sigma)}{r}\,ds\,d\sigma;$$

where one integration is taken round σ, and the other round the whole system of s circuits. Then any electromotive-force induced in circuit σ by a change in the configuration or strength of currents in the system may be written as

$$e = -\frac{d\,W}{d\,t}.$$

Now, confining our attention to one "section" only of the rotating circuit, and considering that the mutual potential between it and the s system varies between two extreme values W' and W'', which it has at the two positions where it passes the brushes, and writing for the time of one rotation τ, we get for the average electromotive-force in any one section

$$\frac{1}{\tfrac{1}{2}\tau}\int_{t}^{t+\tfrac{1}{2}\tau} e\,dt = \frac{1}{\tfrac{1}{2}\tau}(W'-W'').$$

Taking the number of sections from brush to brush as $\tfrac{1}{2}n$, and multiplying up, we get as the average total electromotive-force in the moving circuit

$$E_1 = \frac{n}{\tau}(W'-W''),$$

or, writing v for the number of revolutions per second (the reciprocal of τ)

$$E_1 = n\,(W'-W'')\,v.$$

Next Clausius considers the possible electromotive-force induced in the fixed circuit by the reaction of the moving circuit, and finds an expression for the electrodynamic potential between the former and one element of the latter carrying current j. But on applying to the expressions the fact that in each section the current j is reversed in sign at each half-revolution, it at once follows that the average electromotive-force induced in the fixed circuit is zero. On further considering the mutual reactions between the separate sections of the moving circuit, Clausius considers that there will result a residual electromotive-force proportional to the current and to the speed, and to a certain constant denominated by the symbol ρ, depending on the construction of the machine, and which will be of a sign opposing the main electromotive-force ; or

$$E_2 = - \rho \, i \, v.$$

This makes the net average electromotive-force

$$E = E_1 + E_2 = n \, (W' - W'') \, v - \rho \, i \, v.$$

The work per second done by this electromotive-force is found by multiplying together E and i;

$$E \, i = n \, (W' - W'') \, i \, v - \rho \, i^2 \, v.$$

But the work done against ponderomotive forces may be written

$$T = n \, (W'' - W') \, i \, v \, ;$$

whence

$$E \, i = - \, T - \rho \, i^2 v.$$

Next Clausius takes up the magnetic question, and writing the Frölich equation for the magnetic moment of an electromagnet in the form

$$M = \frac{A \, i}{1 + a \, i},$$

he proceeds to write expressions for the two magnetising forces which act on the armature core, namely

$$\frac{C \, M}{1 + \gamma \, M}$$

for the force due to the field-magnet, and

$$\frac{C \, N}{1 + \gamma \, N}$$

for the force at right angles due to the current in the armature, N being proportional to i and to another constant called B. The resultant magnetising force may be written

$$P = \frac{C \sqrt{M^2 + N^2}}{1 + \beta \, i},$$

where β is another constant. This makes the angle of lead such that

$$\cos \phi = \frac{M}{\sqrt{M^2 + N^2}} \; ; \; \sin \phi = \frac{N}{\sqrt{M^2 + N^2}}.$$

He then resolves the magnetic moment P into two components P_1 and P_2, the former parallel to the magnetic moment of the fixed magnet, and the latter at right angles to it. Then he considers the ponderomotive force which the fixed magnet exerts upon the coils carrying the armature currents, and the work per second done by this force he writes as

$$- h\,\mathrm{M\,N}\,v,$$

h being a coefficient, and the negative sign signifying that in driving the machine the work has in reality to be done against this force. To the above work he adds that done by the ponderomotive force due to the magnetic movement of the iron core, or rather to that component of it named P_1. This gives as the work performed by the ponderomotive forces in unit time

$$\mathrm{T} = - h\,\mathrm{M\,N}\,v - \mathrm{K\,N\,P_1}\,v;$$

where K is another constant.

This equation when compared with previous expressions gives

$$\mathrm{E}\,i = h\,\mathrm{M\,N}\,v + \mathrm{K\,N\,P_1}\,v - \rho\,i^2\,v;$$

$$\mathrm{E}\,i = \mathrm{M\,N}\left(h + \frac{k\,\mathrm{C}}{1 + \beta\,i}\right)v - \rho\,i^2\,v.$$

The author considers that this equation which is true for the case where the armature core is stationary, would be true if the rotation were slow, but if rotation were quick would require to be modified; the angle which the magnetic axis of the armature core forms with that of the fixed electromagnet being now written as

$$\phi' = \phi + \epsilon\,v,$$

where ϵ is a small constant; the magnetic moment P must also be changed to

$$\mathrm{P}' = \mathrm{P}\cos\epsilon\,v,$$

and then P_1 becomes

$$\mathrm{P_1}' = \frac{\mathrm{C}}{1 + \beta\,i}\,(\mathrm{M} - \epsilon\,v\,\mathrm{N}).$$

The effect of eddy-currents in the iron core is taken into account by writing for their magnetic moment the product

$$\eta\,v\,\sqrt{\mathrm{M^2 + N^2}},$$

where η is a small constant; giving a new value for P

$$\eta\, v \sqrt{\text{M}^2 + \text{N}^2}\left(\text{I} + \frac{\text{D}}{\text{I} + \beta\, i}\right),$$

where D is a new constant, corresponding to C, for this case; then writing

$$\epsilon' = \epsilon + \frac{\text{D}}{\text{C}}\,\eta,$$

one finally gets as the fundamental equation

$$\text{E} = \left[\frac{\text{A}\,\text{B}}{\text{I} + a\, i}\left(h + \frac{k\,\text{C}}{\text{I} + \beta\, i}\right) v - \rho\, v - k\,\text{B}^2\left(\eta + \frac{\epsilon'\,\text{C}}{\text{I} + \beta\, i}\right) v^2\right] i.$$

At this point Clausius suddenly changes all his symbols, substituting for them the following set :—

$$a = \frac{\text{I}}{a};\; b = \frac{\text{I}}{\beta};\; p = \frac{h\,\text{A}\,\text{B}}{a};\; q = \frac{k\,\text{A}\,\text{B}\,\text{C}}{a\,\beta};\; l = \frac{\text{A}}{\text{B}\,a};$$

$$\sigma = k\,\text{B}^2\,\eta;\; \lambda = \frac{k\,\text{B}^2\,\text{C}\,\epsilon'}{\beta};\; \text{R} = \frac{\text{E}}{i};\; w = \frac{v}{\text{R} + \rho\, v + \sigma\, v^2};$$

the substitution of which leads to a quadratic equation, the positive root of which is taken. Then by further writing

$$w' = w\left(\text{I} - \frac{\lambda}{p}\, v\right),\; \text{and } c = q - p\, a + p\, b,$$

the equation of the machine finally becomes

$$i = \tfrac{1}{2}\,(p\, w' - a - b) + \tfrac{1}{2}\sqrt{(p\, w' + a - b)^2 + 4\, c\, w}.$$

The new symbols relate, however, to groups of quantities rather than to definite physical entities, and the simplicity of this last expression is unreal. It is impossible to attach a physical significance to the terms which appear in the equation. Clausius has, however, attempted to apply the investigation to the problem of electric transmission of power. See *Wiedemann's Annalen*, xxi. 385, 1884. Stern has further developed the ideas of Clausius in the *Elektrotechnische Zeitschrift*, vii. 14, 1886.

APPENDIX IX.

Rücker's Theory.

PROFESSOR RÜCKER employs a very convenient generalised diagram
of the dynamo, Fig. 375.   Three points A X Y are joined by

FIG. 375.

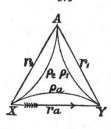

GENERALISED DIAGRAM
OF DYNAMO.

three straight lines representing resistances,
and three curves representing magnetising coils.   X Y represents the armature of
the machine, the current flowing from
X to Y.   Y is the positive brush or terminal.
Either Y A or A X may represent the external resistance.   Let the difference of
potential between Y and A be called e_1 and
that between A and X e_2.   The currents in
r_1 and ρ_1 may be represented by i_1 and γ_1;
and similarly for the other currents.   Let the
resistances of the multiple arcs between Y A
and A X be called respectively R'_1 and R'_2.

All existing forms of dynamo can then be represented by leaving
out some of the conductors of this generalised system.

For example, the ordinary series machine (with the notation used
in Chapters XV. to XXII.) will be as shown in Fig. 376.   There is an

FIG. 376.   FIG. 377.   FIG. 378.

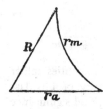

SERIES MACHINE.

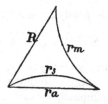

COMPOUND MACHINE.
SHORT SHUNT.

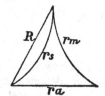

COMPOUND MACHINE.
LONG SHUNT.

advantage in this generalised diagram and in Rücker's notation in
one respect.   The shunt in the compound dynamo may be either
a shunt to the armature alone ρ_a, Fig. 375 [r_s of this book], or it may
be a shunt to the external circuit ρ_2 of Fig. 375 [r_s' of Fig. 378, the
accent being used here to distinguish it from the short shunt].   If it
is agreed that r_2 of Fig. 375 shall always represent the external resist-

ance (r_1 being in any practical case infinite), then ρ_1 represents the series coils [r_m of this work] and ρ_2 the long-shunt coil.

Adopting Frölich's equation for magnetism (see Appendix VII.), Rücker then writes:

$$E = \frac{n\,M\,(s_a\,\gamma_a + s_1\,\gamma_1 + s_2\,\gamma_2)}{1 + \sigma\,(s_a\,\gamma_a + s_1\,\gamma_1 + s_2\,\gamma_2)},$$

where M is a constant of the machine $[= C\,\bar{N}\,/\,(S\,i)'$ of this book], s_a, s_1, s_2, the numbers of turns of wire in the respective magnetising spirals, and σ the usual saturation constant (see p. 307). The products $s_a\,\gamma_a$, &c., are numbers of ampère-turns due to the magnetising currents. Were there initial permanent magnetism, an additional fictitious term $s_0\,\gamma_0$ might be added.

The reader is referred to Prof. Rücker's paper for the steps of the argument by which (neglecting residual magnetism and reactions of armature) he arrives at the following expressions for potential and current in one of the two kinds of compound dynamo, viz. the short shunt.

$$e = \frac{1}{\sigma}\left\{\frac{n\,M\,R\,r_s}{R\,(r_a + r_s) + r_m(r_a + r_s) + r_a r_s} - \frac{R\,r_s}{R\,Z + (Z\,r_m + S\,r_s)}\right\}$$

$$i = \frac{1}{\sigma}\left\{\frac{n\,M\,r_s}{R\,(r_a + r_s) + r_m(r_a + r_s) + r_a r_s} - \frac{r_s}{R\,Z + (Z\,r_m + S\,r_s)}\right\};$$

Now simplify by grouping as follows :—

$$P_1 = \frac{n\,M\,r_s}{\sigma\,\{r_m(r_a + r_s) + r_a r_s\}}, \qquad A_1 = \frac{r_a + r}{r_m(r_a + r_s) + r_a r_s}$$

$$Q_1 = \frac{r_s}{\sigma\,(Z\,r_m + S\,r_s)}, \qquad B_1 = \frac{Z}{Z\,r_m + S\,r_s}, \qquad x_1 = \frac{1}{R}$$

$$P_2 = \frac{n\,M\,r_s}{\sigma\,(r_a + r_s)}, \qquad A_2 = \frac{r_m(r_a + r_s) + r_a r_s}{r_a + r_s}$$

$$Q_2 = \frac{r_s}{\sigma\,Z}, \qquad B_2 = \frac{Z\,r_m + S\,r_s}{Z}, \qquad x_2 = R\,;$$

and we get

$$e = \frac{P_1}{A_1 + x_1} - \frac{Q_1}{B_1 + x_1}$$

$$i = \frac{P_2}{A_2 + x_2} - \frac{Q_2}{B_2 + x_2}.$$

It will be noted that x_1 which appears in the expression for potential is the conductivity, and x_2 which appears in that for current is the resistance of the external circuit.

The equations for long-shunt machines are similarly treated by Prof. Rücker, their symbols grouped and generalised, and expressions of similar form are deduced.

It is therefore clear that we may represent in a general equation of the form

$$\phi = P/(A + x) - Q/(B + x),$$

either the current or the potential at terminals of any compound dynamo; the quantities A, B, P, and Q being independent of the variable external resistance, and x being either the resistance or else the conductivity (according whether we are expressing current or potential) of the external part of the circuit.

The constant A is always either the resistance (or else the reciprocal of the resistance) that the machine would offer when at rest, to the current of a battery placed in the external circuit. B is also either a resistance or the reciprocal of one. P is always either a current or an electromotive-force, and is proportional to the speed and to the maximum magnetism of the field-magnet. Q is also either a current or an electromotive-force, and depends upon the currents requisite to circulate in the magnetising coils in order to magnetise them to the [diacritical] point of half-saturation.

Starting from this generalised formula, Rücker then investigates the conditions of efficiency, of maximum power, and of self-regulation. It is impossible to give an abstract of the mathematical argument; but certain results may be stated.

If ϕ represent potential at terminals, then $P/Q >$ or < 1 according as the speed is greater or less than the critical speed for the resistance of the circuit when R = o.

If ϕ represent external current, then $P/Q >$ or < 1 according as the speed is greater or less than the critical speed when R = ∞.

A large value of A is favourable to good self-regulation.

For a given usual value of R, a high maximum efficiency is favourable to a large value of A in the case where ϕ represents e; but is favourable to a small value of A where ϕ represents i.

Hence it is more difficult to combine high efficiency of a machine with an approximately constant current than with an approximately constant potential.

APPENDIX X.

On the Proper Size of Armature and Field-magnet Windings.

Professors Ayrton and Perry have recently investigated[1] the rule for the best thickness of the conductors on armatures. Using the same notation as in the rest of this book, their argument is as follows :—The power developed by the armature may be written

$$w = E\,i_a = \frac{n\,C\,N}{10^8}\,i_a;$$

or writing a for the number of ampères per square centimetre in the cross-section of the armature, r for the external radius of armature core, and t for thickness of external winding on armature,

$$C\,i_a = 2\,a \times 2\,\pi r t;$$

whence

$$w = \frac{4\,\pi r n\ N.t\,a}{10^8}.$$

Now, writing ρ as the resistance of the armature winding per centimetre cube (including copper and insulation), the watts wasted per cubic centimetre in heating will be $= \rho\,a^2$; and the watts wasted in heating the external conductors on the armature will be $= 2\,\pi r t l\,\rho\,a^2$; where l is the length of the external conductors on the armature. Now let the rise of temperature above that of the surrounding air be called θ (degrees C.), and let z stand for the amount of heat (expressed in watts) emitted per square centimetre of surface, for an excess of temperature of $1°$ C. above the surrounding air. Then the watts emitted as heat from the whole external surface of the armature will be $= z\,\theta\,2\,\pi r l$. Equating the watts generated and the watts emitted at the surface, we have

$$2\,\pi r t l\,\rho\,a^2 = z\,\theta\,2\,\pi r l,$$

$$t\,\rho\,a^2 = z\,\theta,$$

$$t\,a^2 = \frac{z\,\theta}{\rho}.$$

Now write a_1, θ_1, w_1, and N_1 as the highest permissible values of a, θ, w, and N respectively. And write q for $\sqrt{z\,\theta_1/\rho}$. It appears

[1] See their paper to be published in *Phil. Mag.* for June 1888.

that the value of q in the best modern machines is about 288. Now $t\,a = q\,\sqrt{t}$. Inserting this value, we get

$$w_1 = \frac{4\pi r n}{10^8}\, N_1 q \sqrt{t}.$$

Now by the principle of the magnetic circuit, p. 319, we have

$$N = \frac{4\pi S i/10}{\dfrac{2(d+t)}{A_2} + \Sigma\dfrac{lv}{A\mu}};$$

where d is the clearance between the surface of the armature windings and the curved face of the polar surface of the magnet, A_2 the effective area of the polar surface, and $\Sigma\dfrac{lv}{A\mu}$ the magnetic resistance of the iron parts of the magnetic circuit. Assume that N is worked up to its highest permissible value N_1, then we may write

$$w_1 = \frac{4\pi r n q}{10^8}\,\sqrt{t}\;\frac{4\pi S i/10}{\dfrac{2(d+t)}{A_2} + \Sigma\dfrac{lv}{A\mu}}.$$

This expression varies in value when the value given to t is varied, and it can be shown to have a maximum when the value of t is such that

$$\frac{2t}{A_2} = \frac{2d}{A_2} + \Sigma\frac{lv}{A\mu}.$$

That is to say, the permissible continuous output of the machine is a maximum when the thickness of the winding on the armature is such that *the magnetic resistance of the space occupied by the winding on the armature is equal to the resistance of the rest of the magnetic circuit.*

Assuming that practically the whole of the gap-space between armature core and pole-piece is filled with armature winding, the above rule amounts to saying that, given the construction of armature, the dynamo ought to be worked at such a degree of excitation that its total magnetic resistance is run up to be twice as great as that of the gap-space alone. [This is indeed no other than the diacritical stage of magnetisation, the permeability in gross of the magnetic circuit—iron and air together—being at this point reduced to half its initial value.]

Modern practice points to the following proportions in ring armatures: the thickness of external armature winding is from 7 to 11 per cent. of the diameter of the iron core; and in drum armatures, from 9 to 13 per cent.

The heating of field-magnet coils has been discussed by Prof. George Forbes, who has given the following rules. If θ be the excess of temperature above that of the surrounding air, and s the surface of the coil, then the watts wasted in heat that are dissipated from the surface are calculated from the equation

$$\cdot 00125\,\theta\,s = w = i^2 r = e\,i = e^2/r.$$

That is to say, a square centimetre, warmed $1°$ C. above the surrounding air, emits heat at the rate of about $\frac{1}{800}$ of a watt. Hence it follows that for a given maximum permissible heating of θ_1 degrees, and a given maximum permissible waste of w_1 watts, the surface to be provided must be at least such that

$$s = 800\,\frac{w_1}{\theta_1}.$$

If r be the resistance (measured hot) of the coil, the maximum permissible current i_1 is given by the equation

$$i_1 = \sqrt{\frac{\cdot 00125 \times \theta\,s}{r}}.$$

[N.B.—The resistance when warmed θ degrees above atmospheric temperature, is ascertained by multiplying by $1 + 0\cdot 004\,\theta$ the resistance r measured at atmospheric temperature.]

If the permissible rise of temperature be 50 degrees (C.), these formulæ become

$$s \text{ (sq. cms.)} = 16\,w_1 = 16\,i_1^2 r = 16\,e_1 i_1 = 16\,e_1^2 \div r,$$

$$i_1 = \cdot 25\sqrt{\frac{s \text{ (sq. cms.)}}{r \text{ (ohms hot)}}},$$

$$i_1 = \cdot 63\sqrt{\frac{s \text{ (sq. inches)}}{r \text{ (ohms hot)}}},$$

For coils to be connected as shunts, the formulæ giving the maximum permissible potential at terminals of coil are:

$$e_1 = \sqrt{\cdot 00125\,\theta\,s\,r};$$

or if θ be taken as 50 degrees (C.),

$$e_1 = \cdot 25\sqrt{s\,r} \text{ if } s \text{ is in sq. centimetres,}$$

and

$$e_1 = \cdot 63\sqrt{s\,r} \text{ if } s \text{ is in sq. inches.}$$

To reach the same limiting temperature with equal-sized bobbins wound with different-sized wire, the cross-section of the wire must vary as the current it is to carry ; or in other words, the current density (ampères per square centimetre) must remain constant.

To raise to the same temperature two similarly shaped coils, differing in size only, and having the gauge of the wire in the same ratio (so that there are the same number of turns on the large coil as on the small one), the currents must be such that the squares of the currents are proportional to the cubes of the linear dimensions.

The following rules are useful for calculating the windings for machines of same type, but of varying size.

Similar iron cores, similarly wound with lengths of wire proportional to the squares of their linear dimensions, will when excited with equal currents, produce equal magnetic forces at points similarly situated with respect to them.   (Sir W. Thomson, *Phil. Trans.*, 1856, p. 287.)

Similar machines must have ampère-turns proportional to the linear dimensions, if they are to be magnetised up to an equal degree of saturation.   (J. & E. Hopkinson, *Phil. Trans.*, 1886, p. 338.)

If two machines are to give same electromotive-force, the diameter of the wire of the coils must vary as the linear dimensions.

If, in altering the field-magnets of a machine of any given capacity, the lengths of the several portions of the magnetic circuit remain the same, but the several areas are altered, then the wire for rewinding must have its cross-sectional area altered in proportion to the periphery of the section of the cores.

The resistance of a coil, the volume of which is known, and which is wound with (round) copper wire of diameter d millimetres, enlarged by insulation to a diameter of D millimetres, can be calculated by the following rule, which is based on the assumption that the partial bedding of the convolutions allows of 10 per cent. more wires being got in than would be the case if they were exactly wound in square order.   This figure can only be approximate, as the amount of bedding varies somewhat with the relative thickness and pliability of the coating of insulating materials, as well as with the gauge of the wire.   If v be the volume in cubic centimetres, the resistance r of the coil in ohms (cold) will be

$$r = 0.0244 \frac{v}{D^2 d^2}.$$

If v be expressed in cubic inches, and D and d in "mils" (1 "mil" = .001 inch), then the approximate formula becomes

$$r = 960,700 \frac{v}{D^2 d^2}.$$

APPENDIX XI.

ON THE LEAST-SPEED OF SELF-EXCITATION.

ALLUSION was made on pp. 405 and 411 to the fact that for all series-wound and shunt-wound machines, when joined to a circuit of given resistance, there is a certain limiting speed below which they refuse to excite themselves. This least-speed of self-excitation (or critical speed) depends, as was pointed out, upon the resistance. It has not, however, been noticed hitherto that a relation exists between this least-speed and the magnetic properties of the dynamo. This relation is easily established in the case of a series-wound machine.

Writing ΣR for the sum of all the electric resistances in the circuit, we have (in C.G.S. units)

$$E = n\,C\,N,$$

$$i = E \div \Sigma R,$$

$$\frac{i}{N} = \frac{n\,C}{\Sigma R} \quad \cdot \quad \cdot \quad \cdot \quad \cdot \quad \cdot \quad \cdot \quad (1)$$

And writing $\Sigma \rho$ for the sum of all the magnetic resistances of the magnetic circuit (p. 319), we have

$$N = \frac{4\,\pi\,S\,i}{\Sigma \rho},$$

$$\frac{i}{N} = \frac{\Sigma \rho}{4\,\pi\,S}. \quad \cdot \quad \cdot \quad \cdot \quad \cdot \quad \cdot \quad 2$$

Equating (1) and (2), we have

$$\cdot \quad \cdot \quad \cdot \quad \Sigma \rho \,.\, \Sigma R = 4\,\pi\,n\,C\,S. \quad \cdot \quad \cdot \quad \cdot \quad \cdot \quad (3)$$

The right-hand side of this equation must be multiplied by 10^{-9} if ΣR is to be expressed in ohms. It will be obvious that if n is maintained constant, the equation might be written

$$\Sigma \rho \,.\, \Sigma R = \text{constant.} \quad \cdot \quad \cdot \quad \cdot \quad \cdot \quad (3\ bis)$$

No assumption has been made about the law of magnetisation, and the truth of the equation is independent of the way in which $\Sigma \rho$ depends upon N. We have here the key to the behaviour of the dynamo, and the explanation why it is that the series-wound machine, though exciting itself up " at a compound-interest rate," stops short of absolute excitation. As its field-magnets get more and more highly magnetised, the magnetic resistance increases (see p. 295). Whatever be the electric resistance thrown into the circuit, if it excites itself at all, it excites itself up to such a degree of magnetisation that the product of magnetic resistance into electric resistance remains constant.

The two factors of the product $\Sigma \rho \cdot \Sigma R$ both are complex quantities, containing constant as well as variable terms. We may write

$$\Sigma \rho = a + \xi;$$

where a relates to the magnetic resistance of the gap between the armature core and the polar surface of the field-magnets, which gap, filled with copper, insulation, and air, possesses, so far as is known, a constant magnetic resistance. The variable part ξ relates to the iron portions of the magnetic circuit, and is in general the sum of a number of terms, such as $l \div \mu A$ (see p. 319).

Similarly we may write

$$\Sigma R = a + x;$$

where a relates to the constant electric resistances (armature coils, &c.) of the circuit, and x to the variable part. Equation (3) then becomes

$$(a + \xi)(a + x) = \text{constant.} \quad . \quad . \quad . \quad (4)$$

Now, at starting, when the magnetisation of the iron is zero, the magnetic resistance of the iron part of the circuit is small, but depends upon its prior history. Assuming that it is negligibly small relatively to a, then there will be a maximum value of x, which we may call x_1, corresponding to $\xi = o$, which will be the critical external resistance of the machine at the given speed, and its value :

$$x_1 = \frac{4 \pi n C S}{a} - a. \quad . \quad . \quad . \quad . \quad . \quad (5)$$

If any greater resistance than x_1 is inserted in the circuit the machine will refuse to excite itself, even though there is small residual magnetisation present. So far as there is feeble magnetisation present the machine will act as a feeble magneto-electric machine. Suppose, now, x be so far decreased that the total electric resistance is reduced to half its limiting value—that is to say, let $x' = \frac{1}{2}x_1 - \frac{1}{2}a$: then the current, and the magnetisation, will run up until the magnetic resistance is equal to double its initial value, or $\xi = a$. The permeability of the magnetic circuit, taken as a whole, air and iron, will have been halved. It may be remarked in passing that this state of things corresponds to the "diacritical point" first investigated by the author in 1884; and, if the Lamont-Frölich formula is true, this point would correspond to the state of semi-saturation of the magnetism. Suppose, now, the variable part of the electric resistance to be reduced to zero: in other words, let the machine be short-circuited. Then ξ will run up to the maximum value it can have at that speed. Calling this value ξ_2, we have

$$\xi_2 = \frac{4\pi n \, C S}{a} - a. \quad . \quad . \quad . \quad . \quad . \quad . \quad (6)$$

It is also clear that with a given value of x the least-speed of self-excitation n_1 is given by the equation

$$n_1 = \frac{a(a+x)}{4\pi \, C S}. \quad . \quad . \quad . \quad . \quad . \quad . \quad (7)$$

This throws some light on the observed fact that very small machines are not self-exciting except at very high speeds. The fixed part a of the magnetic resistance varies in similar machines of different size, inversely as the linear dimensions; also the fixed part a of the electric resistance varies inversely as the linear dimensions of the machine. Hence the least-speed of self-excitation in similar machines on short-circuit varies inversely as the square of the linear dimensions.

Another important point is that the least-speed of self-excitation of a machine while short-circuited is a measure of the goodness of the magnetic circuit of the machine; for the lower a is, the smaller will the critical speed be.

For a shunt-wound dynamo the equation becomes

$$(a + \xi)\left(r_a + r_s + \frac{r_s r_a}{R}\right) = 4\pi n \, C Z, \quad . \quad . \quad . \quad . \quad (8)$$

where R is the variable external resistance and Z the number of shunt windings. Excitation does not take place at a given speed n unless R has as its minimum value

$$R = \frac{a \, r_a \, r_s}{4\pi n \, C Z - a(r_a + r_s)}. \quad . \quad . \quad . \quad . \quad (9)$$

2 T

There is a still more general way of establishing the relation (3). Every dynamo may be regarded as an arrangement of two circuits, a primary (or field-magnet) circuit and a secondary (or induced, or armature) circuit, between which there is a coefficient of mutual induction M (see pp. 489 and 605), which coefficient is caused to vary by mechanically altering the configuration of the system. If the currents in the two circuits be called respectively i_1 and i_2, then the work done in starting or stopping either current in the presence of the other is equal to $M\,i_1 i_2$. Now in each revolution of the armature the half-current is twice started and twice stopped in each of the sections of the armature coil. Hence in each quarter-revolution the work to be mechanically performed is $M\,i_1 i_2$; or if, as in the series machine, the current in the two circuits is the same, the work to be done is $M\,i^2$. Now M is (see p. 489) equal to $4\pi S_1 S_2 \div \Sigma\rho$; and S_2 is (see p. 339) for this purpose equal to $\frac{1}{4}$ C. But the work done by the machine in the electric circuit during a quarter-revolution is $i^2\,\Sigma R \div 4\,n$. Equating this to the work done on the machine, we get, as before,

$$\Sigma\rho \,.\, \Sigma R = 4\,\pi\,n\,C\,S.$$

APPENDIX XII.

THE DYNAMO IN TELEGRAPHY.

THE early telegraph of Gauss and Weber, of 1832, was carried out by means of a magneto-electric apparatus, as also was that of Steinheil, 1837, in both of which a coil was moved by hand in a magnetic field, for the purpose of sending signals. After this many special magneto-electric telegraphic instruments were made, the best known of which are those of Henley, of Wheatstone, and of Siemens and Halske.

In recent years, since the introduction of modern types of machine, dynamos have been used in telegraphy to replace the batteries at central stations, a single dynamo in some cases being substituted for some thousands of battery cells. The first attempts in this direction were made by L. Schwendler[1] at Calcutta in 1879 and 1880, using a series-wound Siemens' dynamo. Since then dynamos have been regularly established in the central stations of

---

[1] *Zeitschrift für angewandte Elektricitätslehre,* iii. 131.

the Imperial Telegraphs in Berlin, and of the Western Union Telegraph Company in New York. In the former was employed a compound dynamo giving constant potential of 40 volts, to which later a battery of twenty accumulator cells was added. In the New York station five dynamos are used. One of these is for separately exciting the field-magnets of the other four. These work at 25 volts each, and are joined in series so that lines can be fed, according to their various conditions, at 25, 50, 75, or 100 volts. In the central postal station in Paris an arrangement due to M. Picard[1] is employed. Two ordinary small series-wound Gramme dynamos (type d'atélier) of about 100 volts each, united in series, send their currents into a long resistance wire, the middle-point of which is joined to the middle-point between the two dynamos and to earth. Currents can be taken of either sign and of any desired potential up to $\pm$ 100 volts by connecting the lines to the corresponding point on the resistance wire.

The most recent arrangement adopted by the Western Union Telegraph Company in its New York station is described in the *Electrical World*, xi. pp. 67 and 79, 1888.

---

[1] *La Lumière Electrique,* xxv. 301, 1887.

INDEX.

—◆—

PAGE

Index. 659

2 U 2

Index.

J.

PAGE

JABLOCHKOFF, P. 44, 260, 485
Jacobi's Motor 13, 509, 514
   ,,   Law of Electromagnet 303
   ,,   Law of Maximum Activity 14, 518, 526
   ,,   Commutator 509
   ,,   Electric Boat 509
   ,,   Magneto-Machine 6
Jamieson's Electric Governor 619
Jehl and Rupp, Disk Dynamo 220
   ,,   ,,   Continuous-current Transformers 505
Joel's Dynamo 163, 327
Joints in Magnetic Circuit 325
Joubert, J., on Self-induction 100, 463
   ,,   ,, on Speed of Machines 351
   ,,   ,, Theory of Alternate Machines 14, 463
Joule, J. P., on Electric Motor 14, 257, 508
   ,,   ,, Form of Electromagnet.. .. .. .. 270, 570
   ,,   ,, Law of 304, 312
Jürgensen's Dynamo 149, 328

K.

KAPP, Gisbert (*see* Crompton and Kapp).
   ,,   ,,   Dynamos 158, 263, 327, 330
   ,,   ,,   on Compound Winding 123, 152, 424
   ,,   ,,   Formula for Dynamo 156, 305, 314
   ,,   ,,   Transformers 487, 496, 500, 504
   ,,   ,,   Graphic Diagram for Motors 543
   ,,   ,,   Pre-determination of Characteristics 15, 314
   ,,   ,,   on Series Motor 572
Kennedy, R., Iron-clad Dynamo 183, 329
   ,,   Alternating Dynamo 262, 279
   ,,   Transformers 486, 499
   ,,   Regulating Alternating Dynamos 279
Kingdon's Dynamo 276
Kohlrausch, W., on Lahmeyer's Dynamo 186
Koosen, J. H., on Theory 14

L.

LACHAUSSÉE's Dynamo 268
Ladd, W., his Machine 10, 115
Lag, alleged Magnetic 88, 607

N.

LONDON: PRINTED BY WILLIAM CLOWES AND SONS, LIMITED,
STAMFORD STREET AND CHARING CROSS.

Printed in the United States
By Bookmasters